Springer Series in Biophysics 8

S. Damjanovich (Ed.)

Biophysical Aspects of Transmembrane Signaling

With 41 Figures, 17 in Color and 5 Tables

Springer

Professor Sandor Damjanovich MD. DSc.
Department of Biophysics and Cell Biology
University of Debrecen
Medical and Health Science Center
H-4012 Debrecen
Nagyerdei krt.98.
Hungary

ISSN 0932-2353
ISBN-10 3-540-25064-6 Springer-Verlag Berlin Heidelberg New York
ISBN-13 978-3-540-25064-7 Springer-Verlag Berlin Heidelberg New York

Library of Congress Control Number: 2005921138

Springer-Verlag is a part of Springer Science+Business Media

springeronline.com

Printed in Germany

Editor: Dr. Sabine Schreck, Heidelberg
Desk Editor: Anette Lindqvist, Heidelberg
Production: ProEdit GmbH, Heidelberg
Typesetting: Satz-Druck-Service, Leimen
Cover Design: Design & Production, Heidelberg

Printed on acid-free paper 39/3152-Re 5 4 3 2 1 0

Preface

Writing or editing a book on a very topical field of science is always a great challenge, since by the time the book is published some of the newest and latest findings might be obsolete compared with results reported in the most current issues of various high-level journals.

Nevertheless, stimulating good books describe complicated systems and interrelationships and provide a broad spectrum of material for talented researchers who may find new inspiration from a joint presentation of correlated data. Of course, an editor cannot rank his own book. However, in this case, the contributing authors have high scientific standards and well-known achievements which speak for themselves.

Transmembrane signaling has many different forms and biochemical and biophysical details. However, the final outcome of transmembrane signaling at the cellular level elucidates some cellular functions, which are at the center of self-defense, alimentation, escape reactions, etc. Transmembrane signaling can be studied best in the immune system, in particular in lymphocytes, the main cellular carriers of immune defense. The different chapters are independent studies, written by well-known experts in their particular fields. Some chapters include the authors' very recent data, generalized, on the one hand, to provide understandable material for those who are interested in the field but not experts. On the other hand, the current data and novel efforts to unify biochemical and biophysical events in a physiological description provide interesting reading for experts as well.

An effort has been made to describe the broad array of methodical approaches in an understandable fashion for those wishing to apply them. For more details, the reader is referred to the References provided at the end of each chapter.

We hope that all readers will profit from this book and we would be grateful for any criticism in those cases where our efforts fall short of our aims.

Debrecen, Hungary, January 2005 Sandor Damjanovich

Contents

CHAPTER 2
Novel Single Cell Fluorescence Approaches in the Investigation of Signaling at the Cellular Level 33

List of Contributors

Abramson, Jakub
Department of Immunology,
The Weizmann Institute of Science,
Wolfson Bldg., 709a, Rehovot 76100,
Israel

Alves, Nuno L.
Institute for Molecular and Cell
Biology, Rua do Campo Alegre, 823,
4150-180 Porto, Portugal

Arosa, Fernando A.
Institute for Molecular and Cell
Biology, Rua do Campo Alegre, 823,
4150-180 Porto, Portugal

Bodnár, Andrea
Cell Biophysics Research Group of the
Hungarian Academy of Sciences,
University of Debrecen,
Nagyerdei krt. 98,
4012 Debrecen, Hungary

Damjanovich, Sándor
Cell Biophysics Research Group
of the Hungarian Academy of Sciences
and Department of Biophysics
and Cell Biology, Research Center
for Molecular Medicine, Medical
and Health Science Center,
University of Debrecen,
Nagyerdei krt. 98,
4012 Debrecen, Hungary

Darvas, Zsuzsa
Department of Genetics, Cell
and Immunobiology,
Semmelweis University,
Nagyvárad tér 4,
1089 Budapest, Hungary

Dubois, Sigrid
Metabolism Branch, Center for Cancer
Research, National Cancer Institute,
National Institutes of Health,
Bethesda, Maryland 20892-1374, USA

Duda, Ernő
Institute of Biochemistry, Biological
Research Center and Institute
of Medical Microbiology
and Immunology,
University of Szeged,
6701 Szeged, Hungary

Günther Eissner
Department of Hematology
and Oncology,
University of Regensburg,
93053 Regensburg, Germany

Falus, András
Department of Genetics, Cell
and Immunobiology,
Semmelweis University,
Nagyvárad tér 4,
1089 Budapest, Hungary

Fonseca, Ana M.
Institute for Molecular
and Cell Biology,
Rua do Campo Alegre, 823,
4150–180 Porto, Portugal

Friedländer, Elza
Department of Molecular Biology,
Max Planck Institute for Biophysical
Chemistry, Am Fassberg 11,
Göttingen, Germany

Fujiwara, Takahiro
Kusumi Membrane Organizer Project,
Exploratory Research for Advanced
Technology Organization
(ERATO/SORST-JST), Department
of Biological Science,
Institute for Advanced Research,
Nagoya University, Nagoya 464-8602,
Japan

Gáspár, Rezső
Department of Biophysics and Cell
Biology, Research Center
for Molecular Medicine, Medical
and Health Science Center,
University of Debrecen,
Nagyerdei krt. 98,
4012 Debrecen, Hungary

Goldman, Carolyn
Metabolism Branch, Center for Cancer
Research, National Cancer Institute,
National Institutes of Health,
Bethesda, Maryland 20892–1374, USA

Hegyesi, Hargita
Department Genetics, Cell and
Immunobiology,
Semmelweis University,
Nagyvárad tér 4,
1089 Budapest, Hungary

Jenei, Attila
Department of Biophysics and Cell
Biology, Research Center
for Molecular Medicine, Medical
and Health Science Center,
University of Debrecen,
Nagyerdei krt. 98,
4012 Debrecen, Hungary

Kusumi, Akihiro
Kusumi Membrane Organizer Project,
Exploratory Research for Advanced
Technology Organization
(ERATO/SORST-JST), Department
of Biological Science, Institute
for Advanced Research,
Nagoya University, Nagoya 464-8602,
Japan

Männel, Daniela N.
Department of Immunology,
University of Regensburg,
93042 Regensburg, Germany

Martínez-A, Carlos
Department of Immunology
and Oncology, Centro Nacional
de Biotecnología/CSIC,
Campus Universitario de Cantoblanco,
28049, Madrid, Spain

Mátyus, László
Department of Biophysics
and Cell Biology, Research Center
for Molecular Medicine, Medical
and Health Science Center,
University of Debrecen,
Nagyerdei krt. 98,
4012 Debrecen, Hungary

Mellado, Mario
Department of Immunology
and Oncology, Centro Nacional
de Biotecnología/CSIC,
Campus Universitario de Cantoblanco,
28049, Madrid, Spain

Müller, Jürgen
Metabolism Branch, Center for Cancer Research, National Cancer Institute, National Institutes of Health, Bethesda, Maryland 20892–1374, USA

Murakoshi, Hideji
Kusumi Membrane Organizer Project, Exploratory Research for Advanced Technology Organization (ERATO/SORST-JST), Department of Biological Science, Institute for Advanced Research, Nagoya University, Nagoya 464–8602, Japan

Murase, Kotono
Kusumi Membrane Organizer Project, Exploratory Research for Advanced Technology Organization (ERATO/SORST-JST), Department of Biological Science, Institute for Advanced Research, Nagoya University, Nagoya 464–8602, Japan

Nagy, Péter
Department of Molecular Biology, Max Planck Institute for Biophysical Chemistry, Am Fassberg 11, Göttingen, Germany

Nagy, Tünde
Institute of Biochemistry, Biological Research Center and SOLVO Biotechnology Ltd., 6701 Szeged, Hungary

Pecht, Israel
Department of Immunology, The Weizmann Institute of Science, Wolfson Bldg., 709a, Rehovot 76100, Israel

Post, Janine N.
Department of Molecular Biology, Max Planck Institute for Biophysical Chemistry, Am Fassberg 11, Göttingen, Germany

Rodríguez-Frade, José Miguel
Department of Immunology and Oncology, Centro Nacional de Biotecnología/CSIC, Campus Universitario de Cantoblanco, 28049, Madrid, Spain

Santos, Susana G.
Institute for Molecular and Cell Biology, Rua do Campo Alegre, 823, 4150–180 Porto, Portugal

Serrano, Antonio
Department of Immunology and Oncology, Centro Nacional de Biotecnología/CSIC, Campus Universitario de Cantoblanco, 28049, Madrid, Spain

Szöllősi, János
Department of Biophysics and Cell Biology, Research Center for Molecular Medicine, Medical and Health Science Center, University of Debrecen, Nagyerdei krt. 98, 4012 Debrecen, Hungary

Thurmond, Robin L.
Johnson and Johnson Pharmaceutical Research and Development, 3210 Merryfield Rd., San Diego, California 92121, USA

Tóth, Katalin
Division of Biophysics of Macromolecules, Deutsches Krebsforschungszentrum, Im Neuenheimer Feld 580, 69120 Heidelberg, Germany

Vámosi, György
Cell Biophysics Research Group of the Hungarian Academy of Sciences, University of Debrecen, Nagyerdei krt. 98, 4012 Debrecen, Hungary

Varga, Zoltán
Department of Biophysics and Cell Biology, Research Center for Molecular Medicine, Medical and Health Science Center, University of Debrecen, Nagyerdei krt. 98, 4012 Debrecen, Hungary

Vereb, György
Department of Biophysics and Cell Biology, Research Center for Molecular Medicine, Medical and Health Science Center, University of Debrecen, Nagyerdei krt. 98, 4012 Debrecen, Hungary

Waldmann, Thomas A.
Metabolism Branch, Center for Cancer Research, National Cancer Institute, National Institutes of Health, Bethesda, Maryland 20892–1374, USA

The Impact of Environmental Signals on the Growth and Survival of Human T Cells

Fernando A. Arosa, Ana M. Fonseca, Susana G. Santos, Nuno L. Alves

1.1 Introduction

Circulating human T cells come into contact with a variety of non-immunological cells and molecules that are constituents of the different internal environments. These encounters may result in effective triggering of plasma membrane receptors and in the initiation of transmembrane signals that drive T cells into different biological processes (e. g., growth, division, cytokine secretion or death), depending on the differentiation state of the T cell and the nature of the signals and receptors involved. The transmembrane signals delivered by the environment may intersect and ultimately modify the biological outcome of specific signals delivered via the TcR/CD3 complex by professional antigen-presenting cells (APC). We propose that the process of T cell differentiation in the periphery is mandatory, is largely influenced by cells and molecules present in the tissue environment that T cells occupy, and is the result of an adaptive process that permits T cells to embark on a fruitful relationship with the neighboring cells. Accordingly, T cells populating or circulating through the different tissues and organs are in a privileged position to regulate the physiological processes of neighboring cells. A clearer understanding of receptor and cell signaling pathways and levels of cross talk induced by environmental cells and molecules on human T cells should provide novel insights into the phenotypic and functional changes that occur during human T cell differentiation. In this chapter we will review accumulated evidence indicating that signals delivered by red blood cells (RBC) and epithelial cells (EC) influence important aspects of the biology of T cells, such as the capacity to grow, divide, differentiate and survive. The role that IL-15 and MHC-class I molecules play in this context will also be reviewed.

1.2 Red Blood Cells: From Carriers of Respiratory Gases to Modulators of T Cell Survival

1.2.1 RBC: Carriers of O_2 and CO_2

Red blood cells (RBC), or erythrocytes, are anucleate and biconcave cells, with an average diameter of 8 μm. There are normally about 3–5 million per microliter

Springer Series in Biophysics
S. Damjanovich (Ed.)
Biophysical Aspects of Transmembrane Signaling

of blood. RBC are produced from pluripotent haemopoietic cells that reside in the bone marrow and, upon appropriate stimulation, undergo a series of maturation events that eventually produce fully functional erythrocytes. During maturation RBC lose their mitochondria, nucleus, ribosomes and many cytoplasmic enzymes. Maturation in bone marrow takes 24–48 h and the average RBC survives in circulation for 120 days. As red cells age, their membranes become rigid and inflexible, and they are removed from circulation by macrophages of the spleen and liver (Klinken 2002).

The primary function of RBC is considered to be oxygen transport from pulmonary capillaries, at high O_2 tension, to tissue capillaries, where it is exchanged for the metabolically generated CO_2. The most abundant intraerythrocytic protein is hemoglobin (Hb), which is the carrier for O_2 and CO_2 throughout the body. Hb contains four globin subunits, normally 2α and 2β chain in adults, each surrounding a core haem-moiety (Klinken 2002). At the center of the haem is iron which is essential for gaseous transport. Although most erythrocyte iron (Fe^{2+}) is coordinately bound to the protoporphyrin moiety of heme, RBC also have some "free" cytosolic Fe^{2+} (Ferrali et al. 1992; Gabay and Ginsburg 1993).

Despite the limited biosynthetic repertoire available to mature RBC, they are resistant to oxidant-induced damage, because they have a very efficient antioxidant defense system (in the form of scavengers and detoxifying enzymes) that combats free radical-induced oxidative damage to cellular macromolecules. Catalase and glutathione-peroxidase (GSH-PX) form an important defense system to combat hydrogen peroxide (H_2O_2) produced by the superoxide dismutase reaction, and these enzyme systems are thought to be the major antioxidant systems in human RBC. Glutathione, GSH, a low molecular weight thiol, is the required substrate for GSH-PX and is important in reversible protein mixed-disulfide formation, a way to prevent irreversible oxidative damage to protein thiol groups. There are studies supporting the idea that catalase and reduced β-nicotinamide (NADPH) protect RBC against acute, high exogenous levels of H_2O_2, whereas GSH is needed to protect against the low levels that are continuously generated endogenously (Eaton 1991; Dumaswala et al. 1999). Membrane-bound proteinases, a secondary antioxidant defense mechanism recently identified as acylpeptide hydrolases, protect RBC by preferentially degrading the oxidatively damaged proteins (Beppu et al. 1994; Fujino et al. 2000).

1.2.2 Novel Functions of RBC

Recently there has been an increasing amount of data produced concerning the role of RBC in regulating the bioactivities of other cell types, such as regulation of vascular homeostasis (Pawloski and Stamler 2002), inhibition of platelet aggregation induced by ADP or collagen (Pawloski et al. 1998; Megson et al. 2000), enhancement of platelet aggregation after collagen-induced activation (Vallés et al. 2002), inhibition of neutrophil apoptosis (Aoshiba et al. 1999) and regulation of T cell rolling and adhesion to the endothelium (Melder et al. 2000). The capacity of RBC to regulate biological processes in neighboring cells is thought to be due to the permeability of their membranes to exogenous reactive oxygen and nitrogen species, namely nitric oxide (NO) (Halliwell and Gutteridge 1999).

The capacity of Hb to reversibly bind NO according to the changes in O_2 tension make RBC potential carriers of NO bioactivity (Pawloski et al. 2001; McMahon et al. 2002). A model proposed by Stamler and colleagues suggested that NO is transported in Hb by binding to the high conserved β-cysteine 93, forming the S-nitrosated Hb (SNO-Hb), which may serve for regulation of blood pressure and facilitates efficient delivery of oxygen to tissues (Jia et al. 1996; Stamler et al. 1997; Gow and Stamler 1998; Bonaventura et al. 1999). They also presented evidence showing that interactions of SNO-Hb with band 3 (also known as anion-exchanger AE1) on the erythrocyte membrane can facilitate NO release (Pawloski et al. 2001). Despite this hypothesis, many questions about the mechanisms by which SNO-Hb is formed and delivered by RBC in vivo remain unanswered (Hobbs et al. 2002; Cosby et al. 2003; Nagababu et al. 2003).

In addition to the role in NO vasodilatation via the formation of SNO-Hb, other RBC functions that have been generally disregarded include the modulation of T cell proliferation and survival. In fact, the capacity of RBC to modulate immunological responses in vitro has been known for more than three decades (Tarnvik 1971; Silva et al. 1981; Ebert 1985; Rugeles et al. 1987; Virela et al. 1988; Shau and Golub 1988; Kalechman et al. 1993). These studies showed that erythrocytes influence a variety of biological processes in T cells, including cell proliferation, cytokine production (IL-2, IL-3, IL-6, TNF, IFN-γ), IL-2 receptor expression, NK-mediated cytotoxicity and antibody synthesis by B cells (reviewed by Arosa et al. 2004).

Work performed in recent years has contributed to the better understanding of the effect of RBC on human T cells. Essentially, it has been established that RBC reduce oxidative stress and apoptosis of activated T cells and favor cell division, thus modulating T cell growth and survival in vitro (Fonseca et al. 2001, 2003b). The RBC effect is directly exerted on the activated T cells, primarily CD8+ T cells, and is not mediated by accessory cells (Fonseca et al. 2001, 2003a; Porto et al. 2001). On the other hand, it is well known that both immune activation (e.g., augmented circulating immunoglobulins and T cell activation) and immunosuppression (e.g., improved graft survival) can occur because of RBC transfusion (Tartter 1995; Claas et al. 2001). These data, together with studies by Gafter et al. providing evidence that RBC transfusions in humans may induce the appearance of suppressor CD8+ T cells (Gafter et al. 1992), underscore the physiological importance of RBC in the regulation of the immunological system. Consequently, it is plausible to think that when RBC come into contact with activated T cells, for instance at inflammatory sites or in the liver, they deliver "signals" that will promote survival and eventually augment T cell effector functions. However, how do RBC deliver these transmembrane signals to the activated T cells?

1.2.3 Are Transmembrane Signals Involved in the Regulation of T Cell Survival Induced by RBC?

The capacity of RBC to modulate human T cell growth and survival is well documented (Arosa et al. 2004). This capacity was thought to be due to the antioxidant capacity of RBC, to RBC membrane proteins that deliver co-stimulatory signals to T cells, such as CD58, and to heme-related products with co-stimulatory activities. In vitro studies performed over the last 5 years have concluded that the positive influence that RBC have on T cells is not the result of a mere radical scavenging activity of RBC (Fonseca et al. 2001). Moreover, our results indicate that the CD2/CD58 interaction and iron-related products "per se" play a minor role, if any, in the regulation of T cell growth and survival and point to other mechanisms (Fonseca et al. 2003a, 2003b). Yet, it must be emphasized that in order to afford optimal protection, the RBC must come into contact with the activated T cell, suggesting that transmembrane cross talk between the RBC and the activated T cell takes place (Fonseca et al. 2001).

Broadly speaking, the erythrocyte membrane is comprised of a cholesterol-rich phospholipid bilayer that is studded by a large number of transbilayer proteins, of glycosylphosphatidylinositol (GPI)-anchored proteins standing outside, and an important protein assembly, the erythrocyte or membrane skeleton, which laminates the inner surface of the bilayer. The most abundant integral membrane proteins in human RBC membranes are the sialoglycoprotein, glycophorin and the already mentioned band 3 which play a structural role in linking the bilayer with the spectrin skeletal network and are specific to red cells (Schrier 1985). Some red cell-surface proteins appear to be receptor or adhesion molecules, either from their structure or from their known functions in other cells, such as members of the immunoglobulin superfamily.

The immunoglobulin superfamily is a large collection of glycoproteins, abundant on leucocytes, but also present on some other cells. CD58, also called lymphocyte function-associated antigen-3 (LFA-3), is one of those that was reported to be present in RBC. This molecule is a GPI-anchored protein in RBC, and is widely expressed by both hematopoietic and non-hematopoietic cells, namely antigen-presenting cells (APC) (Krensky et al. 1983). The ligand for CD58 is CD2 which is expressed on virtually all T cells, thymocytes and NK cells, and this interaction provides a potent co-stimulus to T cells by means of molecular associations with intracellular kinases. Besides the fact that interaction between CD2 (on T cells) and its receptor CD58 on opposing cells is critical for cell adhesion and, consequently, for an optimal immune recognition, it is important to note that this interaction alone has no stimulating effect on T cells and that only activated T cells can bind and form rosettes with human RBC (Springer 1990). Rosetting of human T cells through the interaction of CD2 with an LFA-3 homologue on sheep erythrocytes is a classic laboratory technique that has been used since the early 1970s by immunologists to purify T lymphocytes (Froland 1972), but only in 1987 was it shown that the interaction between CD2 and CD58 was mediating this phenomenon (Makgoba et al. 1987; Plunkett et al. 1987; Selvaraj and Dustin 1987).

To ascertain the possible involvement of CD58 in the expansion of mitogen-activated T cells induced by RBC, experiments where the CD2/CD58 interaction was blocked by using anti-CD58 antibodies (clone TS2/9) were performed. This treatment did not reverse the increased T cell proliferation induced by the presence of RBC (Fonseca et al. 2003a). Besides, RBC treatment with phosphatidylinositol-specific phospholipase C, which removes GPI-linked proteins such as CD58 and CD59 from the cell surface, did not induce any relevant change in the RBC effect (Fonseca et al. 2003a). In view of these results, we examined the expression of CD58 by RBC. Surprisingly, we were not able to detect significant expression of LFA-3/CD58 on human RBC either by immunoprecipitation or by flow cytometry (Fig. 1.1).

This result disagrees with previous reports (Dustin et al. 1987; Makgoba et al. 1987; Plunkett et al. 1987) despite the fact that TS2/9 antibodies were used in all these studies. The lack of expression of LFA-3/CD58 on human RBC cannot be considered an artifact because TS2/9 antibodies labeled and immunoprecipitated LFA-3/CD58 from the cell surface of an EBV cell line (Fig. 1.1). However, the possibility that the levels of LFA-3/CD58 on the surface of human RBC are below the detection limits of these sensitive techniques cannot be ruled out. In any case, these levels of expression would be inappropriate in terms of the delivery of co-stimulatory signals to activated T cells and most likely do not play a relevant role in the RBC effect. These data, together with results indicating that the RBC effect could be observed even when RBC are added to the cultures of activated T cells 24 and 48 h after the initial stimulation, suggest a lack of involvement of co-stimulatory signals delivered by RBC during the early stages of T cell activation (Fonseca et al. 2003a).

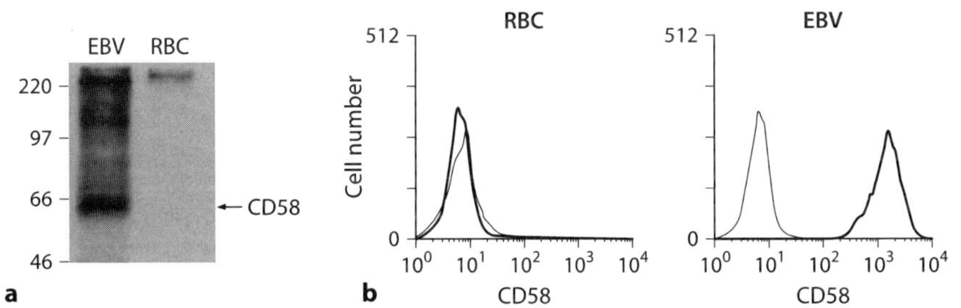

Fig. 1.1. Human RBC lack detectable expression of CD58. **a** 5×10^6 EBV cells and 250×10^6 RBC were surface biotinylated and lysed in 1% Triton X-100. After pre-clearing, the cell lysates were immunoprecipitated with anti-CD58 antibodies (TS2/9) and protein-A Sepharose beads as described (Fonseca et al. 2003b). Immunoprecipitates were resolved in a 10% SDS-PAGE under non-reducing conditions and transferred to nitrocellulose filters. Proteins were visualized after incubation with ExtrAvidin-HRP followed by ECL. The band corresponding to CD58 is indicated. **b** RBC and EBV cells were cell-surface labeled with saturating concentrations of TS2/9 followed by rabbit anti-mouse Igs conjugated with FITC, and acquired in a FACSCalibur. Results show CD58 expression (*thick lines*) in RBC (*left histogram*) and EBV cells (*right histogram*). Background staining with mouse Igs (*thin lines*) is shown

Although these data suggest that transmembrane signals via CD2 are unlikely to play a role in the immunoregulatory effect of RBC, the involvement of other RBC receptors cannot be ruled out. Indeed, the protection afforded by RBC (e. g., inhibition of apoptosis, enhancement of cell division, upregulation of cytoprotective proteins, etc.) conforms with the existence of transmembrane signals that intersect and strengthen intracellular pathways involved in the control of cell division and survival. Importantly, these signals are turned on when RBC come into contact with metabolically active, i.e., activated and or/dividing, T cells. Because RBC markedly reduce the levels of intracellular oxidative stress within the activated T cells (Fonseca et al. 2003b), it is plausible to think that RBC proteins and/or cellular components intersect with pathways regulated by oxygen and nitrogen radicals (Ginn-Pease and Whisler 1998). Furthermore, since RBC favor primarily the expansion of CD8+ T cells (Porto et al. 2001; Fonseca et al. 2003a), transmembrane signals delivered by RBC are also likely to modulate cell cycle-related pathways.

1.2.4 Are RBC-Derived Vesicles Carriers of the RBC Bioactivity?

Although RBC are considered inert to regulatory signals from other cells, they are equipped with the machinery required for intercellular communication (Minetti and Low 1997), and modulation of RBC has become a pharmacological tool for the treatment of certain diseases (Brugnara et al. 1993; Solerte et al. 1997). Although the responsiveness of RBC to a variety of environmental regulators and signals is becoming apparent, the mechanisms and intermediates of the response have not received much attention. One of the most common intermediates in mediating RBC responsiveness is the secretion of vesicles.

The normal shape of RBC is a biconcave disk, but under different chemical and physical conditions they may undergo shape alterations which finally result in the release of two types of vesicles: microvesicles of approximately 150 nm ∅ and nanovesicles of approximately 60 nm ∅ (Allan et al. 1980). A rise in cytosolic Ca^{2+} or ATP depletion in RBC induces biochemical and morphological changes in the RBC that are closely linked to the formation and secretion of vesicles containing cytoplasmic and membrane anchored proteins (Pascual et al. 1993; Dervillez et al. 1997; Salzer et al. 2002). Low pH also induces shape changes and vesiculation in intact human RBC (Gros et al. 1996). Upon a rise in cytosolic Ca^{2+} or ATP depletion, RBC change shape from biconcave disks to smaller echinocytic spheres (Allan and Thomas 1981). Some of the biochemical changes include exposure of phosphatidylserine and efflux of potassium on the outer leaflet of the plasma membrane (Allan and Thomas 1981; de Jong et al. 1997; Andrews et al. 2002). RBC shed vesicles in vivo (Dumaswala and Greenwalt 1984; Bevers et al. 1992) and anomalies in the vesiculation process have been described in bleeding disorders and hemolitic anemias (Bevers et al. 1992; Turner et al. 2003), and are thought to occur in systemic lupus erythematosus (SLE) and HIV patients (Cohen et al. 1989; Dervillez et al. 1997). Although the physiological target(s) and functions of these RBC-derived vesicles remain largely unknown, recent studies have shown that vesicles derived from B cells bind to the cell surface of follicular dendritic cells and regulate their activity (Denzer et al. 2000).

In addition to the classical antioxidant molecules superoxide dismutase, catalase and glutathione (Halliwell and Gutteridge 1999), RBC also contain another family of antioxidant molecules designated as peroxiredoxins (Prx). Prx are thiol-specific antioxidant enzymes that belong to the family of peptidyl prolyl isomerases and that exert their protective effect in cells through their peroxidase activity (ROOH+2e$^-$→ROH+H_2O), whereby hydrogen peroxide, peroxynitrite and a wide range of organic hydroperoxides are reduced and detoxified. Recently, a range of other cellular roles have also been ascribed to mammalian Prx family members, including the modulation of cytokine-induced hydrogen peroxide levels which have been shown to mediate signaling cascades, leading to cell proliferation, differentiation and apoptosis (Fujii and Ikeda 2002; Hofmann et al. 2002; Lee et al. 2003; Neumann et al. 2003; Wood et al. 2003).

Six subclasses of Prx are produced at high levels in mammalian cells. Notably, PrxII (also known as NKEF-B, for natural killer enhancing factor B) together with PrxI (also known as NKEF-A) are among the second or third most abundant proteins in RBC (Moore et al. 1991; Wood et al. 2003). Early studies on the effect of RBC on immune functions showed enhancement of natural killer activity (Shau and Golub, 1988). Afterwards, the same group identified the natural-killer-enhancing factor of RBC as a major antioxidant protein that turned out to be a peroxiredoxin (Shau et al. 1993; Shau and Kim 1994; Wood et al. 2003). Interestingly, a recent study has reported that recombinant NKEF-A and NKEF-B inhibited HIV-1 replication when exogenously added to acutely infected T cells (Geiben-Lynn et al. 2003). Besides Prx, other peptidyl prolyl isomerases present in RBC have been shown to play important roles in signal transduction pathways involved in the regulation of cell cycle progression (Lu et al. 1996; Galat 2003). The likelihood that the modulation of (CD8+) T cell growth and survival induced by RBC results from vesicles containing Prx and/or other peptidyl prolyl isomerases, which are secreted from the RBC upon contact or proximity with metabolically active T cells, is an open question that deserves further investigation (Fig. 1.2).

1.3 Epithelial Cells and IL-15: The Perfect Combination for CD8+ T Cell Differentiation

1.3.1 Interleukin-15: Major Features of a Multifunctional Cytokine

IL-15 is a cytokine belonging to the 4 α-helix-bundle cytokine family which exerts a pivotal role in NK and T cell development and homeostasis. IL-15 binds to a high-affinity receptor (IL-15R) consisting of three distinct membrane components – α, β and γ_c chains – the latter shared by IL-2, IL-4, IL-7, IL-9, IL-13, and IL-21. The interaction of IL-15 with its receptor is followed by a series of intracellular signaling events that have been reviewed elsewhere (Waldmann and Tagaya 1999).

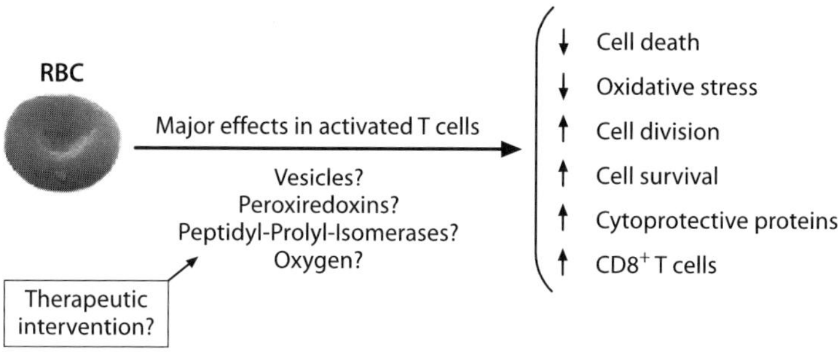

Fig. 1.2. Outline of the influence that human RBC have on activated T cells and the possible mechanisms involved

Most of our knowledge on IL-15 comes from studies in genetically targeted mice for IL-15 and IL-15R genes. IL-15–/– and IL-15Rα–/– mice lack NK cells, implying a crucial effect of IL-15 in murine NK cell ontogeny (Lodolce et al. 1998; Kennedy et al. 2000). In accordance with the murine data, in vitro studies have shown that IL-15 is an essential factor for human NK cell differentiation and maturation (Mrozek et al. 1996; Yu et al. 1998). On the other hand, several reports have shown that IL-15 also controls CD8+ T cells in both mice and humans. Studies in humans showed that naïve CD8+ T cells expand in response to IL-15 in an antigen-independent manner, although a higher responsive capacity to this growth factor is found in memory-type CD8+ T cells. In contrast, the response of CD4+ T cells is much lower (Kanegane and Tosato 1996; Alves et al. 2003; Cookson and Reen 2003; Schonland et al. 2003). Strikingly, when the IL-15R is analysed in the different subsets, naïve CD8+ T cells show, in comparison with memory-type cells, a lower expression of IL-15Rβ chain and a comparable level of IL-15Rγ (constitutively expressed in lymphocytes). In both subsets, the level of IL-15Rα is barely detectable by flow-cytometric analysis. Nonetheless, the capacity of naïve CD8+ T cells to respond to IL-15 is not limited by the lower expression of the IL-15Rβ chain. Rather, naïve CD8+ T cells upregulated the IL-15Rβ chain after contact with IL-15 (Alves et al. 2003). Moreover, the capacity to respond to IL-15 is not precluded by the absence of the IL-15Rα chain (Dubois et al. 2002). These data indicate that in contrast to mice, human naïve CD8+ T cells acquire the ability to respond to IL-15.

Besides the aforementioned effects on CD8+ T cell proliferation, IL-15 also induces or initiates a CD8+ T cell differentiation program. In naïve CD8+ T cells, IL-15 induces a cytotoxic T lymphocyte (CTL)-phenotype, characterized by the loss of CD28 and CD45RA and acquisition of CD56. Concomitant with these phenotypic changes, naïve cells acquire functional properties typical of CTL, such as the ability to produce interferon γ (IFN-γ) and tumor necrosis factor α (TNF-α), the expression of perforin and granzyme B, and cytotoxic capacities (Alves et al. 2003; Cookson and Reen 2003). In concert with its effects on naïve T cells, IL-15 contributes to an augmented effector function of memory CD8+ T cells (Liu et al. 2002; Alves et al. 2003; Mueller et al. 2003).

The regulation of IL-15 transcription, translation and secretion involves multiple complex mechanisms (reviewed by Tagaya et al. 1996), and the posttranscriptional regulation seems to be the most important one. Three main checkpoints negatively regulate IL-15 mRNA translation: (1) multiple start codons in the 5′ untranslated region which may diminish the efficiency of translation. Bamford and coworkers demonstrated that in COS cells the deletion of this region increases the production of IL-15 (Bamford et al. 1998); (2) the short and the long signal peptides (LSP and SSP, respectively) present in the IL-15 precursor protein which give rise to two IL-15 isoforms. Both IL-15 isoforms encode an identical mature IL-15 protein, differing only in the signal sequence. Immunofluorescence studies revealed that LSP-IL-15 localizes in the secretory pathway, while SSP-IL-15 is localized in the cytoplasm and nucleus (Waldmann and Tagaya 1999); and (3) a negative element seems to exist in the C-terminus of the coding sequence. It was demonstrated that modifying the 3′ end of the IL-15 protein increases its production five- to tenfold (Bamford et al. 1998). This complex and tight regulation of the IL-15 gene is uncommon for most cytokines characterized hitherto and may indicate that its overproduction is harmful to the host.

1.3.2 Physiological Expression of IL-15: The Epithelial Cell Connection

SSP-IL-15 mRNA is produced by multiple non-lymphoid tissues including the heart, thymus and testis, whereas LSP-IL-15 is produced in the placenta, skeletal muscle, kidney, lung, heart, fibroblast, and liver (Grabstein et al. 1994; Meazza et al. 1996; Tagaya et al. 1997; Cookson and Reen 2003). IL-15 protein has been detected in a wide range of cells including bone marrow stroma cells, human spleen-derived fibroblasts and activated monocytes, macrophages and dendritic cells (Carson et al. 1995; Reinecker et al. 1996; Briard et al. 2002). Notably, IL-15 is also produced by epithelial cells of the intestine and the liver and its production is upregulated during stress or infection (Reinecker et al. 1996; Fehniger and Caligiuri 2001; Man et al. 2003; Golden-Mason et al. 2004). It has been proposed that an abnormal expression of IL-15 could be involved in the pathogenesis of autoimmune disorders such as rheumatoid arthritis (RA) and inflammatory bowel diseases (IBD). Increased levels of IL-15 have been detected in synovial fluids and in inflamed mucosa of patients with RA and IBD, respectively (McInnes et al. 1996; Liu et al. 2000). It is possible that the high levels of IL-15 found under these conditions might contribute to the maintenance and enhancement of the chronic inflammation, and therefore IL-15 might modulate T cell function in the inflamed environment (Mention et al. 2003).

In this context, IL-15 appears to regulate co-stimulatory pathways in human T cells, such as upregulation of NKG2D in CD8+ T cells lacking the expression of CD28 (Roberts et al. 2001). Interestingly, NKG2D is a ligand for a set of unconventional MHC-class I molecules expressed by epithelial cells that act as potent costimulatory receptors for CD8+ T cells, enhancing their function and survival (see below). These findings suggest that the functions and phenotype induced by IL-15 on CD8+ T cells mimic the ones triggered by TCR. In line with this, recent work by Liu and coworkers highlighted that IL-15 and TCR stimulation induces a remarkable similarity in the gene expression, proliferation, effector molecule production and cytotoxicity of CD8+ T cells (Liu et al. 2002). It would be interesting to investigate thoroughly whether these two parallel signals, inducing similar molecular, cellular and functional events, will converge in a common pathway.

Despite the multiple biological activities exerted by IL-15, it is noteworthy that free IL-15 is not detectable either in biological fluids of healthy individuals or in supernatants of activated cells that express high levels of IL-15 mRNA. Thus, the role of IL-15 in cells of the immune system, particularly CD8+ T cells, is an important point that remains to be elucidated. Recently, Dubois et al. (2002) have elegantly demonstrated that IL-15 can be presented at the cell surface of activated monocytes bound to the IL-15Rα, activating in *trans*-target cells expressing either intermediate- (IL-15Rβ and γ_c) or high- (IL-15Rα, β and γ_c) affinity IL-15 receptors. Furthermore, Briard et al. (2002) have shown that IL-15 is bioactively expressed at the cell surface of human spleen-derived fibroblasts. Hence, these studies indicate a novel mechanism by which cells expressing the IL-15Rα chain form complexes with IL-15 at their cell surface, thereby *trans*-presented to neighboring cells (CD8+ T cells) expressing intermediate-affinity IL-15R. These novel findings place IL-15 as a likely bridge in the communication between T cells and epithelial cells (Fig. 1.3).

1.3.3 Epithelial Tissues as Sites for CD8+ T Cell Positioning and Survival

The majority of T cells populating epithelial tissues (e.g., the small intestine and liver parenchyma) are phenotypically and functionally memory CD8+ T cells (Cerf-Bensussan and Gruy-Grand 1991; Lundquist et al. 1996; Hamann et al. 1997; Norris et al. 1998; Ishihara et al. 1999). Doherty and collaborators (Doherty et al. 1999; Kenna et al. 2003) demonstrated that a large fraction of human hepatic CD8+ T cells express NK receptors. These CD8+ T cells, called NKT cells, are also present in peripheral blood under a number of conditions and are characterized by lack of CD28 expression (Mingari et al. 1996; Speiser et al. 1999; Pittet et al. 2001; Arosa 2002). Although evidence for the expression of NK receptors by human intestinal intraepithelial lymphocytes (iIEL) is scarce, a number of studies suggest that T cells bearing NK receptors exist within the human intestine. Thus, BY55/CD160 is an NK receptor expressed by both circulating CD8+CD28– T cells and iIEL (Anumanthan et al. 1998). First described in circulating NK cells, CD160 is expressed by cord blood and bone marrow CD8+ T cells, and a large fraction of circulating CD8+ T cells in HIV patients (Bensussan et al. 1993, 1994; Maiza et al. 1993). Moreover, it is generally accepted that iIEL and hepatic CD8+ T cells lack cytotoxic activity but gain cytotoxicity after culture in IL-2 (Chott et al. 1997; Doherty et al. 1999), a feature also shared by memory CD8+ T cells. More recently, expression of NKG2D has been described in human intraepithelial CD8+ T cells in celiac disease (Hue et al. 2004). These phenotypic and functional similarities suggest that circulating memory CD8+CD28– T cells originate from

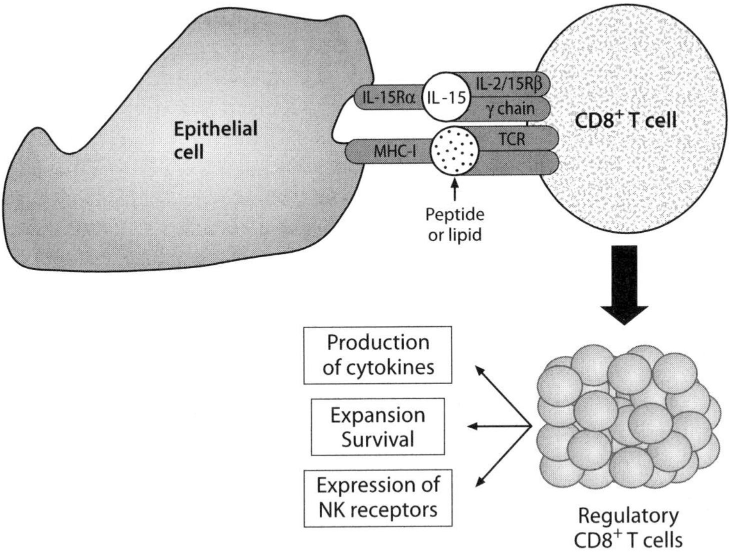

Fig. 1.3. Schematic model illustrating the effects of epithelial cells and IL-15 on human CD8+ T cells. Epithelial cells may activate CD8+ T cells through a variety of MHC-class I molecules presenting either peptides or lipids. In addition, under stressful conditions epithelial cells may produce IL-15 that binds to the IL-15Rα that presents IL-15 in *trans* to CD8+ T cells bearing intermediate-affinity IL-15Rβ/γc receptors. In combination, the transmembrane signals delivered by MHC-I and IL-15 play a pivotal role in activation, homeostasis and function of CD8+ T cells

CD8+CD28+ T cell precursors that have acquired typical NK receptors during the differentiation process in epithelial environments. Considering recent developments (see below), it is plausible to think that thymus and/or bone marrow-derived CD8+ T cells, in the context of signals provided by environments such as the liver parenchyma or the intestine, differentiate, lose expression of CD28 and express cell-surface NK receptors, thus becoming NKT cells (Arosa 2002).

1.3.4 Epithelial Cells Deliver Activation and Survival Signals Preferentially to CD8+ T Cells

Epithelial tissues are composed of sheets of epithelial cells that cover the external and internal body surface and line many important internal organs. In the intestine and the liver, enterocytes and hepatocytes are the most abundant epithelial cells, respectively. While enterocytes play a major role in protection and nutrient absorption, hepatocytes are the major metabolic achievers in the body (Kmiec 2001). Until recently both organs were considered to be inert from an immunological point of view. Evidence accumulated during recent years has changed this view, as gastroenterologists and immunologists have noted that enterocytes and hepatocytes possess immunostimulatory properties (Hershberg and Mayer 2000; Mackay 2002).

Both the intestine enterocytes and the liver hepatocytes are in close contact with a variety of lymphocyte subsets, from NK cells to TcRαβ and TcRγδ cells, and they display features of memory/differentiated T cells. Studies conducted in the mid-1980s unveiled the capacity of intestinal epithelial cells (iEC) to activate human peripheral blood T cells in vitro, namely, CD8+ T cells (Mayer and Shlien 1987). In elucidating the molecular mechanisms involved in the preferential activation of CD8+ T cells by ex vivo iEC, Mayer and collaborators demonstrated that CD1d and a heavy glycosylated protein of 180 kDa (gp180) are the receptors involved in the activation of human CD8+ T cells through interaction with the TCR and CD8, respectively (Panja et al. 1993; Li et al. 1995; Yio and Mayer 1997; Allez et al. 2002; Van de Wal et al. 2003).

CD1d is a non-classical MHC-class I molecule specialized in binding lipids which is expressed by intestinal epithelial cells, hepatocytes, and stromal cells (Blumberg et al. 1991; Canchis et al. 1993), and interaction of the CD1d/gp180 complex with the TCR/CD8 activates the Src tyrosine kinase Lck (Panja et al. 1993; Li et al. 1995; Hershberg and Mayer 2000). The capacity of intestinal epithelial cells to activate human CD8+ T cells in vitro has been reported by other groups and has been observed with parenchymal hepatic cells and cell lines (Bumgardner et al. 1992; Volpes et al. 1992; Arosa et al. 1998), airway epithelial cells (Kalb et al. 1997), endothelial cells (Dengler et al. 2000; Choi et al. 2004) and fibroblasts (Roberts et al. 1997). Therefore, it appears that regardless of their origin, epithelial cells have the intrinsic capacity to deliver signals to human CD8+ T cells. Yet, this intercellular communication may not solely involve classical receptor–ligand interaction. Hence, cytokines such as IL-15 as well as RBC present in the tissue microenvironments might influence T cell growth and survival, and, as mentioned above, contribute to the amplification of the T cell response in inflammatory conditions (Briard et al. 2002; Dubois et al. 2002; Mention et al. 2003; Arosa et al. 2004; Golden-Mason et al. 2004; Hue et al. 2004). Finally, molecules of the MHC-class I molecules other than CD1d may also contribute in unexpected ways to the regulation of T cell growth and survival in the tissue environment, as discussed in the next section.

1.4 MHC-Class I Molecules: Pleiotropic Transmembrane Signaling Structures

1.4.1 The MHC Locus

The major histocompatibility complex (MHC) forms an extended genomic cluster, with an elevated number of genes coding for proteins that accomplish extremely important immunological functions. Two main classes of MHC proteins are encoded within the MHC locus, and in humans are designated as HLA (Gruen and Weissman 1997). The high polymorphism of MHC-encoded proteins defines the unique genetic fingerprint of an individual's immune system, explaining the basis of tissue rejection in transplantation, and defining the repertoire of antigenic determinants to which each one of us is capable of responding (Howard et al. 1999). The MHC region is known to influence the population penetrance of a variety of autoimmune and non-autoimmune disorders, such as multiple sclerosis, systemic lupus erythematosus, myasthenia gravis, rheumatoid arthritis, ankylosing spondyloarthritis and hemochromatosis (Gruen and Weissman 1997; Cardoso and de Sousa 2003).

1.4.2 Folding and Assembly of MHC-Class I Molecules

The MHC-class I protein is a highly polymorphic, membrane-spanning heterotrimeric complex, formed of a transmembrane glycoprotein, named heavy chain (HC) with approximately 46 kDa, non-covalently associated with a light chain, beta 2 microglobulin (β_2m), with approximately 12 kDa and with a small peptide of eight to ten amino acids. The heavy chain contains an extracellular portion divided into three domains – α1, α2 and α3; in contrast to the highly conserved α3 domain, α1 and α2 domains have two polymorphic α helices with interconnected β sheets that form the peptide binding groove. The structure of the groove is believed to be maintained by the non-covalent association with β_2m (Bjorkman and Parham 1990).

Earlier events in the folding of MHC-class I molecules display many similarities with the assembly of other multisubunit glycoproteins. After entering the endoplasmic reticulum (ER), MHC-class I heavy chains are monoglucosylated and associate with the general chaperone calnexin. The soluble chaperone calreticulin comes to replace calnexin and, with the thiol-dependent oxidoreductase ERp57, promotes the formation of the disulfide bonds inside the heavy chain and the assembly with the β_2m (Parodi 2000; Dick et al. 2002). At a later stage, the process becomes more specific and complex, with the association of the accessory molecules tapasin and the transporter associated with antigen processing (TAP), leading to the formation of the peptide-loading complex. This complex is responsible for loading MHC-class I molecules with high-affinity peptides (Rudd et al. 2001; Williams et al. 2002). Correctly folded MHC-class I molecules are transported to the Golgi and then to the cell surface. MHC-class I molecules not properly folded are believed to be retrotranslocated to the cytosol and targeted for degradation (Hughes et al. 1997; Cresswell et al. 1999). The role of the various chaperones and accessory molecules in the folding and assembly of MHC-class I molecules has been recently reviewed (Antoniou et al. 2003).

1.4.3 Cell-Surface MHC-Class I Molecules: *trans*-Interactions with the TCR and NK Receptors

After leaving the ER, folded MHC-class I molecules enter the secretory pathway and egress, via the Golgi, to the plasma membrane of all nucleated cells, where their main function is to present peptides to CD8+ T cells. The T cell receptor (TCR) complex comprises α/β subunits that recognize the peptide-MHC, noncovalently associated with the intracellular signal transduction subunits TCRς, and CD3γ, δ and ε (Pitcher et al. 2003). Activation takes place through immunoreceptor tyrosine-based activation motifs (ITAM) present in the cytoplasmic domains of the CD3 chains. The first step in antigen receptor signal transduction is the activation of cytosolic protein tyrosine kinases of the Src, Syk and Tec families, which phosphorylate the ς chain of the CD3 complex and several adaptor molecules, initiating signaling cascades that lead to calcium rises, proliferation, cytokine secretion and ultimately the generation of differentiated CD8+ T cells (Cantrell 2002; Lanzavecchia and Sallusto 2002). By restricting the peptides recognized by CD8+ T cells, MHC-class I molecules are involved not only in the selection of CD8+ T cells in the thymus, but also in the maintenance of the CD8+ T cell pool in the periphery, which differs according to the differentiation state of the CD8+ T cell (Seddon and Zamoyska 2003).

In addition to the TCR, MHC-class I molecules can interact with a variety of receptors present on human T cells and collectively designated as NK receptors. These receptors include members of the killer Ig-like receptor (KIR) family, both stimulatory and inhibitory, members of the CD94–NKG2 family, and NKG2D (Moretta et al. 2001; Vivier and Anfossi 2004). While the ligands for the inhibitory KIR members include classical MHC-class I molecules as well as non-classical MHC-class I such as HLA-G, the identity of the ligands for the stimulatory KIR remains uncertain (Wagtmann et al. 1995; Dohring and Colonna 1996; Moretta et al. 2001). Members of the CD94–NKG2 family transmit negative signals, resulting in inhibition of target cell lysis upon recognizing the non-classical MHC-class I molecule HLA-E (Baker et al. 1998; Borrego et al. 2002). Finally, the ligands for NKG2D are the unconventional MHC-class I molecules MICA, MICB and ULBP (Bauer et al. 1999; Cosman et al. 2001; Raulet 2003). Importantly, expression of NK receptors is restricted, with the exception of NKG2D, to highly differentiated CD8+ T cells that have lost expression of CD28 (Anumanthan et al. 1998; Baars et al. 2000; Arosa 2002; Snyder et al. 2004). Regarding signal transduction, the various inhibitory receptors share a common regulatory sequence of amino acids in their cytoplasmic tails, named the immunoreceptor tyrosine-based inhibitory motifs (ITIM), which, upon phosphorylation, confer the capacity to recruit SH2-containing tyrosine phosphatase 1 and inhibit cell-mediated cytotoxicity (Baker et al. 1998; Moretta et al. 2001; Vivier and Anfossi 2004).

The immunological role of MHC-class I molecules in presenting peptides to CD8+ T cells through their TCR and interacting with a variety of NK receptors, both through *trans*-associations, is presently undeniable (Yewdell et al. 2003; Snyder et al. 2004). In the 1970s, however, two Danish immunologists put forward a hypothesis whereby MHC-class I molecules were involved in the regulation of ligand–receptor interactions not directly implicated in immune reactions through *cis* associations with those receptors (Svejgaard and Ryder 1976). This hypothesis was further refined and extended in the next decade (Simonsen and Olsson 1983; Simonsen et al. 1985), and from time to time there have been reports on functions of MHC-class I molecules that go beyond immunology (reviewed by Stagsted 1998). The remainder of this section will focus on unconventional aspects of

MHC-class I molecules, with emphasis on their impact on human T cell growth and survival.

1.4.4 Cell-Surface MHC-Class I Molecules: *cis*-Interactions with Growth Receptors

MHC-class I molecules have been associated with the function of many different cell-surface receptors responsible for growth and differentiation. Examples are receptors for glucagon, the epidermal growth factor, epinephrine, thromboxane A and endorphin, among others. In addition, a number of reports have suggested a direct structural association between MHC-class I molecules and hormone receptors, including receptors for luteinizing and gonadotropic hormones and insulin (Cremaschi and Sterin-Borda 1994). Interestingly, the major histocompatibility complex is the only major susceptibility locus that has been identified in association with diabetes mellitus and in animal models for this disease.

In 1986, Olsson and coworkers defined the MHC-class I heavy chain as a structural subunit of the human insulin receptor (IR). They showed that monoclonal antibodies against some of the monomorphic determinants of MHC-class I molecules reduced insulin binding and immunoprecipitated insulin receptors, and suggested that the association between MHC-class I molecules and the IR occurred as a result of displacement of β_2m by the IR (Due et al. 1986). Subsequent studies in humans and mice have contributed to defining the molecular basis for the association between MHC-class I molecules and the IR, as well as characterizing the biological impact of those interactions on IR-mediated intracellular signaling events (Samson et al. 1986; Cousin et al. 1987; Edidin and Reiland 1990). Elegant studies by Edidin and coworkers demonstrated that physical association between MHC-class I molecules and IR and its impact on IR signaling cascades, such as kinase activity, is a function of the HLA-class I allele and the ratio between HLA-class I molecules and insulin receptors (Kittur et al. 1987; Reiland and Edidin 1993; Ramalingam et al. 1997). Notably, as postulated by Olsson and coworkers (Due et al. 1986), cell-surface β_2m-free MHC-class I heavy chains are required for functional association with the IR (Ramalingam et al. 1997). However, whether the dissociation of β_2m unmasks an IR binding site in the MHC-class I heavy chain or induces a conformational change that allows IR binding remains uncertain.

Despite the body of evidence implicating MHC-class I molecules in regulatory roles beyond immunology, the novelty of the findings and some discrepancies between studies led some authors to question these discoveries (Liegler et al. 1990, 1991). To overcome that problem and address this topic, Olsson and coworkers used a novel approach. In a series of pioneering studies, they showed that peptides derived from the α1 domain of MHC-class I molecules regulated IR functions, namely, inhibition of the tyrosine kinase activity of the IR, resulting in inhibition of IR internalization, augmented steady-state levels of IR signaling and finally enhanced glucose uptake (Hansen et al. 1989; Stagsted et al. 1990, 1993; Stagsted 1998). Importantly, it was shown that the bioactive peptides attached to

MHC-class I molecules on the cell surface, apparently to the $\alpha 1$ domain (Olsson et al. 1994). This innovative work shed light on the influence of MHC-class I molecules in receptor internalization and function. Considering that the interaction between MHC-class I molecules and the insulin receptor was shown to depend on $\beta 2$ m-free class I molecules, these studies strengthen the possible importance of misfolded MHC-class I molecules and their biochemical modifications upon association with the IR, or receptors for other growth factors such as the IL-2 receptor, on the regulation of T cell function (see below).

1.4.5 Cell-Surface MHC-Class I Molecules in T Cells: Are Misfolding and *cis*-Association Related Events?

MHC-class I molecules physically associate in *cis* on the cell surface of human T cells with receptors whose engagement influences cell activation and proliferation, including receptors for cytokines such as IL-2 (Pedersen et al. 1999). Work performed in the 1980s is crucial to envision the importance of these *cis* associations. In 1983, Monos and Cooper reported that MHC-class I proteins are the most prominently synthesized and rapidly turned over of all the cell proteins in quiescent human T lymphocytes, changing that pattern when the T cell is activated to a lower increase in synthesis. Subsequent studies showed that cell-surface MHC-class I molecules undergo spontaneous internalization via coated vesicles in resting human and mice T cells (Tse and Pernis 1984; Machy et al. 1987). Moreover, using monocytic cells, Dasgupta and coworkers (1988) showed that the endocytosed molecules are recycled back to the cell surface via trans-Golgi cisternae. These data revealed the capacity of MHC-class I molecules to behave like receptor molecules and traffic through endolysosomal compartments, a capacity that has been associated with their intrinsic ability to become misfolded and associate with receptors that may regulate T cell signaling, growth and survival, depending on the stage of the cell cycle and the differentiation status of the T cell.

Indeed, a remarkable but overlooked feature of cell-surface MHC-class I molecules is that they may dissociate from the peptide and the β_2m, becoming misfolded MHC-class I heavy chains that are detected by antibodies recognizing specific peptide sequences within the polymorphic region of the $\alpha 1$ domain (Madrigal et al. 1991; Perosa et al. 2003). The activation and/or differentiation state of the lymphocyte was found to be the most important factor determining the MHC-class I/peptide dissociation and the generation of misfolded MHC-class I molecules at the cell surface of normal human T cells (Schnabl et al. 1990; DeMaria et al. 1992; Arosa et al. 1999). On the other hand, two separate studies demonstrated that dissociation of the MHC-class I heavy chain from the peptide and β_2m involves conformational changes and partial dimerization of the heavy chains, as well as recycling of MHC-class I molecules through endosomal compartments (Little et al. 1995; Pickl et al. 1996). Finally, studies in mice have suggested that dissociation of MHC-class I heavy chains and β_2m and formation of MHC-class I dimers are related events that occur as a result of loss or unavailability of β_2m (Capps et al. 1993).

The self-association of MHC-class I molecules at the cell surface of a variety of transformed cell lines, including lymphoblastoid cells, has been shown by a number of groups and is likely the result of the activation state of these cells (Matko et al. 1994; Damjanovich et al. 1995 Jenei et al. 1997; Triantafilou et al. 1999). Indeed, misfolded MHC-class I molecules as well as MHC-class I dimers are seldomly observed in resting human T cells collected from peripheral blood, their expression increases sharply upon activation, and the ratio between folded and misfolded MHC-class I molecules changes according to the activation status of the T cell, with misfolded molecules increasing during active cell division and decreasing with a decline in proliferation (Arosa et al. 1999; Santos et al. 2004). Moreover, misfolded MHC-class I molecules present at the cell surface of activated normal human T cells are associated with chaperones such as calreticulin and ERp57, suggesting that cell-surface MHC-class I molecules may also be subjected to protein quality control (Arosa et al. 1999; Santos et al. 2004).

The biological importance of the misfolded MHC-class I molecules and the MHC-class I clusters present at the cell surface has several interpretations. Capps and coworkers, as well as other groups, regarded MHC-class I misfolding and clustering as a mechanism to remove immunologically dysfunctional molecules (Capps et al. 1993; Pickl et al. 1993; DeMaria et al. 1994). In contrast, Damjanovich and collaborators interpret the MHC-class I clustering as a mechanism to strengthen *trans*-interactions between professional APC and T cells (Jenei et al. 1997; Bodnar et al. 2003). Finally, Schnabl and colleagues (1990) envisioned MHC-class I misfolding as a mechanism whereby loss of β_2m and peptide allows the MHC-class I heavy chain to associate in *cis* with a new structure . This *cis*-association may be functionally important to the activated T cells depending on the nature of the new molecules with which the "free" heavy chains can associate (Schnabl et al. 1990; Arosa et al. 1999; Santos et al. 2004). Studies performed during the last two decades on the impact that cell-surface MHC-class I ligation has on parameters of T cell activation, proliferation and survival strongly support this view.

1.4.6 MHC-Class I *cis*-Associations and the Regulation of T Cell Growth and Survival

Early reports using soluble anti-MHC-class I antibodies showed inhibition of T cell proliferation induced by T cell mitogens (Taylor et al. 1986; De Felice et al. 1987; Dasgupta et al. 1988; Smith et al. 1994). In contrast, cross-linked anti-MHC-class I antibodies induced proliferation of stimulated T cells (Geppert et al. 1988; Gilliland et al. 1989; Wacholtz et al. 1989). It was soon demonstrated that T cell-surface MHC-class I molecules were involved in the regulation of biological processes in T lymphocytes, from calcium fluxes, tyrosine phosphorylation and cytokine secretion to cell division and death (Houlden et al. 1991; Skov et al. 1995; Genestier et al. 1997, 1998). More recent studies have shown that engagement of T cell-surface MHC-I molecules has different outputs depending on the domain engaged and the activation status of the cells. Thus, whereas cross-linking of anti-β_2m antibodies induces T cell apoptosis of activated human T cells, a process paralleled by the activation of intracellular kinases such as Lck, ZAP-70, PI-3 and JNK, cross-linking of anti-MHC-class I heavy chain antibodies induces proliferation of resting normal human T cells (Bregenholt et al. 1996; Skov et al. 1997). These data reinforce the role played by T cell-surface MHC-I molecules in the regulation of transmembrane signaling on human T cells, and indicate that MHC-class I ligation has opposite effects in activated and non-activated T cells (Pedersen et al. 1999).

Because some of the signaling events modulated by MHC-class I ligation, such as calcium fluxes and IL-2 secretion, do not require the cytoplasmic domain of the molecule, the regulation is likely achieved by *cis*-associations with other cell-surface molecules (Gur et al. 1990). Indeed, a number of reports have shown that MHC-class I molecules are physically and functionally associated on the surface of normal and transformed human T cells with molecules such as CD3, CD8, CD45 and the IL-2R, all directly or indirectly associated with a signaling machinery (Bushkin et al. 1986, 1988; Blue et al. 1988; Sharon et al. 1988; Pedersen et al. 1999). More recent studies suggest that misfolded MHC-class I molecules preferentially associate with the CD8β chain and also with intracellular Src kinases such as Lck (Santos et al. 2004). The MHC-class I *cis* associations with signaling molecules may reflect timely regulation of the function mediated by those receptors as demonstrated for growth receptors (Due et al. 1986; Olsson et al. 1994, Ramalingam et al. 1997). In this context, the negative modulatory effects observed on activated human T cells when cell-surface MHC-class I molecules are ligated could be a consequence of the interference with ongoing biological processes in the activated T cells regulated by MHC-class I molecules. The recent demonstration that MHC-class I molecules associate in *cis* with the natural killer receptor Ly49A and restrict natural killer cell inhibition is a further illustration of the pleiotropic nature of MHC-class I molecules (Doucey et al. 2004), and has renewed interest is this overlooked characteristic of MHC-class I molecules.

1.5 Concluding Remarks and Future Prospects

The preferential proliferation and differentiation of CD8+ T cells induced by erythrocytes, epithelial cells and IL-15 argue in favor of the existence of a specific receptiveness of CD8+ T cells to signals from the internal environment delivered by non-professional APC, such as epithelial cells from the gut and liver, as opposed to CD4+ T cells which are more receptive to signals from professional APC, such as dendritic cells and macrophages. The privileged communication that takes place between CD8+ T cells and their environment is a unique feature that may help to shed light on the phenotypic and functional changes that occur within the human T cell pool in a variety of situations (Arosa 2002).

Regarding RBC, we suggest that the modulation of CD8+ T cell growth and survival induced by RBC results from secreted vesicles. After binding to the plasma membrane of the activated T cells, these RBC vesicles initiate and/or intersect with intracellular signals that will ultimately modulate fundamental processes such as the entrance into the cell cycle, resistance to oxidative stress and apoptosis, and cell survival. Yet, considering that RBC deliver oxygen, and that hypoxia can affect T cell function and fate (Naldini and Carraro 1999; Conforti et al. 2003), the possibility that RBC contribute to the increase in T cell growth and survival because of the more efficient oxygen transport into T cells deserves consideration. In any case, identification of the RBC molecules responsible may provide a valuable tool to develop therapeutic protocols aimed at reversing pathological situations linked to excessive or faulty T cell growth (Fig. 1.2).

Concerning the impact that epithelial cells and IL-15 have on CD8+ T cells, it is anticipated that this interaction will have reciprocal effects, i. e., the activated CD8+ T cells would be involved in the regulation of physiological processes of the neighboring epithelial cells. The likelihood of the activation of CD8+ T cells by IL-15 presented by liver epithelial cells, though not proven yet, has to be considered. In this scenario, molecules with the MHC-fold will likely play a critical and multifunctional role in the relationship between epithelial tissues and T cells. Depending on their folding status, cell-surface MHC-class I molecules may intervene in a variety of processes that will impact both the T cell and the neighboring cell (Fig. 1.4).

The study and characterization of novel functions for this family of molecules (e. g., regulation of receptor internalization) will contribute to a better understanding of the biology of human T lymphocytes.

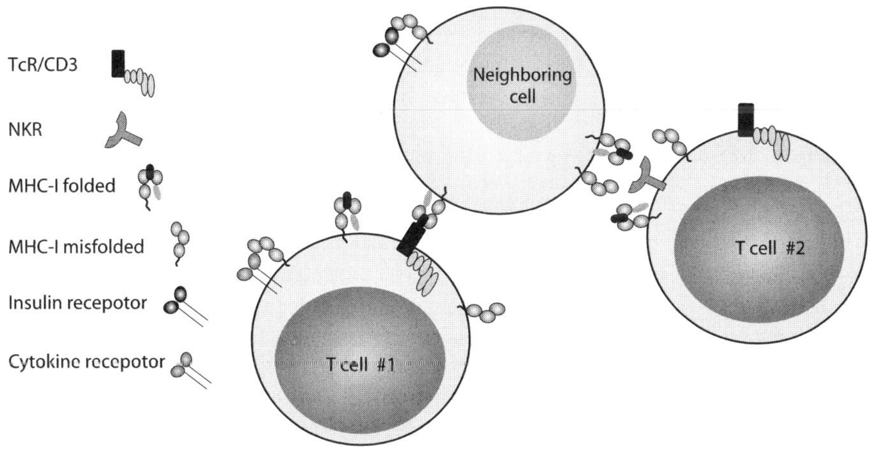

Fig. 1.4. Schematic model illustrating the role of folded and misfolded MHC-class I molecules on human T cell function. While folded MHC-class I molecules present on neighboring cells may interact in *trans* with the TCR/CD3 complex (*T Cell #1*) and with NK receptors (*T Cell #2*), misfolded MHC-class I molecules present in either neighboring cells or T cells (T cells #1 and #2) may interact in *cis* with receptors for growth factors (e.g., insulin) and cytokines (e.g., IL-15). While the first type of interaction may result in T cell activation and proliferation, the second type may result in survival signals that facilitate expansion and differentiation. Ligation of cell-surface MHC-class I molecules in T cells may affect the balance between folded and misfolded MHC-class I molecules as well as *cis* interactions with signaling and/or growth receptors, thus interfering with the process of growth and survival

Acknowledgements. This work was supported by grants from FEDER and the Fundação para a Ciência e a Tecnologia (POCTI/38264/MGI/2001), from the Gulbenkian Foundation (Prémio Estímulo à Investigação 2001) and from the American Portuguese Biomedical Research Fund.

References

Allan D, Thomas P (1981) Ca^{2+}-induced biochemical changes in human erythrocytes and their relation to microvesiculation. Biochem J 198:433–440

Allan D, Thomas P, Limbrick AR (1980) The isolation and characterization of 60 nm vesicles ('nanovesicles') produced during ionophore A23187-induced budding of human erythrocytes. Biochem J 188:881–887

Allez M, Brimnes J, Dotan I, Mayer L (2002) Expansion of CD8+ T cells with regulatory function after interaction with intestinal epithelial cells. Gastroenterology 123:1516–1526

Alves NL, Hooibrink B, Arosa FA, van Lier RA (2003) IL-15 induces antigen-independent expansion and differentiation of human naive CD8+ T cells in vitro. Blood 102:2541–2546

Andrews DA, Yang L, Low PS (2002) Phorbol ester stimulates a protein kinase C-mediated agatoxin-TK-sensitive calcium permeability pathway in human red blood cells. Blood 100:3392–3399

Antoniou AN, Powis SJ, Elliott T (2003) Assembly and export of MHC class I peptide ligands. Curr Opin Immunol 15:75–81

Anumanthan A, Bensussan A, Boumsell L, Christ AD, Blumberg RS, Voss SD, Patel AT, Robertson MJ, Nadler LM, Freeman GJ (1998) Cloning of BY55, a novel Ig superfamily member expressed on NK cells, CTL, and intestinal intraepithelial lymphocytes. J Immunol 161: 2780–2790

Aoshiba K, Nakajima Y, Yasui S, Tamaoki J, Nagai A (1999) Red blood cells inhibit apoptosis of human neutrophils. Blood 93:4006–4010

Arosa FA (2002) CD8+CD28– T cells: certainties and uncertainties of a prevalent human T-cell subset. Immunol Cell Biol 80:1–13

Arosa FA, Irwin C, Mayer L, de Sousa M, Posnett DN (1998) Interactions between peripheral blood CD8 T lymphocytes and intestinal epithelial cells (iEC). Clin Exp Immunol 112: 226–236

Arosa FA, Jesus O, Porto G, Carmo AM, de Sousa M (1999) Calreticulin is expressed on the cell surface of activated human peripheral blood T lymphocytes in association with major histocompatibility complex class I molecules. J Biol Chem 274:16917–16922

Arosa FA, Pereira CF, Fonseca AM (2004) Red blood cells as modulators of T cell growth and survival. Curr Pharm Des 10:191–201

Baars PA, Ribeiro Do Couto LM, Leusen JH, Hooibrink B, Kuijpers TW, Lens SM, van Lier RA (2000) Cytolytic mechanisms and expression of activation-regulating receptors on effector-type CD8+CD45RA+CD27– human T cells. J Immunol 165:1910–1917

Bakker ABH, Phillips JH, Figdor CG, Lanier LL (1998) Killer cell inhibitory receptors for MHC class I molecules regulate lysis of melanoma cells mediated by NK cells, $\gamma\delta$T cells, and antigen-specific CTL. J Immunol 160:5239–5245

Bamford RN, DeFilippis AP, Azimi N, Kurys G, Waldmann TA (1998) The 5′ untranslated region, signal peptide, and the coding sequence of the carboxyl terminus of IL-15 participate in its multifaceted translational control. J Immunol 160:4418–4426

Bauer S, Groh V, Wu J, Steinle A, Phillips JH, Lanier LL, Spies T (1999) Activation of NK cells and T cells by NKG2D, a receptor for stress-inducible MICA. Science 285:727–729

Bensussan A, Rabian C, Schiavon V, Bengoufa D, Leca G, Boumsell L (1993) Significant enlargement of a specific subset of CD3+CD8+ peripheral blood leukocytes mediating cytotoxic T-lymphocyte activity during human immunodeficiency virus infection. *Proc Natl Acad Sci USA 90:9427–9430*

Beppu M, Inoue M, Ishikawa T, Kikugawa K (1994) Presence of membrane-bound proteinases that preferentially degrade oxidatively erythrocyte membrane proteins as secondary antioxidant defense. Biochem Biophys Acta 1196:81–87

Bevers EM, Wiedmer T, Comfurius P, Shattil SJ, Weiss HJ, Zwaal RF, Sims PJ (1992) Defective $Ca^{(2+)}$-induced microvesiculation and deficient expression of procoagulant activity in erythrocytes from a patient with a bleeding disorder: a study of the red blood cells of Scott syndrome. Blood 79:380–388

Bjorkman PJ, Parham P (1990) Structure, function, and diversity of class I major histocompatibility complex molecules. Annu Rev Biochem 59:253–288

Blue ML, Craig KA, Anderson P, Branton KR Jr, Schlossman SF (1988) Evidence for specific association between class I major histocompatibility antigens and the CD8 molecules of human suppressor/cytotoxic cells. Cell 54:413–421

Blumberg RS, Terhorst C, Bleicher P, McDermott FV, Allan CH, Landau SB, Trier JS, Balk SP (1991) Expression of a nonpolymorphic MHC class I-like molecule, CD1D, by human intestinal epithelial cells. J Immunol 147:2518–2524

Bodnar A, Bacso Z, Jenei A, Jovin TM, Edidin M, Damjanovich S, Matko J (2003) Class I HLA oligomerization at the surface of B cells is controlled by exogenous beta(2)-microglobulin: implications in activation of cytotoxic T lymphocytes. Int Immunol 15:331–339

Bonaventura C, Ferruzzi G, Tesh S, Stevens RD (1999) Effects of S-nitrosation on oxygen binding by normal and sickle cell hemoglobin. J Biol Chem 274:24742–24748

Borrego F, Kabat J, Sanni TB, Coligan JE (2002) NK cell CD94/NKG2A inhibitory receptors are internalized and recycle independently of inhibitory signaling processes. J Immunol 169: 6102–6111

Bregenholt S, Ropke M, Skov S, Claesson MH (1996) Ligation of MHC class I molecules on peripheral blood T lymphocytes induces new phenotypes and functions. J Immunol 157:993–999

Briard D, Brouty-Boye D, Azzarone B, Jasmin C (2002) Fibroblasts from human spleen regulate NK cell differentiation from blood CD34+ progenitors via cell surface IL-15. J Immunol 168:4326–4332

Brugnara C, de Franceschi L, Alper SL (1993) Inhibition of $Ca^{(2+)}$-dependent K^+ transport and cell dehydration in sickle erythrocytes by clotrimazole and other imidazole derivatives. J Clin Invest 92:520–526

Bumgardner GL, Chen S, Almond PS, Bach FH, Ascher NL, Matas AJ (1992) Cell subsets responding to purified hepatocytes and evidence of indirect recognition of hepatocyte major histocompatibility complex class I antigen. II. In vitro-generated 'memory' cells to class I+ class II– hepatocytes. Transplantation 53:863–868

Bushkin Y, Posnett DN, Pernis B, Wang CY (1986) A new HLA-linked T cell membrane molecule, related to the beta chain of the clonotypic receptor, is associated with T3. J Exp Med 164:458–473

Bushkin Y, Demaria S, Le JM, Schwab R (1988) Physical association between the CD8 and HLA class I molecules on the surface of activated human T lymphocytes. Proc Natl Acad Sci USA 85:3985–3989

Canchis PW, Bhan AK, Landau SB, Yang L, Balk SP, Blumberg RS (1993) Tissue distribution of the non-polymorphic major histocompatibility complex class I-like molecule, CD1d. Immunology 80:561–565

Cantrell DA (2002) T-cell antigen receptor signal transduction. Immunology 105:369–374

Capps GG, Robinson BE, Lewis KD, Zúñiga MC (1993) In vivo dimeric association of class I MHC heavy chains. J Immunol 151:159–169

Cardoso CS, de Sousa M (2003) HFE, the MHC and hemochromatosis: paradigm for an extended function for MHC class I. Tissue Antigens 61:263–275

Carson WE, Ross ME, Baiocchi RA, Marien MJ, Boiani N, Grabstein K, Caligiuri MA (1995) Endogenous production of interleukin 15 by activated human monocytes is critical for optimal production of interferon-gamma by natural killer cells in vitro. J Clin Invest 96:2578–2582

Cerf-Bensussan N, Guy-Grand D (1991) Intestinal intraepithelial lymphocytes. *Gastroenterol Clin N Am 20:549–576*

Choi J, Enis DR, Koh KP, Shiao SL, Pober JS (2004) T lymphocyte–endothelial cell interactions. Annu Rev Immunol 22:683–709

Chott A, Gerdes D, Spooner A, Mosberger I, Kummer JA, Ebert EC, Blumberg RS, Balk SP (1997) Intraepithelial lymphocytes in normal human intestine do not express proteins associated with cytolytic function. Am J Pathol 151:435–442

Claas FHJ, Roelen DL, van Rood JJ, Brand A (2001) Modulation of the alloimmune response by blood transfusions. Transfus Clin Biol 8:315–317

Cohen JH, Caudwell V, Levi-Strauss M, Bourgeois P, Kazatchkine MD (1989) Genetic analysis of CR1 expression on erythrocytes of patients with systemic lupus erythematosus. Arthritis Rheum 32:393–397

Coleman JW (2001) Nitric oxide in immunity and inflammation. Immunopharmacology 1:1397–1406

Conforti L, Petrovic M, Mohammad D, Ma Q, Barone S, Filipovich AH (2003) Hypoxia regulates expression and activity of Kv1.3 channels in T lymphocytes: a possible role in T cell proliferation. J Immunol 170:695–702

Cookson S, Reen D (2003) IL-15 drives neonatal T cells to acquire CD56 and become activated effector cells. Blood 102:2195–2197

Cosby K, Partovi KS, Crawford JH, Patel RP, Reiter CD, Martyr S, Yang BK, Waclawiw MA, Zalos G, Xu X, Huang KT, Shields H, Kim-Shapiro DB, Schechter AN, Cannon RO, Gladwin MT (2003) Nitrite reduction to nitric oxide by deoxyhemoglobin vasodilates the human circulation. Nat Med 9:1498–1505

Cosman D, Mullberg J, Sutherland CL, Chin W, Armitage R, Fanslow W, Kubin M, Chalupny NJ (2001) ULBPs, novel MHC class I-related molecules, bind to CMV glycoprotein UL16 and stimulate NK cytotoxicity through the NKG2D receptor. Immunity 14:123–133

Cousin JL, Samson M, Pilch PF, Fehlmann M (1987) Internalization of insulin receptors and HLA antigens in human hepatoma cells. Biochem J 242:403–410

Cremaschi GA, Sterin-Borda L (1994) MHC class I-hormone receptor associations: still a physiological enigma? Acta Physiol Pharmacol Ther Lat Am 44:55–64

Cresswell P, Bangia N, Dick T, Diedrich G (1999) The nature of the MHC class I peptide-loading complex. Immunol Rev 172:21–28

Damjanovich S, Vereb G, Schaper A, Jenei A, Matko J, Starink JP, Fox GQ, Arndt-Jovin DJ, Jovin TM (1995) Structural hierarchy in the clustering of HLA class I molecules in the plasma membrane of human lymphoblastoid cells. Proc Natl Acad Sci USA 92:1122–1126

Dasgupta JD, Watkins S, Slayter H, Yunis EJ (1988) Receptor-like nature of class I HLA: endocytosis via coated pits. J Immunol 141:2577–2580

De Felice M, Turco MC, Giarrusso PC, Corbo L, Pizzano R, Martinelli V, Ferrone S, Venuta S (1987) Differential regulatory role of monomorphic and polymorphic determinants of histocompatibility leukocyte antigen class I antigens in monoclonal antibody OKT3-induced T cell proliferation. J Immunol 139:2683–2689

De Jong K, Geldwerth D, Kuypers FA (1997) Oxidative damage does not alter membrane phospholipid asymmetry in human erythrocytes. Biochemistry 36:+6768–6776

DeMaria S, Schwab R, Bushkin Y (1992) The origin and fate of β2m-free MHC class I molecules induced on activated T cells. Cell Immunol 142:103–113

DeMaria S, Schwab R, Gottesman SR, Bushkin Y (1994) Soluble beta 2-microglobulin-free class I heavy chains are released from the surface of activated and leukemia cells by a metalloprotease. J Biol Chem 269:6689–6694

Dengler TJ, Pober JS (2000) Human vascular endothelial cells stimulate memory but not naive CD8+ T cells to differentiate into CTL retaining an early activation phenotype. J Immunol 164:5146–5155

Denzer K, Kleijmeer MJ, Heijnen HF, Stoorvogel W, Geuze HJ (2000) Exosome: from internal vesicle of the multivesicular body to intercellular signaling device. J Cell Sci 19:3365–3374

Dervillez X, Oudin S, Libyh MT, Tabary T, Reveil B, Philbert F, Bougy F, Pluot M, Cohen JH (1997) Catabolism of the human erythrocyte C3b/C4b receptor (CR1, CD35): vesiculation and/or proteolysis? Immunopharmacology 38:129–140

Dick TP, Bangia N, Peaper DR, Cresswell P (2002) Disulfide bond isomerization and the assembly of MHC class I-peptide complexes. Immunity 16:87–98

Doherty DG, Norris S, Madrigal-Estebas L, McEntee G, Traynor O, Hegarty JE, O'Farrelly C (1999) The human liver contains multiple populations of NK cells, T cells, and CD3+CD56+ natural T cells with distinct cytotoxic activities and Th1, Th2, and Th0 cytokine secretion patterns. J Immunol 163:2314–2321

Dohring C, Colonna M (1996) Human natural killer cell inhibitory receptors bind to HLA class I molecules. Eur J Immunol 26:365–369

Doucey MA, Scarpellino L, Zimmer J, Guillaume P, Luescher IF, Bron C, Held W (2004) *Cis* association of Ly49A with MHC class I restricts natural killer cell inhibition. Nat Immunol 5:328–336

Dubois S, Mariner J, Waldmann TA, Tagaya Y (2002) IL-15Ralpha recycles and presents IL-15 in *trans* to neighboring cells. Immunity 17:537–547

Due C, Simonsen M, Olsson L (1986) The major histocompatibility complex class I heavy chain as a structural subunit of the human cell membrane insulin receptor: implications for the range of biological functions of histocompatibility antigens. Proc Natl Acad Sci USA 83:6007–6011

Dumaswala UJ, Greenwalt TJ (1984) Human erythrocytes shed exocytic vesicles in vivo. Transfusion 24:490–492

Dumaswala UJ, Zhuo L, Jacobsen DW, Jain SK, Sukalski KA (1999) Protein and lipid oxidation of banked human erythrocytes: role of glutathione. Free Rad Biol Med 27:1041–1049

Dustin ML, Sanders ME, Shaw S, Springer TA (1987) Purified lymphocyte function-associated antigen 3 binds to CD2 and mediates T-lymphocyte adhesion. J Exp Med 165:677–692

Eaton J (1991) Catalases and peroxidases and glutathione and hydrogen peroxide: mysteries of bestiary. J Lab Clin Med 118:3–4

Ebert EC (1985) Sheep red blood cells enhance T-lymphocyte proliferation. Clin Immunol Immunopathol 37:203–212

Edidin M, Reiland J (1990) Dynamic measurements of the associations between class I MHC antigens and insulin receptors. Mol Immunol 27:1313–1317

Fehniger TA, Caligiuri MA (2001) Interleukin 15: biology and relevance to human disease. Blood 97:14–32

Ferrali M, Signorini C, Ciccoli L, Comporti M (1992) Iron release and membrane damage in erythrocytes exposed to oxidizing agents, phenilhydrazine, divicine and isouramil. Biochem J 285:293–301

Fonseca AM, Porto G, Uchida K, Arosa FA (2001) Red blood cells inhibit activation induced cell death and oxidative stress in human peripheral blood T lymphocytes. Blood 97:3152–3160

Fonseca AM, Porto G, Arosa FA (2003a) Red blood cells promote survival and cell cycle progression of dividing human peripheral blood T cells: role of CD58/LFA-3 and heme compounds. Cell Immunol 224:17–28

Fonseca AM, Porto G, Arosa FA (2003b) Red blood cells upregulate cytoprotective proteins and the labile iron pool in dividing human T cells despite a reduction in oxidative stress. Free Radic Biol Med 35:1404–1416

Freitas AA, Rocha BB (1993) Lymphocyte lifespans: homeostasis, selection and competition. Immunol Today 14:25–29

Froland SS (1972) Binding of sheep erythrocytes to human lymphocytes. A probable marker of T lymphocytes. Scand J Immunol 1:269–280

Fujii J, Ikeda Y (2002) Advances in our understanding of peroxiredoxin, a multifunctional, mammalian redox protein. Redox Rep 7:123–130

Fujino T, Watanabe K, Beppu M, Kikugawa K, Yasuda H (2000) Identification of oxidized protein hydrolase of human erythrocytes as acylpeptide hydrolase. Biochim Biophys Acta 1478:102–112

Gabay T, Ginsburg H (1993) Hemoglobin denaturation and Fe release in acidified red blood cell lysate – a possible source of Fe for intraerythrocytic malaria parasites. Exp Parasitol 77:261–272

Gafter U, Kalechman Y, Sdreni B (1992) Induction of a subpopulation of suppressor cells by a single blood transfusion. Kidney Int 41:143–148

Galat A (2003) Peptidyl *cis/trans* isomerases (immunophilins): biological diversity–targets–functions. Curr Top Med Chem 3:1315–1347

Geiben-Lynn R, Kursar M, Brown NV, Addo MM, Shau H, Lieberman J, Luster AD, Walker BD (2003) HIV-1 antiviral activity of recombinant natural killer cell enhancing factors, NKEF-A and NKEF-B, members of the peroxiredoxin family. J Biol Chem 278:1569–1574

Genestier L, Paillot R, Bonnefoy-Berard N, Waldmann H, Revillard JP (1997) T cell sensitivity to HLA class I-mediated apoptosis is dependent on interleukin-2 and interleukin-4. Eur J Immunol 27:495–499

Genestier L, Fournet S, Flacher M, Assossou O, Revillard J, Bonnefoy-Berad N (1998) Induction of Fas (Apo-1, CD95)-mediated apoptosis of activated lymphocytes by polyclonal antithymocyte globulins. Blood 92:2360–2368

Geppert TD, Wacholtz MC, Davis LS, Lipsky PE (1988) Activation of human T4 cells by cross-linking class I MHC molecules. Immunology 140:2155–2164

Gilliland LK, Norris NA, Grosmaire LS, Ferrone S, Gladstone P, Ledbetter JA (1989) Signal transduction in lymphocyte activation through crosslinking of HLA class I molecules. Hum Immunol 25:269–289

Ginn-Pease ME, Whisler RL (1998) Redox signals and NF-kappaB activation in T cells. Free Radic Biol Med 25:346–361

Golden-Mason L, Kelly AM, Doherty DG, Traynor O, McEntee G, Kelly J, Hegarty JE, O'Farrelly C (2004) Hepatic interleukin 15 (IL-15) expression: implications for local NK/NKT cell homeostasis and development. Clin Exp Immunol 138:94–101

Gow AJ, Stamler JS (1998) Reactions between nitric oxide and haemoglobin under physiological conditions. Nature 391:169–173

Grabstein KH, Eisenman J, Shanebeck K, Rauch C, Srinivasan S, Fung V, Beers C, Richardson J, Schoenborn MA, Ahdieh M (1994) Cloning of a T cell growth factor that interacts with the beta chain of the interleukin-2 receptor. Science 264:965–968

Gros M, Vrhovec S, Brumen M, Svetina S, Zeks B (1996) Low pH induced shape changes and vesiculation of human erythrocytes. Gen Physiol Biophys 15:145–163

Gruen JR, Weissman SM (1997) Evolving views of the major histocompatibility complex. Blood 90:4252–4265

Gur H, el-Zaatari F, Geppert TD, Wacholtz MC, Taurog JD, Lipsky PE (1990) Analysis of T cell signaling by class I MHC molecules: the cytoplasmic domain is not required for signal transduction. J Exp Med 172:1267–1270

Halliwell B, Gutteridge JMC (1999) Free radicals in biology and medicine. Oxford University Press, New York

Hamann D, Baars PA, Rep MH, Hooibrink B, Kerkhof-Garde SR, Klein MR, van Lier RA (1997) Phenotypic and functional separation of memory and effector human CD8+ T cells. J Exp Med 186:1407–1418

Hansen T, Stagsted J, Pedersen L, Roth RA, Goldstein A, Olsson L (1989) Inhibition of insulin receptor phosphorylation by peptides derived from major histocompatibility complex class I antigens. Proc Natl Acad Sci USA 86:3123–3126

Hershberg RM, Mayer LF (2000) Antigen processing and presentation by intestinal epithelial cells – polarity and complexity. Immunol Today 21:123–128

Hobbs AJ, Gladwin MT, Patel RP, Williams DLH, Butler AR (2002) Haemoglobin: NO transporter, NO inactivator or none of the above? Trends Pharmacol Sci 23:406–411

Hofmann B, Hecht HJ, Flohe L (2002) Peroxiredoxins. Biol Chem 383:347–364

Houlden BA, Widacki SM, Bluestone JA (1991) Signal transduction through class I MHC by a monoclonal antibody that detects multiple murine and human class I molecules. J Immunol 146:425–430

Howard MC, Spack EG, Choudhury K, Greten TF, Schneck JP (1999) MHC-based diagnosis and therapeutics – clinical applications for disease-linked genes. Immunol Today 20:161–165

Hue S, Mention JJ, Monteiro RC, Zhang S, Cellier C, Schmitz J, Verkarre V, Fodil N, Bahram S, Cerf-Bensussan N, Caillat-Zucman S (2004) A direct role for NKG2D/MICA interaction in villous atrophy during celiac disease. Immunity 21:367–377

Hughes EA, Hammond C, Cresswell P (1997) Misfolded major histocompatibility complex class I heavy chains are translocated into the cytoplasm and degraded by the proteasome. Proc Natl Acad Sci USA 94:1896–1901

Ishihara S, Nieda M, Kitayama J, Osada T, Yabe T, Ishikawa Y, Nagawa H, Muto T, Juji T (1999) CD8+NKR-P1A+ T cells preferentially accumulate in human liver. Eur J Immunol 29: 2406–2413

Jenei A, Varga S, Bene L, Matyus L, Bodnar A, Bacso Z, Pieri C, Gaspar R Jr, Farkas T, Damjanovich S (1997) HLA class I and II antigens are partially co-clustered in the plasma membrane of human lymphoblastoid cells. Proc Natl Acad Sci USA 94:7269–7274

Jia L, Bonaventura C, Bonaventura J, Stamler JS (1996) S-nitrosohaemoglobin: a dynamic activity of blood involved in vascular control. Nature 380:221–226

Kalb TH, Yio XY, Mayer L (1997) Human airway epithelial cells stimulate T-lymphocyte Lck and Fyn tyrosine kinase. Am J Respir Cell Mol Biol 17:561–570

Kalechman Y, Herman S, Gafter U, Sredni B (1993) Enhancing effects of autologous erythrocytes on human or mouse cytokine secretion and IL-2R expression. Cell Immunol 148:114–129

Kanegane H, Tosato G (1996) Activation of naive and memory T cells by interleukin-15. Blood 88:230–235

Kenna T, Golden-Mason L, Porcelli SA, Koezuka Y, Hegarty JE, O'Farrelly C, Doherty DG, Mason LG (2003) NKT cells from normal and tumor-bearing human livers are phenotypically and functionally distinct from murine NKT cells. J Immunol 171:1775–1779

Kennedy MK, Glaccum M, Brown SN, Butz EA, Viney JL, Embers M, Matsuki N, Charrier K, Sedger L, Willis CR, Brasel K, Morrissey PJ, Stocking K, Schuh JC, Joyce S, Peschon JJ (2000) Reversible defects in natural killer and memory CD8 T cell lineages in interleukin 15-deficient mice. J Exp Med 191:771–780

Kittur D, Shimizu Y, DeMars R, Edidin M (1987) Insulin binding to human B lymphoblasts is a function of HLA haplotype. Proc Natl Acad Sci USA 84:1351–1355

Klinken SP (2002) Red blood cells. Int J Biochem Cell Biol 34:1513–1518

Kmiec Z (2001) Cooperation of liver cells in health and disease. Adv Anat Embryol Cell Biol 161:1–151

Krammer PH (2000) CD95's deadly mission in the immune system. Nature 407:789–795

Krensky AM, Sanchez-Madrid F, Robbins E, Nagy TA, Springer T, Burakoff SJ (1983) The functional significance, distribution, and structure of LFA-1, LFA-2, and LFA-3: cell surface antigens associated with CTL–target interactions. J Immunol 131:611–616

Lanzavecchia A, Sallusto F (2002) Progressive differentiation and selection of the fittest in the immune response. Nat Rev Immunol 2:982–987

Lee TH, Kim SU, Yu SL, Kim SH, Park do S, Moon HB, Dho SH, Kwon KS, Kwon HJ, Han YH, Jeong S, Kang SW, Shin HS, Lee KK, Rhee SG, Yu DY (2003) Peroxiredoxin II is essential for sustaining life span of erythrocytes in mice. Blood 101:5033–5038

Li Y, Yio XY, Mayer L (1995) Human intestinal epithelial cell-induced CD8+ T cell activation is mediated through CD8 and the activation of CD8-associated p56lck. J Exp Med 182:1079–1088

Liegler T, Alexander J, Cresswell P, Goldfine I, Goodenow RS (1990) An analysis of insulin receptor expression and binding on MHC class I positive and negative human lymphoblastoid cells. J Immunol 145:1788–1793

Liegler T, Szollosi J, Hyun W, Goodenow RS (1991) Proximity measurements between H-2 antigens and the insulin receptor by fluorescence energy transfer: evidence that a close association does not influence insulin binding. Proc Natl Acad Sci USA 88:6755–6759

Little AM, Nöâner E, Parham P (1995) Dissociation of β2-microglobulin from HLA class I heavy chains correlates with acquisition of epitopes in the cytoplasmic tail. J Immunol 154:5205–5215

Liu K, Catalfamo M, Li Y, Henkart PA, Weng NP (2002) IL-15 mimics T cell receptor crosslinking in the induction of cellular proliferation, gene expression, and cytotoxicity in CD8+ memory T cells. Proc Natl Acad Sci USA 99:6192–6197

Liu Z, Geboes K, Colpaert S, D'Haens GR, Rutgeerts P, Ceuppens JL (2000) IL-15 is highly expressed in inflammatory bowel disease and regulates local T cell-dependent cytokine production. J Immunol 164:3608–3615

Lodolce JP, Boone DL, Chai S, Swain RE, Dassopoulos T, Trettin S, Ma A (1998) IL-15 receptor maintains lymphoid homeostasis by supporting lymphocyte homing and proliferation. Immunity 9:669–676

Lu KP, Hanes DS, Hunter T (1996) A human peptidyl-propyl isomerase essential for regulation of mitosis. Nature 380:544–547

Lundquist C, Baranov V, Hammarstrom S, Athlin L, Hammarstrom ML (1995) Evidence for regional specialization and extrathymic T cell maturation in the human gut epithelium. Int Immunol 7:1473–1478

Machy P, Truneh A, Gennaro D, Hoffstein S (1987) Major histocompatibility complex class I molecules internalized via coated pits in T lymphocytes. Nature 328:724–726

Mackay IR (2002) Hepatoimmunology: a perspective. Immunol Cell Biol 80:36–44

Madrigal JA, Belich MP, Benjamin RJ, Little AM, Hildebrand WH, Mann DL, Parham P (1991) Molecular definition of a polymorphic antigen (LA45) of free HLA-A and -B heavy chains found on the surfaces of activated B and T cells. J Exp Med 174:1085–1095

Maiza H, Leca G, Mansur IG, Schiavon V, Boumsell L, Bensussan A (1993) A novel 80-kD cell surface structure identifies human circulating lymphocytes with natural killer activity. *J Exp Med 178:1121–1126*

Makgoba MW, Shaw SW, Gugel EA, Sanders ME (1987) Human T cell rosetting is mediated by LFA-3 on autologous erythrocytes. J Immunol 138:3587–3589

Man K, Lo CM, Lee TK, Li XL, Ng IO, Fan ST (2003) Intragraft gene expression profiles by cDNA microarray in small-for-size liver grafts. Liver Transpl 9:425–432

Matko J, Bushkin Y, Wei T, Edidin M (1994) Clustering of class I HLA molecules on the surfaces of activated and transformed human cells. J Immunol 152:3353–3360

Mayer L, Shlien R (1987) Evidence for function of Ia molecules on gut epithelial cells in man. J Exp Med 166:1471–1483

McInnes IB, al Mughales J, Field M, Leung BP, Huang FP, Dixon R, Sturrock RD, Wilkinson PC, Liew FY (1996) The role of interleukin-15 in T-cell migration and activation in rheumatoid arthritis. Nat Med 2:175–182

McMahon TJ, Moon RE, Luschinger BP, Carraway MS, Stone AE, Stolp BW, Gow AJ, Pawloski JR, Watke P, Singel DJ, Piantadosi CA, Stamler JS (2002) Nitric oxide in the human respiratory cycle. Nat Med 8:711–717

Meazza R, Verdiani S, Biassoni R, Coppolecchia M, Gaggero A, Orengo AM, Colombo MP, Azzarone B, Ferrini S (1996) Identification of a novel interleukin-15 (IL-15) transcript isoform generated by alternative splicing in human small cell lung cancer cell lines. Oncogene 12:2187–2192

Megson IL, Sogo N, Mazzei FA, Butler AR, Walton JC, Webb DJ (2000) Inhibition of human platelet aggregation by a novel S nitrosothiol is abolished by haemoglobin and red blood cells in vitro: implications for anti-thrombotic therapy. Br J Pharmacol 131:1391–1398

Melder RJ, Yuan J, Munn LL, Jain RK (2000) Erythrocytes enhance lymphocyte rolling and arrest in vivo. Microvasc Res 59:316–322

Mention JJ, Ben Ahmed M, Begue B, Barbe U, Verkarre V, Asnafi V, Colombel JF, Cugnenc PH, Ruemmele FM, McIntyre E, Brousse N, Cellier C, Cerf-Bensussan N (2003) Interleukin 15: a key to disrupted intraepithelial lymphocyte homeostasis and lymphomagenesis in celiac disease. Gastroenterology 125:730–745

Minetti G, Low PS (1997) Erythrocyte signal transduction pathways and their possible functions. Curr Opin Hematol 4:116–121

Mingari MC, Schiavetti F, Ponte M, Vitale C, Maggi E, Romagnani S, Demarest J, Pantaleo G, Fauci AS, Moretta L (1996) Human CD8+ T lymphocyte subsets that express HLA class I-specific inhibitory receptors represent oligoclonally or monoclonally expanded cell populations. Proc Natl Acad Sci USA 93:12433–12438

Monos DS, Cooper HL (1983) Rapid turnover of HLA proteins in quiescent lymphocytes: proposed connection with immunological surveillance. J Immunol 131:341–346

Moore RB, Mankad MV, Shriver SK, Mankad VN, Plishker GA (1991) Reconstitution of $Ca^{(2+)}$-dependent K^+ transport in erythrocyte membrane vesicles requires a cytoplasmic protein. J Biol Chem 266:18964–18968

Moretta A, Bottino C, Vitale M, Pende D, Cantoni C, Mingari MC, Biassoni R, Moretta L (2001) Activating receptors and coreceptors involved in human natural killer cell-mediated cytolysis. Annu Rev Immunol 19:197–223

Mrozek E, Anderson P, Caligiuri MA (1996) Role of interleukin-15 in the development of human CD56+ natural killer cells from CD34+ hematopoietic progenitor cells. Blood 87:2632–2640

Mueller YM, Bojczuk PM, Halstead ES, Kim AH, Witek J, Altman JD, Katsikis PD (2003) IL-15 enhances survival and function of HIV-specific CD8+ T cells. Blood 101:1024–1029

Nagababu E, Ramasamy S, Abernethy DR, Rifkind J (2003) Active nitric oxide produced in the red cell under hypotoxic conditions by deoxyhemoglobin mediated nitrite reduction. J Biol Chem 278:46349–46356

Naldini A, Carraro F (1999) Hypoxia modulates cyclin and cytokine expression and inhibits peripheral mononuclear cell proliferation. J Cell Physiol 181:448–454

Neumann CA, Krause DS, Carman CV, Das S, Dubey DP, Abraham JL, Bronson RT, Fujiwara Y, Orkin SH, Van Etten RA (2003) Essential role for the peroxiredoxin Prdx1 in erythrocyte antioxidant defence and tumour suppression. Nature 424:561–565

Norris S, Collins C, Doherty DG, Smith F, McEntee G, Traynor O, Nolan N, Hegarty J, O'Farrelly C (1998) Resident human hepatic lymphocytes are phenotypically different from circulating lymphocytes. *J Hepatol 28:84–90*

Olsson L, Goldstein A, Stagsted J (1994) Regulation of receptor internalization by the major histocompatibility complex class I molecule. Proc Natl Acad Sci USA 91:9086–9090

Panja A, Blumberg RS, Balk SP, Mayer L (1993) CD1d is involved in T cell–intestinal epithelial cell interactions. J Exp Med 178:1115–1119

Parodi AJ (2000) Protein glycosylation and its role in protein folding. Annu Rev Biochem 69:69–93

Pascual M, Lutz HU, Steiger G, Stammler P, Schifferli JA (1993) Release of vesicles enriched in complement receptor 1 from human erythrocytes. J Immunol 151:397–404

Pawloski JR, Stamler JS (2002) Nitric oxide in RBCs. Transfusion 42:1603–1609

Pawloski JR, Swaminathan RV, Stamler JS (1998) Cell-free and erythrocytic S-nitrosohemoglobin inhibits human platelet aggregation. Circulation 97:263–267

Pawloski JR, Hess DT, Stamler JS (2001) Export by red blood cells of nitric oxide bioactivity. Nature 409:622–626

Pedersen AE, Skov S, Bregenholt S, Ruhwald M, Claesson MH (1999) Signal transduction by the major histocompatibility complex class I molecule. Acta Pathol Microbiol Immunol Scand (APMIS) 107:887–895

Perosa F, Luccarelli G, Prete M, Favoino E, Ferrone S, Dammacco F (2003) Beta 2-microglobulin-free HLA class I heavy chain epitope mimicry by monoclonal antibody HC-10-specific peptide. J Immunol 171:1918–1926

Pickl WF, Majdic O, Fae I, Reuschel R, Holter W, Knapp W (1993) The soluble pool of beta 2-microglobulin free HLA class I alpha-chains. Qualitative and quantitative characterization. J Immunol 151:2613–2622

Pickl WF, Holter W, Stöckl J, Majdic O, Knapp W (1996) Expression of β2-microglobulin-free HLA class I α-chains on activated T cells requires internalization of HLA class I heterodimers. Immunology 88:104–109

Pitcher LA, Young JA, Mathis MA, Wrage PC, Bartók B, van Oers NSC (2003) The formation and functions of the 21- and 23-KDa tyrosine-phosphorylated TCR ς subunits. Immunol Rev 191:47–61

Pittet MJ, Speiser DE, Valmori D, Cerottini JC, Romero P (2001) Cutting edge: cytolytic effector function in human circulating CD8+ T cells closely correlates with CD56 surface expression. J Immunol 164:1148–1152

Plunkett ML, Sanders ME, Selvaraj P, Dustin ML, Springer TA (1987) Rosetting of activated human T lymphocytes with autologous erythrocytes. Definition of the receptor and ligand molecules as CD2 and lymphocyte function-associated antigen 3 (LFA-3). J Exp Med 165:664–676

Porto B, Fonseca AM, Godinho I, Arosa FA, Porto G (2001) Human red blood cells have an enhancing effect on the relative expansion of CD8+ T lymphocytes in vitro. Cell Prolif 34:359–367

Ramalingam TS, Chakrabarti A, Edidin M (1997) Interaction of class I human leukocyte antigen (HLA-I) molecules with insulin receptors and its effect on the insulin-signaling cascade. Mol Biol Cell 8:2463–2474

Raulet DH (2003) Roles of the NKG2D immunoreceptor and its ligands. Nat Rev Immunol 3:781–790

Reiland J, Edidin M (1993) Chemical cross-linking detects association of insulin receptors with four different class I human leukocyte antigen molecules on cell surfaces. Diabetes 42:619–625

Reinecker HC, MacDermott RP, Mirau S, Dignass A, Podolsky DK (1996) Intestinal epithelial cells both express and respond to interleukin 15. Gastroenterology 111:1706–1713

Roberts AI, Nadler SC, Ebert EC (1997) Mesenchymal cells stimulate human intestinal intraepithelial lymphocytes. Gastroenterology 113:144–150

Roberts AI, Lee L, Schwarz E, Groh V, Spies T, Ebert EC, Jabri B (2001) NKG2D receptors induced by IL-15 costimulate CD28-negative effector CTL in the tissue microenvironment. J Immunol 167:5527–5530

Rudd PM, Elliott T, Cresswell P, Wilson IA, Dwek RA (2001) Glycosylation and the immune system. Science 291:2370–2376

Rugeles MT, La Via M, Goust JM, Kilpatrick JM, Hyman B, Virella G (1987) Autologous red blood cells potentiate antibody synthesis by unfractionated human mononuclear cell cultures. Scand J Immunol 26:119–127

Salzer U, Hinterdorfer P, Hunger U, Borken C, Prohaska R (2002) $Ca^{(++)}$-dependent vesicle release from erythrocytes involves stomatin-specific lipid rafts, synexin (annexin VII), and sorcin. Blood 99:2569–2577

Samson M, Cousin JL, Fehlmann M (1986) Cross-linking of insulin receptors to MHC antigens in human B lymphocytes: evidence for selective molecular interactions. J Immunol 137: 2293–2298

Santos SG, Powis SJ, Arosa FA (2004) Misfolding of MHC-class I molecules in activated T cells allows *cis*-interactions with receptors and signaling molecules and is associated with tyrosine phosphorylation. J Biol Chem 279:53062–53070

Schnabl E, Stockinger H, Majdic O, Gaugitsch H, Lindley IJD, Maurer D, Hajek-Rosenmayr A, Knapp W (1990) Activated human T lymphocytes express MHC class I heavy chains not associated with β2-microglobulin. J Exp Med 171:1431–1442

Schonland SO, Zimmer JK, Lopez-Benitez CM, Widmann T, Ramin KD, Goronzy JJ, Weyand CM (2003) Homeostatic control of T-cell generation in neonates. Blood 102:1428–1434

Schrier SL (1985) Red cell membrane biology – introduction. Clin Haematol 14:1–12
Seddon B, Zamoyska R (2003) Regulation of peripheral T-cell homeostasis by receptor signaling. Curr Opin Immunol 15:321–324
Selvaraj P, Dustin ML (1987) Rosetting of human T lymphocytes with sheep and human erythrocytes. Comparison of human and sheep ligand binding using purified E receptor. J Immunol 138:2690–2695
Sharon M, Gnarra JR, Baniyash M, Leonard WJ (1988) Possible association between IL-2 receptors and class I HLA molecules on T cells. J Immunol 141:3512–3515
Shau H, Golub SH (1988) Modulation of natural killer-mediated lysis by red blood cells. Cell Immunol 116:60–72
Shau H, Kim A (1994) Identification of natural killer enhancing factor as a major antioxidant in human red blood cells. Biochem Biophys Res Commun 199:83–88
Shau H, Gupta RK, Golub SH (1993) Identification of a natural killer enhancing factor (NKEF) from human erythroid cells. Cell Immunol 147:1–11
Silva A, Lopez-Botet M, Alvarez J, de Landazuri MO (1981) Enhancement of the functional activities of human T cells after their interaction with SRBC. J Immunol 126:393–397
Simonsen M, Olsson L (1983) Possible roles of compound membrane receptors in the immune system. Ann Immunol (Paris) 134:85–92
Simonsen M, Skjodt K, Crone M, Sanderson A, Fujita-Yamaguchi Y, Due C, Ronne E, Linnet K, Olsson LL (1985) Compound receptors in the cell membrane: ruminations from the borderland of immunology and physiology. Prog Allergy 36:151–176
Skov S, Odum N, Claësson MH (1995) MHC class I signaling in T cells leads to tyrosine kinase activity and PLC-γ1 phosphorylation. J Immunol 154:1167–1176
Skov S, Bregenholt S, Claësson MH (1997) MHC class I ligation of human T cells activates the ZAP70 and p56lck tyrosine kinases, leads to an alternative phenotype of the TCR/CD3ς-chain, and induces apoptosis. J Immunol 158:3189–3196
Smith DM, Bluestone JA, Jeyarajah DR, Newberg MH, Engelhard VH, Thistlethwaite JR Jr, Woodle ES (1994) Inhibition of T cell activation by a monoclonal antibody reactive against the alpha 3 domain of human MHC class I molecules. J Immunol 153:1054–1067
Snyder MR, Weyand CM, Goronzy JJ (2004) The double life of NK receptors: stimulation or costimulation? Trends Immunol 25:25–32
Solerte SB, Fioravanti M, Cerutti N, Severgnini S, Locatelli M, Pezza N, Rondanelli M, Trecate L, Balza G, Ferrari E (1997) Retrospective analysis of long-term hemorheologic effects of pentoxifylline in diabetic patients with angiopathic complications. Acta Diabetol 34:67–74
Speiser DE, Valmori D, Rimoldi D, Pittet MJ, Lienard D, Cerundolo V, MacDonald HR, Cerottini JC, Romero P (1999) CD28-negative cytolytic effector T cells frequently express NK receptors and are present at variable proportions in circulating lymphocytes from healthy donors and melanoma patients. Eur J Immunol 29:1990–1999
Springer TA (1990) Adhesion receptors of the immune system. Nature 346:425–434
Stagsted J (1998) Journey beyond immunology. Regulation of receptor internalization by major histocompatibility complex class I (MHC-I) and effect of peptides derived from MHC-I. Acta Pathol Microbiol Immunol Scand (APMIS) Suppl 85:1–40
Stagsted J, Reaven GM, Hansen T, Goldstein A, Olsson L (1990) Regulation of insulin receptor functions by a peptide derived from a major histocompatibility complex class I antigen. Cell 62:297–307
Stagsted J, Olsson L, Holman GD, Cushman SW, Satoh S (1993) Inhibition of internalization of glucose transporters and IGF-II receptors. Mechanism of action of MHC class I-derived pep-

tides which augment the insulin response in rat adipose cells. J Biol Chem 268: 22809–22813

Stamler JS, Jia L, Eu JP, McMahon TJ, Demchenko IT, Bonaventura J, Gernert K, Piantadosi CA (1997) Blood flow regulation by S-nitrosohemoglobin in the physiological oxygen gradient. Science 276:2034–2037

Svejgaard A, Ryder LP (1976) Interaction of HLA molecules with non-immunological ligands as an explanation of HLA and disease associations. Lancet 2:547–549

Tagaya Y, Bamford RN, De Filippis AP, Waldmann TA (1996) IL-15: a pleiotropic cytokine with diverse receptor/signaling pathways whose expression is controlled at multiple levels. Immunity 4:329–336

Tarnvik A (1971) Lymphocyte stimulation by a complex of phytohaemagglutinin and red cell membrane. Acta Pathol Microbiol Scand (Sect B) 79:588–594

Tartter PI (1995) Immunologic effects of blood transfusion. Immunol Invest 24:277–288

Taylor DS, Nowell PC, Kornbluth J (1986) Functional role of HLA class I cell-surface molecules in human T-lymphocyte activation and proliferation. Proc Natl Acad Sci USA 83: 4446–4450

Triantafilou K, Triantafilou M, Wilson KM, Cherry RJ, Fernandez N (1999) Intracellular and cell surface heterotypic associations of human leukocyte antigen-DR and human invariant chain. Hum Immunol 60:1101–1112

Tse DB, Pernis B (1984) Spontaneous internalization of class I major histocompatibility complex molecules in T lymphoid cells. J Exp Med 159:193–207

Turner EJ, Jarvis HG, Chetty MC, Landon G, Rowley PS, Ho MM, Stewart GW (2003) ATP-dependent vesiculation in red cell membranes from different hereditary stomatocytosis variants. Br J Haematol 120:894–902

Vallés J, Santos MT, Aznar J, Martinez M, Moscardo A, Pinon M, Broekman MJ, Marcus AJ (2002) Platelet–erythrocyte interactions enhance alpha(IIb)beta(3) integrin receptor activation and P-selectin expression during platelet recruitment: down-regulation by aspirin ex vivo. Blood 99:3978–3984

Van de Wal Y, Corazza N, Allez M, Mayer LF, Iijima H, Ryan M, Cornwall S, Kaiserlian D, Hershberg R, Koezuka Y, Colgan SP, Blumberg RS (2003) Delineation of a CD1d-restricted antigen presentation pathway associated with human and mouse intestinal epithelial cells. Gastroenterology 124:1420–1431

Virella G, Rugeles MT, Hyman B, La Via M, Goust JM, Frankis M, Bierer BE (1988) The interaction of CD2 with its LFA-3 ligand expressed by autologous erythrocytes results in enhancement of B cell responses. Cell Immunol 116:308–319

Vivier E, Anfossi N (2004) Inhibitory NK-cell receptors on T cells: witness of the past, actors of the future. Nat Rev Immunol 4:190–198

Volpes R, van den Oord JJ, Desmet VJ (1992) Can hepatocytes serve as 'activated' immunomodulating cells in the immune response? J Hepatol 16:228–240

Wacholtz MC, Patel SS, Lipsky PE (1989) Patterns of costimulation of T cell clones by cross-linking CD3, CD4/CD8, and class I MHC molecules. J Immunol 142:4201–4212

Wagtmann N, Rajagopalan S, Winter CC, Peruzzi M, Long EO (1995) Killer cell inhibitory receptors specific for HLA-C and HLA-B identified by direct binding and by functional transfer. Immunity 3:801–809

Waldmann TA, Tagaya Y (1999) The multifaceted regulation of interleukin-15 expression and the role of this cytokine in NK cell differentiation and host response to intracellular pathogens. Annu Rev Immunol 17:19–49

Williams AP, Peh CA, Purcell AW, McCluskey J, Elliott T (2002) Optimization of the MHC class I peptide cargo is dependent on tapasin. Immunity 16:509–520
Wood ZA, Schroder E, Robin Harris J, Poole LB (2003) Structure, mechanism and regulation of peroxiredoxins. Trends Biochem Sci 28:32–40
Yewdell JW, Reits E, Neefjes J (2003) Making sense of mass destruction: quantitating MHC class I antigen presentation. Nat Rev Immunol 3:952–961
Yio XY, Mayer L (1997) Characterization of a 180 kDa intestinal epithelial cell membrane glycoprotein, gp180. A candidate molecule mediating T cell–epithelial interactions. J Biol Chem 272:12786–12792

Novel Single Cell Fluorescence Approaches in the Investigation of Signaling at the Cellular Level

Péter Nagy, György Vereb, Janine N. Post, Elza Friedländer, János Szöllősi

2.1 Introduction

Treatment of human diseases relies on a clear understanding of the molecular mechanisms and cellular events behind pathophysiological processes. In most cancers, one or more steps in the transduction of signals from the extracellular environment to the nucleus are subverted, leading directly to the formation and evolution of the tumor. Consequently, drugs targeting transmembrane signaling proteins can be used as an effective and selective treatment approach (Sausville et al. 2003). Classical molecular biological methods (blotting, immunoprecipitation, chemical cross-linking) have given important insight into the details of transmembrane signaling processes. Most of these approaches include the disruption of the integrity of the cell (e.g. by detergent treatment), thereby precluding the study of the intricate details of signal transduction in its native, unperturbed environment. Additionally, these methods are inherently unable to reveal cell-to-cell differences caused by heterogeneous, endogenous or ectopic expression of proteins. Biochemical methods often lack the quantitative information needed to describe the physico-chemical processes accurately. In situ labeling of biomolecules by fluorescence offers a valuable alternative to biochemical methods. Detection of fluorescently labeled molecules is (1) sensitive: under favorable conditions even single molecules can be detected; (2) not harmful to the cell or to the environment because the use of ionizing radiation can be avoided; and (3) flexible and specific: fluorescently labeled antibodies or genes engineered to code for proteins with a fluorescent tag accurately report on the subcellular distribution of proteins. In this chapter we give a brief overview of the application of fluorescence in quantitative cell biological studies, with special emphasis on Förster-type fluorescence resonance energy transfer (FRET) and intracellular trafficking, and address the advantages and potential pitfalls of different detection and labeling strategies from the standpoint of a cell or molecular biologist.

2.2 Subcellular and Temporal Resolution Versus Statistical Reliability

Spatial and temporal organization of signal transduction events is important for the maintenance of the integrity of a complex biological system. Although both the temporal and spatial resolutions of fluorescence microscopy are limited

Springer Series in Biophysics
S. Damjanovich (Ed.)
Biophysical Aspects of Transmembrane Signaling

(discussed in detail later), the complexity of a large fraction of cell biological questions can be adequately addressed using this technique. From a practical point of view, the spatial resolution of conventional fluorescence microscopy is about 200–500 nm; thus it provides subcellular resolution and is able to interrogate subcellular organelles separately with a temporal resolution in the millisecond to second time domain. However, even the simplest biological parameters or phenomena usually show a large population variance, e.g. transient transfection often yields a two to three orders of magnitude difference in the cellular expression of the transfected gene. Consequently, sampling of a relatively large number of cells is necessary to obtain a statistically reliable representation of the population. Fluorescence microscopy is usually limited to the measurement of tens or hundreds of cells. The use of flow cytometry, where single cells cross a laser beam at a rate of hundreds or thousands of cells per second, provides statistically accurate data about fluorescence intensity, i.e. the concentration of fluorescent molecules in cells (Ormerod 1994; Shapiro 2002). Unlike microscopy, flow cytometry yields very limited information about morphology by measuring light scatter parameters sensitive to the shape and internal structure of cells (Jovin et al. 1976). Additionally, commercial flow cytometers provide no information about the subcellular distribution of fluorescent molecules. This can be a problem if one wishes to study, e.g., the properties of green fluorescent protein (GFP)-tagged membrane proteins which are present not only in the plasma membrane, but also in organelles along the secretory pathway (e.g. the Golgi complex). However, if less than ~10% of the protein is outside the plasma membrane, the contribution of these proteins to the measured signal will be negligible. In addition, attached cells have to be removed from the culture dish. Therefore, flow cytometric measurement of suspended cells may not faithfully reflect the properties of proteins in an attached cell. Given the disadvantages of microscopic and flow cytometric approaches, a combination of these two techniques is sometimes necessary (Lidke et al. 2003). However, both microscopy and flow cytometry are able to discriminate dead cells from live ones, an achievement not shared by conventional biochemical or fluorometric approaches.

Attempts have been made to combine the statistical reliability of flow cytometry with the subcellular and morphological resolution of fluorescence microscopy. This less widespread technique is called laser scanning cytometry (LSC) (Tarnok and Gerstner 2002). Fluorochrome-labeled cells are imaged under a conventional fluorescence microscope equipped with special software capable of identifying single cells on the microscope slide. Not only the fluorescence intensity but also parameters describing the morphology of the cell and the distribution of fluorescence are recorded in addition to the location of each cell. This latter feature makes repeated measurement of the same cell possible.

2.3 Labeling of Cells with Fluorescent Molecules

2.3.1 Monoclonal Antibodies and Their Derivatives

Monoclonal antibodies (mAb) are bivalent molecules with an approximate molecular weight (MW) of 160 kDa. They contain two antigen-binding fragments (Fab) and a constant fragment (Fc) (Hudson and Souriau 2003). Fluorescent dyes can be covalently conjugated to them, creating one of the most versatile tools in the hands of molecular biologists. Since the size of fluorescent molecules (Cy dyes, Alexa fluorophores, fluorescein, etc.) is small (MW ~0.5 kDa), they generally do not alter the binding affinity of the antibody. Similarly, the Ab usually does not deteriorate the fluorescent properties of the dye, especially if it is conjugated to the Ab through a spacer. In addition to labeling strategies with fluorescent primary antibodies, secondary fluorescent Abs can be used to label a non-fluorescent primary antibody. Alternatively, mAbs can be biotinylated in vitro, and fluorescent avidin can be used as a secondary label. Increases the signal to noise ratio, because several secondary Abs bind to a single primary antibody and they obviate the need to fluorescently modify every primary antibody. However, secondary labeling strategies have some disadvantages: (1) an increase in non-specific labeling [especially when using avidin for labeling intracellular antigens (Hollinshead et al. 1997)]. However, saturation of intracellular endogenous biotin with non-labeled avidin, and then – since avidin is tetravalent – saturation of the biotin binding sites of avidin with biotin can significantly reduce the non-specific background. In addition, high background due to binding of antibodies to Fc receptors can be prevented by using Fab or $F(ab')_2$ fragments; (2) due to the multivalent nature of the secondary Ab, cross-linking of the primary antibody and consequently the labeled protein may occur. This can result in activation, e.g. in the case of membrane proteins. Even the bivalent primary antibodies themselves can sometimes induce cross-linking and receptor activation (Park et al. 1995; Le et al. 2000; Cardarelli et al. 2002). Several strategies have been developed to reduce labeling-induced receptor activation: (1) monovalent Fab fragments can be used either as a primary or as a secondary label; (2) the secondary Fc specific Fab can be conjugated to the primary antibody in vitro (Zenon kits from Molecular Probes), an approach that also reduces non-specific binding; (3) single chain variable fragments (scFv) of an antibody can also be used for primary labeling (Hudson and Souriau 2003); (4) camels, llamas and sharks naturally produce single chain antibodies (VHH) binding monovalently to their epitope (van der Linden et al. 1999; Desmyter et al. 2001; Hudson and Souriau 2003); and (5) small oligonucleotides called aptamers have been developed for strong and specific binding of various cellular entities (Jayasena 1999).

2.3.2
Visible Fluorescent Proteins

The discovery of green fluorescent protein (GFP) and its subsequent development revolutionized fluorescence microscopy because researchers could follow trafficking of proteins specifically in living cells by creating chimeras of the studied protein with GFP. Originally, GFP was isolated from the jellyfish *Aequorea victoria*. Several naturally occurring variants of GFP and other evolutionarily unrelated fluorescent proteins exist (Chalfie 1995; Matz et al. 1999). In addition, mutagenesis of the original proteins generated a large variety of fluorescent proteins with improved properties and different colors. We collectively refer to them as visible fluorescent proteins (VFPs, Table 2.1).

The original GFP from *Aequorea victoria* displayed two absorption peaks at 395 and 470 nm corresponding to the neutral and anionic forms of the chromophore, respectively, and a dim emission peak at 510 nm originating from the anionic form. The chromophore is formed by cyclization and oxidation of residues 65–67 (Ser-Tyr-Gly). The chromophore is buried in a β-barrel structure (Ormo et al. 1996; Yang et al. 1996; Youvan and Michel-Beyerle 1996), and its spectral properties are substantially modified by interactions with neighboring amino acids (Heim et al. 1994). An S65T mutation practically eliminated the UV absorption peak. Enhanced GFP (EGFP) contains an additional mutation (F64L). EGFP fluoresces 35 times brighter then wild-type GFP and as opposed to wild-type GFP the rate of chromophore maturation at 37 °C is not significantly lower than at 28 °C (Patterson et al. 1997). A number of different spectral variants were developed (cyan – CFP, 434/477; yellow – YFP, 514/527; blue – BFP, 380/440; the numbers refer to the excitation/emission maxima) (Tsien 1998; Clontech 2001). Although often referred to as CFP, YFP or BFP, these spectral variants are derivatives of EGFP, and are more correctly called ECFP, EYFP and EBFP. YFP was notorious for its sensitivity to pH, chloride concentration and photobleaching. Second-generation YFP displayed slightly better pH resistance. The perfect YFP has not yet been developed, but two third-generation YFPs have promising properties: citrine has much reduced chloride and pH sensitivity and higher photostability (Griesbeck et al. 2001) and Venus is the brightest and fastest maturing YFP so far (Nagai et al. 2002; Rekas et al. 2002).

Long-wavelength fluorescent proteins provide a valuable alternative to GFP family members, because their red fluorescence shows negligible overlap with cellular autofluorescence and with short-wavelength VFPs. The first such protein, DsRed, was isolated from a coral (Matz et al. 1999). Its widespread application has been severely limited by its slow maturation and the strong oligomerization tendency required for the generation of a functional chromophore (Sacchetti et al. 2002). A large number of other VFPs, including short and long wavelength types, and VFPs originating from different species are still being developed. These variants display brighter fluorescence, less oligomerization, lower photobleaching rate and faster maturation (Labas et al. 2002).

Most VFPs show at least a weak tendency to dimerize. In most cases, this phenomenon does not cause problems, because it takes place only at relatively high concentrations (K_d ~0.1 mM). YFP and CFP have slightly higher dimerization

Table 2.1. Photophysical properties of visible fluorescent proteins

Name	Mutations from wtGFP	Synonym	λ_{ex} (nm)	λ_{em} (nm)	Extinction coefficient (at λ_{ex}, max.)	Quantum yield (QY)	Notes	References
wtGFP		Wild type	395–397 and 470–475	508 (540)	36,500?	0.79		Shimomura et al. (1962)
EGFP	F64L, S65T		488	507–509	55,000	0.6	Possible FRET acceptor of EBFP	Heim et al. (1995)
PAGFP	V163A, T203H	WEGFPH	400 (488)	507–509			Fluorescence increases >100 times upon radiation at 400 nm	Patterson and Lippincott-Schwartz (2002)
YFP	S65G, S72A, T203Y	Gfp-10c	513	527				Ormo et al. (1996)
EYFP	YFP + V68L, Q69L		513	527	67,000	0.67	Possible FRET acceptor of ECFP	(Ormo et al. 1996)
ECFP	F64L, S65T, Y66W, N146I, M153T, V163A	W7	433 (453)	476	26,000	0.37	Possible FRET donor of EYFP	Heim and Tsien 1996
Cerulean	ECFP+ S72A, Y145A, H148D		433	475	43,000	0.62	Better possible FRET donor of EYFP, because of higher QY and single component lifetime	Rizzo et al. 2004

Table 2.1. Photophysical properties of visible fluorescent proteins (*Continued*)

Name	Mutations from wtGFP	Synonym	λ_{ex} (nm)	λ_{em} (nm)	Extinction coefficient (at λ_{ex}, max.)	Quantum yield (QY)	Notes	References
EBFP	F64L, Y66H, Y145F	P4–3	380–383	440–447			Possible FRET donor of EGFP	Patterson et al. (1997)
Citrine	YFP + V68L, Q69M		516	529	77,000	0.76	Acid resistant	Griesbeck et al. 2001
Venus	YFP + F46L, F64L, M153T, V163A, S175G		515	528	92,200	0.57	Faster maturation, less acid sensitive	Nagai et al. (2002)
mYFP	YFP + A206K		513	527	79,000	0.62	Monomeric, K_d=74 mM (compared to YFP K_d=0.1 mM)	Zacharias et al. (2002)
DsRed		drFP583	558	583	57,000	0.79	From *Discosoma* coral, tetrameric	Matz et al. 1999
mRFP			584	607	44,000	0.25	Monomeric DsRed	Campbell et al. (2002)
KFP1		asCP	532 (580)	600	123,000 (initial) 59,000 (kindled)	<0.001 (initial) 0.07 (kindled)	Kindling of this protein can be induced by 532 nm light. Fluorescence intensity increases 30-fold when kindling is irreversible	Chudakov et al. (2003)

Table 2.1. Photophysical properties of visible fluorescent proteins (*Continued*)

Name	Mutations from wtGFP	Synonym	λ_{ex} (nm)	λ_{em} (nm)	Extinction coefficient (at λ_{ex}, max.)	Quantum yield (QY)	Notes	References
Kaede			508 (480) 572	518 (green) 582 (red)			Forms a homotetrameric complex	Ando et al. (2002)
BiFC	YN and YC, GN and GC, CN and CC + combinations thereof	Bimolecular fluorescence complemen tation-	Dependent on basis fluorophore				Two nonfluorescent fluorophores complement each other to form one fluorescent molecule	Hu et al. (2002); Hu and Kerppola (2003)

potency, and DsRed shows a strong tendency to form tetramers. This may result in the formation of large, nonfunctional protein aggregates especially if the fusion partner also shows a propensity to oligomerize (Zhang et al. 2002). Even if the oligomerized proteins are not inactivated, VFP-driven association results in the formation of artificial oligomers, hindering studies of protein associations. YFP- or CFP-containing chimeras have been shown to aggregate (Clayton et al. 2002; Zacharias et al. 2002; Lidke et al. 2003). In these VFPs, the dimerization tendency was practically eliminated by replacing a hydrophobic alanine with a positively charged lysine (A206 K) (Zacharias et al. 2002; Lidke et al. 2003). It required 33 mutations in DsRed to eliminate its tetramerization tendency (Campbell et al. 2002). Even if correct functioning of a protein is not disrupted by VFP-driven association, artificial aggregation may result in mistargeting of the protein. This was shown for DsRed-thymidine kinase (DsRed-TK) chimeras, which are not correctly imported to the nucleus despite the presence of several nuclear localization signals in TK. Coexpression of DsRed-TK with GFP-TK or untagged TK allowed these proteins to enter the nucleus by inhibiting the formation of DsRed-TK oligomers (Soling et al. 2002).

2.3.3 Novel Applications of VFPs in Cell Biology

Studying intracellular trafficking of VFP chimeras is made difficult by the presence of the VFP fusion protein in many cellular organelles. The presence of a VFP fusion protein in endocytotic and secretory vesicles due to the continuous recycling of membrane proteins is a confounding factor in studies of its internalization. The solution could be a VFP that can be activated selectively at a certain location of the cell (i.e. the cell membrane). Such a VFP is the T203H mutant of wild-type GFP (PA-GFP-64F-65S, photoactivatable GFP), which undergoes a ~100-fold increase in its fluorescence excited at 488 nm after irradiation with blue light at ~400 nm (Fig. 2.1; Patterson and Lippincott-Schwartz 2002; Post et al. 2005).

Non-activated PA-GFP has hardly any fluorescence, making it difficult to identify cells expressing it. Therefore Kaede, a GFP variant that changes color from green to red by irradiation at 350–400 nm, is even more suitable for these kinds of studies, because both the red and the green versions can be separately detected with excitation at 475 and 550 nm without any further change in color (Ando et al. 2002). Another variation on the topic has been recently published. Kindling fluorescent protein (KFP1) shows a 30-fold increase in its fluorescence (λ_{ex}=580 nm, λ_{em}=600 nm) when illuminated with green light. When the intensity of green light is low, photoactivation is reversible, and photoactivated KFP1 is spontaneously and reversibly quenched, which can be accelerated by illumination with blue light. When intense green light is used, photoactivation of KFP1 is irreversible (Chudakov et al. 2003).

VFP fused to protein domains interacting with special messengers can be used to visualize the spatial and temporal changes in the concentration of a messenger in living cells. Such an approach has already been successfully used to localize different phosphoinositides (Balla and Varnai 2002), diacylglycerol (Oancea et

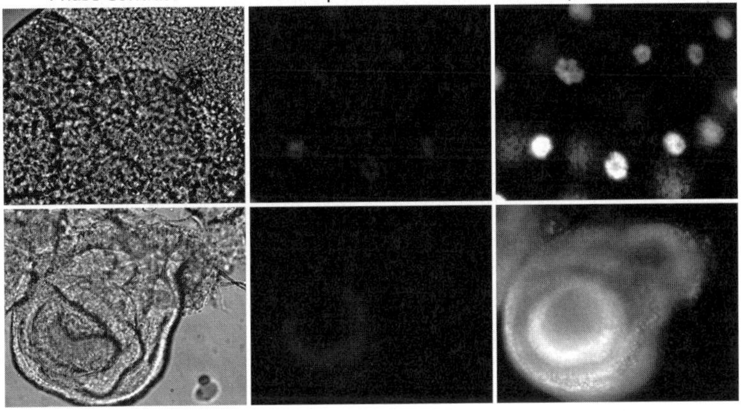

Fig. 2.1. Photoactivation of PA-GFP in *Drosophila* larvae. A transgenic *Drosophila melanogaster* fly line expressing PA-GFP-histone 2AvD was generated. PA-GFP was activated using a mercury-HBO 50-W lamp and an Ex 436/12 filter. For imaging, the following filters were used: Ex450–490, DC510FT, Em520 LP. Activation time was ~1 s when using a 40× 1.3NA oil Plan-NeoFluar objective. Fluorescence intensity increased after activation in different tissues dissected from third instar larvae. This image shows a salivary gland (*above*) and an imaginal disk (*below*)

al. 1998), phosphatidic acid (Rizzo et al. 2000) and calcium (Teruel and Meyer 2002).

FRET has found widespread application in the study of molecular conformations and associations (Jares-Erijman and Jovin 2003). We will give a detailed introduction to basic concepts and measuring methods of FRET in a separate section of this chapter. Now, we would like to highlight the benefit of combining VFPs with FRET experiments. FRET is usually measured between spectroscopically distinct molecules (hetero-FRET), less frequently between identical molecules (homo-FRET). The most widely used donor–acceptor VFP pair for hetero-FRET measurements is CFP-YFP. The BFP-GFP pair, which was used earlier, had unfavorable properties because of the dimness and fast photobleaching of BFP (Zhang et al. 2002). FRET between conventional fluorophores has been frequently used to study conformational changes in molecules. In such a case, both the donor and the acceptor are part of the same molecule. In one such application a calcium-sensitive protein called cameleon is engineered, consisting of CFP and YFP fused together through CaM, a peptide linker and M13. In the presence of Ca^{2+} the interaction between CaM and M13 brings CFP and YFP together, resulting in increased FRET (Miyawaki et al. 1997, 1999). Constructs sensitive to the activity of tyrosine kinases have also been reported: a phosphotyrosine-binding SH2 domain and a substrate peptide for the kinase of interest have been sandwiched between CFP and YFP. Kinase activation phosphorylates the substrate peptide in the reporter that will be recognized and bound by the SH2 domain, inducing a conformational change, and will consequently alter the FRET efficiency (Kurokawa et al. 2001; Sato et al. 2002). Another described reporter is sensitive to caspase activity. In this case, CFP and YFP are fused together through a caspase-sensitive linker. Upon caspase activation, cleavage of the linker separates the donor and the acceptor, thus abolishing FRET (Rehm et al. 2002). In another type of application the donor and the acceptor constitute an intermolecular FRET pair reporting on the association state of the proteins they are fused to. EGF stimulation of cells expressing epidermal growth factor receptor (EGFR) fused to CFP and Grb2 fused to YFP resulted in a large increase in FRET efficiency, demonstrating recruitment of Grb2 to EGFR. Additionally, subcellular distribution of FRET showed that EGFR and Grb2 interact in endosomal compartments, indicating that signaling can occur in endosomes (Sorkin et al. 2000). Molecular associations can also be measured by homo-FRET. If homo-FRET takes place, the fluorescence anisotropy of VFP decreases. The high intrinsic anisotropy (r_0) of GFP and its variants (Volkmer et al. 2000) and the large R_0 values for VFP homo-FRET pairs (Patterson et al. 2000) make them suitable for such applications (for a detailed explanation of homo-FRET see Sect. 2.5.5.6). Time-resolved anisotropy measurements demonstrated homo-aggregation of herpes simplex virus thymidine kinase in transfected cells (Gautier et al. 2001). Steady-state anisotropy measurements have been successfully used to detect the homoassociation of MHC class I molecules (Rocheleau et al. 2003) and EGFR (Lidke et al. 2003).

Hu et al. (2002) described a method named bimolecular fluorescence complementation (BiFC), which is based on the formation of a fluorescent protein from two fragments of YFP brought together by the association of two interacting proteins fused to the two YFP fragments (Hu et al. 2002). Since in the absence of complementation no fluorescence is present (no background), this approach is exquisitely suited to monitoring the interaction between two lowly expressed proteins in living cells. However, both BiFC and FRET are limited in their applicability to measuring the association of a single protein pair. Multicolor fluorescence complementation has been developed from BiFC (Hu and Kerppola 2003). The authors realized that fragments from different VFP variants form fluorescent proteins with different colors. They have identified 12 different pairs of VFP fragments corresponding to 7 distinct spectral classes and used the method to simultaneously visualize the interaction of Fos with Jun ATF2 in live cells.

2.3.4 Labeling of Tetracysteine Motifs with Biarsenical Derivatives of Fluorophores

Fluorescent chimeras produced by fusing VFP to a protein offer a limited set of available colors and can have undesirable effects due to the large size of the VFP (~25 kDa). Griffin et al. proposed introducing a tetracysteine motif (usually Cys-Cys-Xaa-Xaa-Cys-Cys, where Xaa is any non-cysteine amino acid), probably forming a hairpin structure, into a recombinant protein. After expression of the protein, the tetracysteine motif can be reversibly labeled by a membrane-permeable biarsenical derivative of fluorescein (FlAsH-EDT_2, fluorescein arsenical helix binder-ethanedithiol), if cysteines are reduced, which is usually the case for intracellular proteins (Griffin et al. 1998; Adams et al. 2002). FlAsH-EDT_2 is non-fluorescent until it binds to the four cysteines in the recombinant target protein. Toxicity and non-specific binding of the biarsenical compound to endogenous thiols are of concern. Both of these effects can be decreased by the simultaneous administration of micromolar concentrations of EDT or disperse blue to block binding of FlAsH-EDT_2 to endogenous cysteines and hydrophobic pockets, respectively (Griffin et al. 2000). However, addition of micromolar concentrations of EDT can also interfere with some cellular processes necessitating properly designed control experiments. Upon removal of excess FlAsH-EDT_2, the biarsenic dye persists on the tetracysteine motif, but newly synthesized proteins are not labeled. Therefore, biarsenic reagents can be used to follow the trafficking of proteins (Ju et al. 2004). Multicolor labeling can be achieved by combining FlAsH-EDT_2 and ReAsH-EDT_2 (resorufin arsenic helix binder). The latter exhibits red fluorescence after binding to a tetracysteine motif and also generates singlet oxygen which can photoconvert diaminobenzidine (DAB) to an electron-dense product, allowing visualization by electron microscopy (Gaietta et al. 2002). Serial labeling with FlAsH-EDT_2 and ReAsH-EDT_2 produced green and red proteins, respectively, because ReAsH-EDT_2 only labeled proteins synthesized after labeling with FlAsH-EDT_2, since tetracysteine motifs in old proteins were occupied by FlAsH-EDT_2. Recently, blue fluorescing biarsenical dyes have also become available (Adams et al. 2002). Under strong illumination the amount of singlet oxygen generated by ReAsH is sufficient to inactivate proteins within 5–50 nm. The approach, termed chromophore-assisted light inactivation (CALI), has been successfully used to abolish the function of gap junctions (Tour et al. 2003). Negatively charged, sulfonated biarsenical derivatives of fluorescent dyes can be used for selective labeling of tetracysteine motifs in the extracellular domain of membrane proteins after cysteines are reduced (Adams et al. 2002). A brief comparison between VFPs and molecules binding to tetracysteine motifs is given in Table 2.2.

2.3.5 Quantum Dots

Conventional fluorescent dyes and VFPs usually suffer from low brightness and photobleaching, i.e., a limited number of allowed excitation-emission cycles. One of the most exciting new approaches in fluorescence technologies has been the development of quantum dots (QDs) for biological applications (Jovin 2003). They are semiconductor nanocrystals displaying molecular properties due to their small size (quantum confinement; see later) and very bright fluorescence. A QD consists of several layers. In the center, a semiconductor core (e.g. CdSe) with a diameter of 3–5 nm is found. The core itself is the source of fluorescence, and its size determines the color (wavelength of the emitted photon) of the QD: larger QDs display red-shifted fluorescence compared to smaller ones (Alivisatos 1996). In semiconductor crystals absorption of a photon generates an excited, negatively charged electron, leaving a positively charged vacancy, a hole, behind. They are coupled to each other and are called an exciton. If the size of a semiconductor crystal is reduced below the characteristic size of an exciton (Bohr radius), quantum confinement takes place, resulting in the appearance of a well-defined band

Table 2.2. Pros and cons of labeling with VFP and tetracysteine motif-binding molecules

	VFP	Labeling of tetracysteine motifs
Molecular weight	~25 kDa	FlAsH: <1 kDa Tetracysteine motif: ~2 kDa
Position	Usually the C- or N-terminus	Can be added practically anywhere in the protein with the restriction that only reduced tetracysteine motifs can be labeled
Readout	Only fluorescence	Depends on the tetracysteine motif-binding molecule, e.g. fluorescence, fluorescent Ca^{2+} sensor, electron microscopy, NMR
Maturation/kinetics of labeling	Depends a lot on the spectral variant, but takes at least an hour	Labeling of the tetracysteine motif is completed within minutes (Adams et al. 2002)
Detection limit and background in live cells	~1 μM (Niswender et al. 1995) Low background	~10 μM (Adams et al. 2002) Higher, cell type-dependent background that can be reduced with EDT
Dynamics of the label	Once synthesized, VFP forms an integral part of the protein	Labeling of the tetracysteine motif is reversible, millimolar concentration of EDT reverse labeling

in the fluorescence emission spectrum. The size of the QD core is approximately equal to the Bohr radius, conferring molecular, quantum properties on it. The fluorescence quantum yield of an isolated semiconductor core is very low (typically in the order of a few percent). Therefore, the core is surrounded by a semiconductor shell (CdS, ZnS), increasing the quantum yield to 30–70% (Peng et al. 1997; Michalet et al. 2001). Core-shell QDs are soluble in non-polar solvents only. Water solubility is conferred by a shell of functionalized silica, amphipathic linkers or a polymer coating that will also allow further functionalization (Bruchez et al. 1998; Jovin 2003). Commercially available QDs (Quantum Dot Corporation, Evident Technologies) are either functionalized with streptavidin, protein A, or protein G for convenient and versatile use, or can be easily conjugated to proteins by the user.

QDs display several properties that make them an almost ideal label for cell biological applications: (1) they are bright primarily due to their very high absorption coefficient, (2) they are very photostable even during extended illumination, and (3) they have a continuous excitation spectrum and a narrow emission spectrum (full width at half-maximum is ~30–40 nm). The emission peak is typically shifted very little from the long wavelength edge of the excitation spectrum (small Stokes shift) and its wavelength is independent of the excitation wavelength. Due to the continuous nature of the excitation spectrum, lower wavelength excitation can be used, so the excitation and emission wavelengths can be very far from each other (large apparent Stokes shift), reducing the background due to light scattering and cellular autofluorescence (typically displaying a low Stokes shift). A further advantage of the continuous excitation spectrum is that several different QDs can be excited at the same wavelength, and they can still be spectrally separated from each other due to their well-defined narrow emission bands. The disadvantages of QDs are their relatively large size (the core and the shell surrounded by the outer layer is 15–20 nm) and their blinking behavior (intermittent fluorescent and nonfluorescent states). The long-term toxicity of QDs is unknown, but short-term incubation of live cells with QDs does not seem to affect their viability (Chan and Nie 1998; Jaiswal et al. 2003; Lidke et al. 2004). A report about in vivo imaging of sentinel lymph nodes with QDs and an apparent lack of toxic effects in live animals also shows that the "therapeutic index" (the ratio between the toxic and therapeutic/diagnostic dose of a drug) is wide enough for live cell and animal experiments (Kim et al. 2003).

QDs have been successfully used as a secondary label on cultured cells and tissue sections demonstrating low non-specific binding. Streptavidin or anti-mouse IgG-coated QDs were specifically bound to primary biotinylated and non-biotinylated Abs (Wu et al. 2003). Multicolor labeling and long-term imaging have also been demonstrated (Jaiswal et al. 2003). Alternatively, QDs coated with streptavidin can be conjugated to biotinylated ligands or antibodies in a tube prior to cell labeling. In this way, QDs can be used to follow the trafficking of receptors in live cells (Dahan et al. 2003; Lidke et al. 2004). Using this labeling approach it was demonstrated that conjugation of QD-streptavidin to biotinylated EGF did not alter the affinity of the ligand to its receptor and that the relatively large size of QDs did not interfere with the ability of two QD ligands to bind to the two monomers of a dimeric receptor (Lidke et al. 2004).

2.3.6 Steric Effects of Fluorescent Labeling

Correct functioning of proteins depends on their interactions with each other. Fluorescent labeling of proteins may interfere with their interactions: the larger the label, the more profound effects can be expected. Additionally, the relatively low permeability of large whole antibodies across the membrane (even in detergent-permeabilized samples) causes them to have more restricted access to their epitope compared to Fab or $F(ab')_2$ fragments. The size of a conventional fluorescent dye is negligible compared to the size of an antibody. However, VFPs (MW ~25 kDa) and QDs (diameter ~15–20 nm) are comparable in size to a protein (Jares-Erijman and Jovin 2003). In some cases, even large fluorescent dyes (APC, PE) can limit how close the labeled proteins can get to each other (Batard et al. 2002). On the other hand, even large labels like QDs can have no effect on the binding properties of the antibody or ligand to which they are conjugated (Lidke et al. 2004). Therefore, one has to carefully consider the pros and cons of labels before deciding which one to use. One cannot predict, a priori, whether the label will perturb the system being investigated. Proper controls must be employed to ensure that the label faithfully reflects the native system.

2.4 Methods to Detect Fluorescence in Cell Biological Applications

2.4.1 Spatial Resolution of Fluorescence Microscopy

The spatial resolution of a conventional fluorescence microscope is limited by the Abbe principle which states that the spatial resolution of a focusing light microscope is related to the size of its focal spot, which is in turn determined by the wavelength, the half aperture of the objective, and the index of refraction of the medium between the objective and the sample (Pawley 1995):

$$r = 0.61\frac{\lambda}{NA} = 0.61\frac{\lambda}{n \cdot \sin\omega} \tag{2.1}$$

where r is the spatial resolution defined as the minimal spacing between two point-like objects that can be separated from each other, λ is the wavelength of the emitted light, NA is the numerical aperture of the objective, n is the index of refraction of the medium between the objective and the sample and ω is the half-angular aperture of the objective. Due to theoretical and practical limitations on these parameters, the spatial resolution of a fluorescence microscope is approximately 200 nm (Hell 2003). Strictly speaking, the above reasoning applies to the lateral resolution of light microscopy, which is the smallest separation of two point-like objects in the X–Y plane that can be separated from each other. Both the temporal and axial resolutions of a microscope influence the apparent lateral resolution. Obviously, if the imaged object moves during the acquisition time, the image gets blurred. This may be a significant problem if a sample of low fluorescent intensity is imaged. In a conventional fluorescence microscope, the image is formed from a mixture of in-focus and a high proportion of out-of-focus photons, causing a loss of contrast and sharpness, because the axial resolution of objectives is three to four times lower than the lateral one (Pawley 1995). Therefore, the contribution of out-of-focus light to image formation decreases the effective lateral resolution. Considering the effect of the above factors, the lateral resolution power of a fluorescence microscope is ~500 nm from a practical point of view.

Although the theoretical limit of the lateral resolution of conventional fluorescence microscopy is ~200 nm, the position of a single object can be determined with far better accuracy (Betzig 1995). The peak in the intensity profile of a single fluorescent molecule corresponds to the location of the molecule. If the quality of the image is good, the position of the peak can be determined with subpixel accuracy (10–30 nm). If two different fluorescent molecules are imaged in separate channels, their colocalization can be determined with a similar accuracy (Koyoma-Honda et al. 2005). QDs are especially good candidates for such nanometer-resolution multicolor colocalization experiments due to their brightness and photostability (Michalet et al. 2001).

2.4.2 Confocal Microscopy

The usefulness of fluorescence microscopy for biological applications can be profoundly increased by excluding out-of-focus light from image formation. In confocal fluorescence microscopy, a pinhole is placed between the objective and the detector, which is usually a photomultiplier tube. The pinhole rejects out-of-focus light, thus endowing a confocal microscope with the capability of making images of a single optical section with a thickness of ~500 nm. A single section is recorded on a pixel-by-pixel basis by raster scanning the illuminating laser beam over the field; therefore it is usually called confocal laser scanning microscopy (CLSM). A stack of X–Y sections can be used for 3D reconstruction of the distribution of fluorescence. The improved axial resolution of CLSM comes at a price: the pinhole rejects a considerable fraction of the fluorescence, hindering the proper imaging of dim samples. The lateral resolution of confocal microscopy is not significantly different from conventional fluorescence microscopy. Confocal microscopes are very widespread and affordable instruments that have revolutionized cell biological applications of microscopy (Pawley 1995).

Recently described improvements in confocal imaging include fast image acquisition with a programmable array microscope (Hanley et al. 1999, 2000; Heintzmann et al. 2001) and tomographic reconstitution of a 3D object imaged from different angles (Heintzmann and Cremer 2002).

2.4.3 Multiphoton Microscopy

Although detection of emitted photons in CLSM is limited to the focal plane, excitation and, consequently, photobleaching are not limited to a single plane. In addition to avoidable photodamage, this phenomenon decreases the effective resolution of the microscope because scattered photons from outside the observation volume get onto the detector. Multiphoton fluorescence microscopy gets around this problem (Zipfel et al. 2003). It was shown both theoretically and experimentally that photons of lower energy can collectively cause an excitation normally produced by a single photon, if they interact with the absorbing molecule at the same time (less than 10^{-16} s difference). In most applications, multiphoton processes are limited to two photons, and the process is called two-photon excitation (TPE). The probability of TPE is proportional to the square of the intensity; therefore it steeply declines away from the focal point (Denk et al. 1990). Consequently, TPE microscopy achieves confocality without a pinhole by essentially eliminating out-of-focus excitation of fluorescence (Sheppard and Gu.

1990). As a good approximation, TPE occurs at twice the wavelength of single photon excitation. Since selection rules governing single- and two-photon excitation processes are different, deviations from the above rule are not uncommon. The probability of TPE is very low; therefore high illumination intensities are necessary, which are usually achieved by pulsed lasers (Xu et al. 1996). This is one of the drawbacks of multiphoton microscopy, because pulsed laser sources are still rather expensive. The axial and lateral resolutions of TPE microscopy are ~500 and ~200 nm, respectively, not significantly different from confocal microscopy (Zipfel et al. 2003).

The advantages of multiphoton microscopy arise from the confinement of the excitation to the focal point. Consequently, all photons generated contribute to image formation and there is less photodamage and no out-of-focus background due to light scattering; therefore emission detection can be efficient and can be accomplished without a pinhole that would reject a valuable fraction of the emitted photons. Thick samples can also be imaged (Oertner 2002), because high wavelength (infrared) excitation used for multiphoton microscopy is scattered to a much lesser extent than visible and UV light, and the scattered infrared photons are not able to cause multiphoton excitation because their intensity reaches a level sufficient for such a process only in the focal point. Even whole animal imaging can be done with multiphoton excitation (Brown et al. 2001; McDonald and Choyke 2003).

2.4.4 Focusing Microscopes Breaking the Abbe Limit

In the past decade several innovative concepts have been developed to increase the resolution power of light microscopes beyond that predicted by the Abbe principle. Currently, these devices are very expensive and available in few institutions only. Therefore we refer to these approaches only briefly.

One of these uses confocal detection through two opposing lenses. This doubles the effective numeric aperture, resulting in ~100 nm axial resolution. This approach is termed 4π microscopy, based on the quest to increase numerical aperture as close to the full 4π steric angle as possible. (Hell and Stelzer 1992; Egner et al. 2002; Hell 2003). I^5M microscopy is based on a similar principle, but uses non-confocal detection, resulting in significantly reduced image acquisition times (Gustafsson et al. 1999; Hell 2003). Stimulated emission depletion (STED) microscopy achieves a lateral resolution superior to any previously described approach. Fluorescent molecules are excited by laser light. Another red-shifted laser beam is used to deplete the excited state by stimulated emission. The intensity of the depleting laser light is spatially modulated such that depletion of the excited state does not occur at locations corresponding to the minima of the depleting laser intensity. Fluorescent molecules are only present inside these spots whose dimension, and consequently the lateral resolution of STED microscopy can be reduced infinitely from a theoretical point of view. The combination of the 4π concept with STED microscopy resulted in subwavelength (~50 nm) resolution in both the axial and X–Y directions (Dyba and Hell 2002; Dyba et al. 2003; Hell 2003).

2.4.5 Scanning Near-Field Optical Microscopy (SNOM)

As we have seen in previous sections, diffraction is a formidable problem in focusing light microscopes. Near-field optics developed in the 1980s breaks the lateral resolution limit by using an aperture with subwavelength diameter as a light source (Lewis et al. 1984; Lewis 1991). The distance of the aperture from the imaged surface is very small (~5–10 nm), and therefore only those fluorescent molecules that are under the aperture are excited. Consequently, the lateral resolution of SNOM is determined by the size of the aperture, which can be as small as 50 nm. The intensity of light decreases exponentially in the near field, i.e., very close to the aperture; therefore molecules far from the aperture in the axial direction are not excited. The penetration of the near-field into the object depends on the aperture size, and in the case of large apertures it can be tens or hundreds of nanometers (Kirsch et al. 1998; Jenei et al. 1999), but usually it is limited to 10–50 nm (Lewis et al. 2003); thus SNOM is a surface imaging device from a practical point of view. It has been used to study single fluorescent molecules (Betzig and Chichester 1993) and phospholipid monolayers (Hwang et al. 1995) and only less frequently for cellular imaging due to the difficulties in maintaining the small separation between the aperture and the cell surface, which is especially problematic if SNOM imaging is carried out on cells in buffer. Therefore, most cellular imaging with SNOM has been carried out on dried cells. SNOM imaging of fluorescently labeled membrane proteins has demonstrated a patchy distribution on cells (Vereb et al. 1997; Hwang et al. 1998; Nagy et al. 1999b).

2.5 Fluorescence Resonance Energy Transfer (FRET)

2.5.1 A Brief Introduction to FRET

FRET is a process in which energy is transferred non-radiatively from an excited donor fluorophore to an acceptor molecule. From a biologist's point of view, the major merit of FRET is its sensitivity to the separation between the interacting molecules. Harnessing this property makes distance measurements between the donor and acceptor molecules possible. This is why FRET was called a "spectroscopic ruler" capable of determining distances in the 1- to 10-nm range (Stryer 1978). Although the theoretical foundation of FRET was laid in the middle of the 20th century by Förster (1946) and Dexter (1953), it was only in the past 10–20 years that we witnessed an exponential growth in the number of papers about the biological applications of FRET. We will first give a brief introduction to the physics of FRET and then describe some approaches to measure it.

Several papers providing in-depth analysis of FRET have been published (Eisinger 1976; Matkó et al. 1988; Clegg 1995; Szöllősi et al. 1998; Vereb et al. 2002; Jares-Erijman and Jovin 2003), so we only give a short overview of the process. An excited fluorescent molecule has several alternative ways to lose energy in order to return to the ground state. In the absence of a suitable acceptor, there are fluorescent and non-fluorescent de-excitation mechanisms whose rate constants are denoted by k_f and k_{nf}, respectively. In the presence of an acceptor, FRET presents itself as an alternative relaxation mechanism for the donor. If k_{fret} denotes the rate constant of FRET, the FRET efficiency is defined by the following formula:

$$E=\frac{k_{fret}}{k_f+k_{nf}+k_{fret}} \tag{2.2}$$

E gives the fraction of excited donor molecules transferring energy to an acceptor. The FRET rate can be described by the following equation:

$$k_{fret}=c\,J\,n^{-4}\,R^{-6}\,\kappa^2 \tag{2.3}$$

where c is a constant, J is the normalized overlap integral between the donor emission spectrum and the acceptor absorption spectrum, n is the index of refraction of the medium, R is the separation distance between the donor and the acceptor and κ is the orientation factor. Equation (2.3) shows that the FRET rate is inversely proportional to the sixth power of the separation distance between the interacting molecules. The orientation factor is a function of the relative orientations of the donor and acceptor molecules, or, more strictly speaking, the function of the relative orientation of the donor emission and acceptor absorption dipoles. Equation (2.3) states that the FRET efficiency is high if the donor emission spectrum overlaps with the acceptor absorption spectrum, the relative orientation of the two molecules is favorable and their separation distance is small. FRET efficiency can be expressed in a concise form:

$$E=\frac{R_o^6}{R_o^6+R^6},\; R_o^6=c\,J\,\kappa^2\,n^{-4}\,Q_d \tag{2.4–2.5}$$

where R_o is a characteristic distance for a given donor–acceptor pair at which the FRET efficiency is 50%, and Q_d is the fluorescence quantum efficiency of the donor in the absence of the acceptor: $Q_d=k_f/(k_f+k_{nf})$, i.e., the fraction of excited molecules emitting fluorescence. R_o is usually 5–10 nm, and this defines the distance range for which FRET is applicable. The higher the R_o, the more efficient the given donor–acceptor pair is regarding their FRET interaction. J, n and Q_d can be relatively easily determined and all of them are usually constant during a FRET experiment. Therefore they and the constant c can be pooled into a new constant, c'. Thus, Eq. (2.4) can be rewritten as:

$$E=\frac{c'\kappa^2}{c'\kappa^2+R^6} \tag{2.6}$$

It follows that a change in FRET efficiency depends not only on the separation distance between the donor and the acceptor, but also on their relative orientation. Fortunately, in most cell biological applications, the rotational freedom of the donor and the acceptor is high, so that they sample all possible relative orientations during the excited state of the donor and an average random orientation can be used. It can be shown that $\kappa^2=2/3$ in this case (Dale et al. 1979), and any change in the FRET efficiency can be attributed to a change in the donor–acceptor separation. If R_0 is known for a given donor–acceptor pair ($\kappa^2=2/3$ is usually assumed), the separation distance between the donor and the acceptor can be calculated from Eq. (2.4).

In cell biological FRET experiments it has to be emphasized that FRET efficiency reflects the *average* separation distance between the donor and the acceptor. In a homogeneous system where the distance between the donor and the acceptor is same for all molecule pairs, an increase in FRET efficiency means that the donor and the acceptor had gotten closer to each other. On the other hand, in a heterogeneous system where only a fraction of the acceptors is within FRET distance from a donor, an increase in FRET efficiency can be caused by an increase in the fraction of acceptors associating with a donor. These considerations have to be taken into account in the interpretation of cell biological FRET experiments.

2.5.2 Different Ways to Measure FRET

FRET manifests itself in the change of a number of physical parameters. Most of these manifestations are associated with a way to detect FRET. These approaches have been recently reviewed (Berney and Danuser 2003; Jares-Erijman and Jovin 2003).

2.5.2.1 Donor Quenching, Sensitized Emission of the Acceptor

FRET competes with all other de-excitation processes of the donor. This results in reduced donor fluorescence, i.e., donor quenching. Since FRET is an additional relaxation mechanism, it takes the donor less time to relax to the ground state, and thus the fluorescence lifetime of the donor is reduced. It can be shown that

$$E=1-\frac{I_{DA}}{I_D}=1-\frac{\tau_{DA}}{\tau_D} \tag{2.7–2.8}$$

where I_{DA} and I_D are the fluorescence intensities, and τ_{DA} and τ_D are the fluorescence lifetimes of the donor in the presence and absence of the acceptor, respectively.

The measurement of fluorescence quenching is probably one of the most straightforward ways to measure FRET. Unfortunately, this approach is probably the most error-prone for cell biological applications. The problem is that one has to compare the fluorescence intensities of two samples, one labeled with donor, another labeled with donor and acceptor. Due to the natural variability of the expression level of proteins and other unknown environmental factors, FRET is not the only process that can change the fluorescence intensity. If a sufficiently large number of cells are measured and averaged, cell-to-cell differences are canceled. Therefore, donor quenching can only be reliably measured with flow cytometry.

The measurement of fluorescence lifetime is less error-prone than intensity measurements, because it does not depend on the number of fluorophores, i.e., the expression level of an antibody-labeled or VFP-tagged protein. Therefore, detection of a decreased fluorescence lifetime of the donor in the presence of an acceptor compared to that of the free donor is a reliable indicator of FRET. However, fluorescence lifetime measurements need sophisticated and expensive instrumentation (Bastiaens and Squire 1999; Haj et al. 2002).

If the acceptor molecule is fluorescent, it emits photons after the donor transfers energy to it. This is called sensitized emission, and constitutes another very basic way to detect FRET. Simple detection of acceptor emission excited at the donor absorption wavelength is error-prone, because the acceptor is usually directly excited at the donor absorption band too, and the intensity of sensitized emission depends not only on FRET, but also on the concentration of the donor. With narrow bandpass filters that minimize direct excitation of the acceptor these measurements are still feasible (Uster and Pagano 1986). A more accurate way to quantitate FRET through sensitized emission is based on the following formula:

$$E=\left(\frac{F_{AD}}{F_A}-1\right)\frac{\varepsilon_A\, c_A}{\varepsilon_D\, c_D} \tag{2.9}$$

where F_{AD} and F_A denote the fluorescence intensities of the acceptor excited at the donor absorption wavelength in the presence and absence of the donor, respectively, and ε and c (D and A indices refer to donor and acceptor, respectively) denote the molar absorption coefficient and the concentration of the molecules respectively.

2.5.5.2
Acceptor Photobleaching, Photochromic FRET

Donor quenching can be turned into a useful experimental approach to measure FRET, if the same sample can be measured with and without acceptor. First, the donor intensity in a donor–acceptor double-labeled sample is measured corresponding to I_{DA} in Eq. (2.7). Then the acceptor is photodestructed by an intense laser light, releasing of the donor from quenching. Therefore, the donor intensity measured afterwards corresponds to I_D in Eq. (2.7). The method called acceptor photobleaching uses the same pixels for the determination of I_D and I_{DA}; therefore pixel-by-pixel determination of FRET is possible (Bastiaens et al. 1996).

A drawback of the above method, similar to any other photobleaching-based methods, is the lack of repeatability: kinetic measurements cannot be made. When diheteroarylethenes are exposed to near-UV light, they undergo photoconversion from an open to a closed form associated with an increase in their absorption at ~540–550 nm. Exposing the closed form to green light induces the reverse photoconversion. The closed form can serve as an acceptor for dyes emitting in the 500–600 nm range, and donor intensity measured in its presence corresponds to I_{DA}. Since the open form shows negligible absorption at ~540–550 nm, measurement of donor intensity in the presence of the open form corresponds to the unquenched donor intensity (I_D). As switching between the open and closed forms can be repeated an unlimited number of times, this approach, termed photochromic FRET (pcFRET), can be used for following temporal changes in FRET (Giordano et al. 2002).

2.5.5.3 Donor Photobleaching

Photobleaching of a dye is initiated from its excited state. Since FRET decreases the fluorescence lifetime of the donor (Eq. 2.9), it spends less time in the excited state if FRET occurs. Consequently, FRET decreases the rate of donor photobleaching. It can be shown that

$$E = 1 - \frac{T_D}{T_{DA}} \tag{2.10}$$

where T_D and T_{DA} are the photobleaching time constants of the donor in the absence and presence of the acceptor, respectively (Jovin and Arndt Jovin 1989). FRET measurements based on donor photobleaching (pbFRET) also need two separate samples (donor-labeled, donor–acceptor double-labeled), but because the photobleaching rate is relatively insensitive to donor concentration (i.e., expression level of the labeled molecule), Eq. (2.10) is less error-prone than Eq. (2.7). Due to its simplicity, pbFRET has found widespread application and its physical background has also been intensely investigated (Young et al. 1994; Song et al. 1995, 1996, 1997; Nagy et al. 1998b).

2.5.5.4 FRET-Sensitized Photobleaching of the Acceptor

As already pointed out, photobleaching occurs only with molecules in the excited state. Since photobleaching is independent of the process by which the molecule got into the excited state, acceptor molecules excited by FRET can also undergo photobleaching. Acceptor molecules that are beyond FRET distance from a donor are insensitive to FRET-sensitized photobleaching. Consequently, the fraction of acceptor molecules within FRET distance from a donor can be determined using this method (Mekler et al. 1997); e.g., if only 80% of acceptor fluorescence can be bleached through the donor, 20% of acceptors do not actually accept energy from donors. Furthermore, the FRET efficiency can also be determined by comparing the rate of acceptor photobleaching in the presence and absence of the donor:

$$E = \frac{k_s(\lambda_D)\,\varepsilon_A(\lambda_D)\,c_a}{k_b(\lambda_D)\,\varepsilon_D(\lambda_D)\,c_D} \tag{2.11}$$

where $k_s(\lambda_D)$ and $k_b(\lambda_D)$ are the rate constants of FRET-sensitized acceptor bleaching and direct bleaching of the acceptor, respectively, excited at the donor excitation wavelength, $\varepsilon_A(\lambda_D)$ and $\varepsilon_D(\lambda_D)$ are the molar absorption coefficients of the acceptor and the donor, respectively, at the donor excitation wavelength, and c_A and c_D are the concentrations of the acceptor and the donor, respectively. In principle, FRET efficiencies as low as 0.1% can be reliably measured with this technique due to the large increase in the rate of acceptor photobleaching in the presence of the donor, i.e., $k_s(\lambda_D)/k_b(\lambda_D) >> 1$) (Mekler 1994).

2.5.5.5 Combined Donor and Acceptor Emission Measurements

For FRET methods based on measuring either only donor or only acceptor emission, two separate measurements on a donor and a donor–acceptor-labeled sample are necessary. Even though two samples are not needed for acceptor photobleaching or pcFRET measurements, the same sample is measured twice sequentially. Therefore any change (e.g., in donor concentration or fluorescence quantum yield) occurring with the sample between the two measurements distorts the calculations. A combined detection of donor and acceptor fluorescence signals makes the calculation of FRET efficiency possible from a single measurement. Such an approach has been implemented for flow cytometry (Trón et al. 1984; Szöllősi et al. 1984, 1987; Sebestyén et al. 2002) and digital imaging microscopy (Gordon et al. 1998; Nagy et al. 1998b; Elangovan et al. 2003). All of these approaches are based on solving a system of linear equations that usually contain equations describing donor emission, sensitized acceptor emission (FRET channel), and direct acceptor fluorescence and correction factors for the spectral overlap between the fluorescent detection channels.

Combined donor and acceptor emission measurements make pixel-by-pixel and cell-by-cell FRET measurements possible in microscopy and flow cytometry, respectively. The accuracy of pixel-by-pixel and cell-by-cell measurements is very sensitive to reliable autofluorescence subtraction. In most applications, an average autofluorescence value is subtracted from the intensity of every pixel or cell. This method is, however, not accurate when either the donor or the acceptor intensity is low. For cell-by-cell (Sebestyén et al. 2002) and pixel-by-pixel (Nagy et al. 1998b) corrections, the autofluorescence emission of cells without donor and acceptor labeling is determined at wavelengths corresponding to the fluorescence channels to be corrected (λ_{donor}, $\lambda_{acceptor}$, λ_{FRET}) and at a separate wavelength ($\lambda_{\mathrm{autofluorescence}}$), where donor and acceptor fluorescence is negligible and spectral correction factors are determined, e.g.:

$$S = \frac{I(\lambda_{donor})}{I(\lambda_{autofluorescence})} \tag{2.12}$$

Then the autofluorescence of labeled samples is measured at $\lambda_{autofluorescence}$, and the contribution of autofluorescence to the donor channel, $AF(\lambda_{donor})$, is calculated and subtracted on a cell-by-cell or pixel-by-pixel basis:

$$AF\left(\lambda_{donor}\right) = S \cdot AF\left(\lambda_{autofluorescence}\right) \tag{2.13}$$

Using the above approach, the homoassociation of the CD45 tyrosine phosphatase could be studied in cells with a very low expression level (Dornan et al. 2002).

2.5.5.6 FRET Between Spectroscopically Identical Molecules

In the previous sections, we considered FRET between a donor and a spectroscopically distinct acceptor molecule (hetero-FRET). However, if the absorption and emission spectra of a molecule show significant overlap, an excited fluorophore can transfer energy nonradiatively to a nearby, identical molecule, if their relative orientation is favorable ($\kappa^2 \neq 0$). The process is termed homo-FRET or emFRET (energy migration FRET). A molecule excited by FRET is not different from a directly excited molecule: it can emit a fluorescent photon or relax nonradiatively, including FRET to a nearby molecule. It can be shown that due to the chain-like FRET interactions, neither the fluorescence intensity nor the fluorescence lifetime nor the photobleaching rate of the molecule changes. Fortunately, there is one single parameter, the fluorescence anisotropy or polarization of the molecule that is sensitive to FRET.

If a fluorescent molecule is excited by polarized light, molecules with absorption dipoles oriented parallel to the polarization of the light will be excited preferentially. During the excited state, lifetime, molecules may rotate randomly, and the more they rotate, the less oriented they will be when they emit. Furthermore, FRET to a spectroscopically identical molecule also depolarizes fluorescence emission, because molecules excited secondarily through FRET are orientationally uncorrelated with the donor. The more FRET takes place, the less polarized the emitted fluorescence is. The feasibility of such an experimental approach has been demonstrated for steady-state and dynamic fluorescence anisotropy measurements (Clayton et al. 2002). However, the fact that besides FRET, rotational mobility also influences the fluorescence anisotropy of a molecule complicates interpretation of the results and has hampered widespread application of this approach. Researchers may be intimidated by the overwhelmingly complex calculations and by the need for structural information about the orientation of the fluorophores (Runnels and Scarlata 1995; Blackman et al. 1998).

Three concepts have been recently applied to get around the above-mentioned problems. (1) This approach is analogous to the original observation of Perrin in the 1920s and 1930s (Perrin 1932). He found that the anisotropy of a concentrated solution is lower than that of a dilute one. The phenomenon termed concentration depolarization is the result of homo-FRET as described above. We used the same principle to interpret the fluorescence anisotropy of cells expressing GFP fused to epidermal growth factor receptor (Lidke et al. 2003). Since the measurements were carried out using flow cytometry, it was possible to correlate cellular anisotropy of GFP fluorescence with the expression level of the protein construct in a statistically reliable way due to the large number of cells measured. It was found that cells expressing a high amount of EGFR-GFP showed lower anisotropy than low expressers even in the absence of ligand (EGF), indicating that the receptors formed homoassociations without ligand stimulation. (2) We and others introduced another method based on gradual photobleaching of GFP in order to decrease the number of GFP molecules. Assuming that the GFP-tagged molecules form clusters and homo-FRET occurs, the decrease in GFP density gradually abolishes homotransfer and increases fluorescence anisotropy, as it was observed for EGFR and class I major histocompatibility complex (MHC-I) molecules (Lidke et al. 2003; Rocheleau et al. 2003). (3) Recently, a novel approach has been introduced in which the contribution of homo-FRET to steady-state anisotropy is discerned from those due to rotational diffusion by exploiting the fact that homo-FRET fails upon excitation at the long wavelength edge of the absorption spectrum (Weber and Shinitzky 1970). An anisotropy value upon red-edge excitation that is higher than that upon excitation at around the absorption peak indicates that homo-FRET takes place (Squire et al. 2004).

Although homo-FRET measurements can only be used to detect homoassociation of the labeled molecule, they have an obvious advantage: the expression of only a single expression probe is needed, obviating the requirement for the balanced expression of a donor and an acceptor-tagged probe, which is often difficult to achieve.

2.6 Cell Biophysical Methods Reveal Molecular Interactions in the ErbB Signaling Pathway

The ErbB family of receptor tyrosine kinases has four members named ErbB1 (or epidermal growth factor receptor, EGFR), ErbB2, ErbB3, and ErbB4. They engage in a large variety of homo- and heteroassociations that undergo substantial reorganization upon binding of peptide growth factors to their extracellular domain. The role of ErbB proteins in physiological processes and in cancer development has been reviewed recently (Nagy et al. 1999a; Yarden and Sliwkowski 2001; Vereb et al. 2002; Holbro et al. 2003). The 284/1 issue of *Experimental Cell Research* was devoted to this topic. Below, we review some of our results concerning the molecular events behind ErbB-mediated signaling.

2.6.1 Small-Scale Associations: Homo- and Heterodimerization in the ErbB Family

Molecular-scale physical associations among ErbB family members have been studied by classical biochemical, molecular biological (Sliwkowski et al. 1994; Tzahar et al. 1996) and biophysical methods (Gadella and Jovin 1995; Nagy et al. 1998a, 2002). We estimated the association state of ErbB2 and assessed how it was affected by EGF treatment in breast tumor cell lines by measuring FRET between fluorescent monoclonal antibodies or Fab fragments using flow cytometry. There was considerable homoassociation of ErbB2 and heteroassociation of ErbB2 with EGFR in quiescent breast tumor cells. ErbB2 homoassociation was enhanced by EGF treatment in SKBR-3 cells and in the BT474 subline BT474M1 with high tumorogenic potential, whereas the original BT474 line was resistant to this effect. These differences correlated well with EGFR expression. Flow cytometric FRET measurements provided only one single FRET efficiency value for each cell analyzed. In order to reveal heterogeneity in the homoassociation pattern of ErbB2 within a single cell, one of the microscopic FRET approaches had to be used. We used donor photobleaching FRET microscopy to visualize FRET efficiency within single cells, with spatial resolution limited only by diffraction in the optical microscope. This allows detailed analysis of the spatial heterogeneity of molecular interactions. First, we applied the donor photobleaching FRET in wide-field microscopy and revealed extensive pixel-by-pixel heterogeneity in ErbB2 homoassociation (Nagy et al. 1998a). In our measurements, ErbB2 homoassociation was also heterogeneous in unstimulated breast tumor cells; and membrane domains with ErbB2 homoassociation had mean diameters of less than 1 μm (Nagy et al. 1998a). It was not clear whether the domain size was imposed by the optical resolution limit of wide-field microscopy in the X–Y plane or whether it originated from the actual size of ErbB2 aggregates. In order to refine the size estimate of domains containing ErbB2 molecules, we combined FRET measurements with confocal microscopy (Nagy et al. 2002).

We implemented acceptor photobleaching FRET in a CLSM equipped with three lasers in order to study the correlation between the association of ErbB2 with ErbB2 and with ErbB3, and the local density of these RTKs. From fluorescence intensities generated in two optical channels, we were able to determine the FRET efficiency values on a pixel-by-pixel basis, while the third laser beam provided the signal for measuring the expression level of ErbB3. The homoassociation of ErbB2 correlated positively with the local concentration of ErbB2, but negatively with ErbB3 local density. This negative correlation suggests that ErbB2–ErbB3 heterodimers compete with ErbB2 homodimers and therefore a high number of ErbB3 molecules can disassemble ErbB2 homodimers (Nagy et al. 2002).

2.6.2 Large-Scale Associations of ErbB Proteins

In addition to the direct molecular association between membrane proteins, another level of clustering has been identified involving the concentration of proteins in domains with a characteristic diameter of several hundred nanometers (Damjanovich et al. 1995; Hwang et al. 1998). We used SNOM and confocal microscopy to image the distribution of ErbB2 on the surface of quiescent and ligand-stimulated cells (Nagy et al. 1999b). Clusters with an average diameter of 0.5 μm containing ~1,000 ErbB2 proteins were identified on several different breast cancer cell lines (Fig. 2.2). The diameter of ErbB2 clusters increased upon stimulation of cells with EGF, heregulin (ErbB3- and ErbB4-specific ligand) or Herceptin [partially agonistic anti-ErbB2 antibody used in the treatment of breast cancer (Baselga et al. 1996)]. The EGF-induced increase in ErbB2 cluster

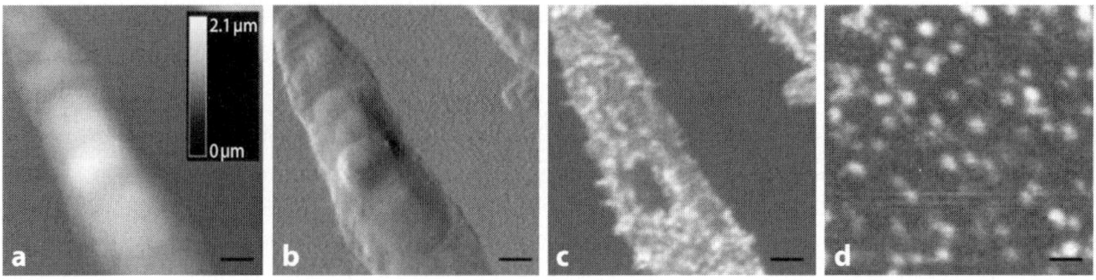

Fig. 2.2. SNOM images of ErbB2-expressing cells. CHO cells transfected with ErbB2 (**a–c**) or MCF-7 breast cancer cells (**d**) were labeled with a fluorescent anti-ErbB2 antibody on ice to prevent antibody-induced artifacts. **a** Topography of ErbB2-transfected CHO cells. **b** Feedback signal proportional to steepness of cell surface. **c**, **d** Fluorescence images showing the clustered distribution of ErbB2 on CHO (**c**) and MCF-7 (**d**) cells. *Scale bars* 4 µm (**a–c**), 1 µm (**d**)

size was dependent on the tyrosine kinase activity of EGFR (Nagy et al. 1999b). Large-scale clustering of receptor tyrosine kinases generates a high local density of signal transducing molecules, probably decreasing the distance receptors have to diffuse to form associations. The role of this level of clustering in signal transduction is underlined by the kinase activity-dependent increase in its size.

Lipid rafts are thought to be key players in transmembrane signal transduction (Simons and Toomre 2000). None of the methods used for raft identification is perfect, and the structures identified as lipid rafts may differ from method to method (Edidin 2001). We used the fluorescently labeled subunit B of cholera toxin (CTX-B) to label lipid rafts in intact cell membranes. We found that clusters of ErbB2 showed a substantial overlap with lipid raft domains. Glycosyl-phosphatydilinositol (GPI)-anchored proteins were reported to co-patch with CTX-B, if cells were incubated at 37 °C to allow the pentavalent CTX-B to cross-link lipid rafts (Harder et al. 1998). The raft localization of ErbB2 was less stable than that of GPI-anchored proteins, because cross-linking of rafts by CTX-B did not induce co-migration of ErbB2. The removal of ErbB2 from lipid rafts by CTX-B was followed by changes in ErbB2-mediated signal transduction, indicating an important role of the correct membrane localization of ErbB2 in signaling (Nagy et al. 2002).

The use of GFP-fused proteins to image the unperturbed distribution of proteins in live cells has become very popular. We used multiphoton 4π microscopy to image the distribution of EGFR-GFP in live CHO cells in buffer (Fig. 2.3). The high lateral and axial resolutions of 4π microscopy let us visualize the nonhomogeneous distribution of EGFR in the membrane which was similar to that of ErbB2 imaged with SNOM (Fig. 2.2), although the speckled distribution in the 4π microscopic image was less conspicuous.

2.6.3 Focal Stimulation of Cells with Magnetic Beads

The question of whether ligands activate receptors all over the cell surface or whether they rather induce a local activation is an intensely investigated topic.

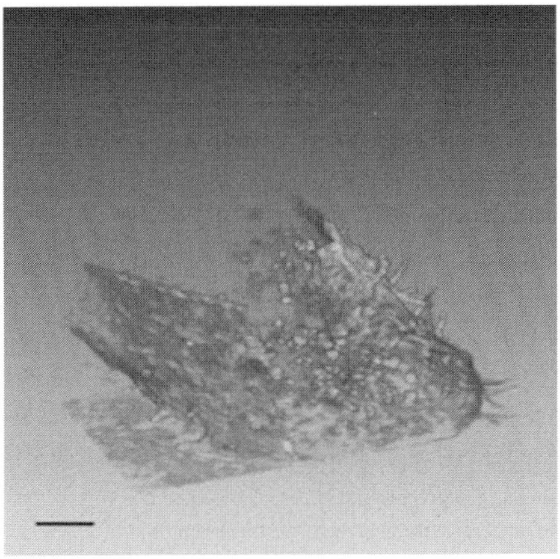

Fig. 2.3. Three-dimensional reconstruction of the distribution of EGFR-GFP in CHO cells. Multiphoton 4π microscopy was used to image live CHO cells expressing EGFR-GFP. A 3D reconstruction of part of a cell is shown. The cytoplasm contains numerous endo- and exocytic vesicles. Several membrane processes are clearly discernible on the membrane on the *right*. The *left* membrane segment shows the inhomogeneous distribution of EGFR-GFP. This clustered feature is reminiscent of what was observed using SNOM imaging ErbB2 shown in Fig. 2.2. *Scale bar* 1 μm

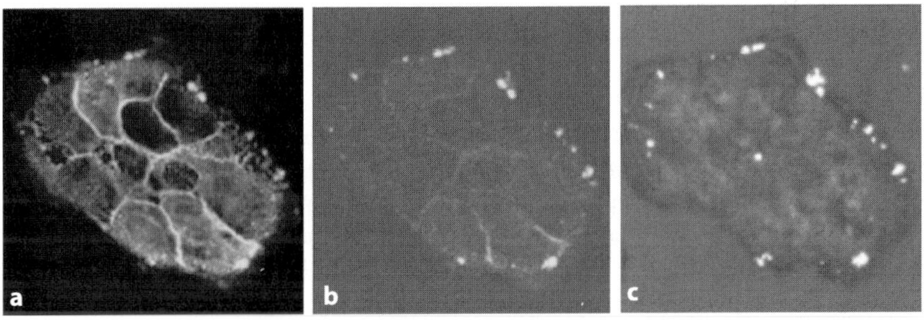

Fig. 2.4. Local activation of ErbB2 by Herceptin-coated microbeads. A431 cells expressing ErbB2 fused to the monomeric mutant of YFP (mYFP) were stimulated with Herceptin-coated magnetic beads and stained with a phosphorylation-specific anti-ErbB2 antibody. Bright spots in the reflection image (**c**) reveal the location of beads. Beads induced recruitment of ErbB2-mYFP (**a**) and localized phosphorylation of ErbB2 (**b**)

Tissues present a highly structured environment for ligand stimulation, and accumulating evidence shows that, at least in some cases, focal stimulation of receptors occurs (Meyer and Birchmeier 1995; Harris et al. 2003; Shilo 2003). Another important question is whether focal receptor stimulation gets generalized through the spread of receptor activation. To this end we used magnetic beads covalently coupled to EGF or Herceptin and showed that receptor activation remained localized to the site where the ligand- or antibody-coated bead bound (Brock and Jovin 2001; Brock and Jovin 2003; Fig. 2.4). Others have found that the extent of signal propagation from the site of receptor stimulation depends on receptor density (Sawano et al. 2002) or the activity of tyrosine phosphatases (Reynolds et al. 2003).

2.6.4 Image Cytometry in Assessing the Activation State of ErbB Receptor Tyrosine Kinases

The "central dogma" related to activation of RTKs postulates ligand-induced dimerization and consequential cross-phosphorylation of the receptors in the dimer. This is frequently followed by receptor internalization and downregulation, leading to the proposition that one mode of action of the anti-ErbB2 therapeutic antibody Herceptin (Trastuzumab) is the partial activation and consequent downregulation of the receptor. In contrast, other antibodies have been produced that do not activate the receptor, but nonetheless enhance internalization (Neve et al. 2001). Hence, experiments that can determine how spontaneous and ligand-induced receptor di- or oligomerization correlate with tyrosine phosphorylation, and how dimerization and/or activation evoked by ligands or antibodies correlates with concomitant internalization, could contribute significantly to our understanding of both signaling through and therapy against members of the ErbB family.

Development of red-shifted fluorescent dyes that allow relatively straightforward implementation of acceptor photobleaching FRET (in particular the indocarbocyanine dye Cy5 paired with either Cy3 or AlexaFluor 546) has opened the possibility to using the conventional laser lines (488, 514, and 543 nm) of confocal microscopes for detecting the emission from one or two additional labels. We used extracellular labeling of ErbB family members on live cells with pairs of antibodies carrying Cy3 and Cy5, respectively, followed by fixation, permeabilization, blocking of intracellular phosphatases, and indirect immunofluorescent labeling of phosphorylated ErbB2 with Ab18 as the primary antibody and AlexaFluor 488-conjugated secondary Fabs. FRET efficiency was measured between ErbB2–ErbB2 and ErbB1–ErbB2 receptor pairs and evaluated with a custom-written algorithm, taking into consideration possible shifts of the sample during imaging, photobleaching of the donor during the protocol, and fluorescence of the photoproducts of the bleached acceptor. The pixel-by-pixel FRET efficiency maps were then correlated with the fluorescence intensity characteristic of activated phospho-ErbB2 using image cross-correlation (Vereb et al. 2000). In SKBR3 breast tumor cells, highest activation of ErbB2 was found in focal adhesion points and at surfaces where adjacent cells were in contact. These areas also exhibited higher ErbB2–ErbB2 FRET values, even though the absolute density of ErbB2 was not higher than average in these pixels. ErbB2–ErbB1 FRET efficiency was neither correlated nor anticorrelated with ErbB2 phosphorylation in resting cells. However, upon stimulation with EGF, the prevalence of ErbB2–ErbB1 dimers or higher oligomers has increased, and ErbB2 phosphorylation also increased in parallel. As opposed to spontaneous ErbB2 homodimers, no preferential subcellular localization of these heterodimers could be observed.

When comparing the Herceptin-sensitive SKBR3 cells with a newly established Herceptin-resistant breast cancer line, it also became evident that basal levels of ErbB2 homoassociation and phosphorylation are both extremely low in the resistant line compared to the sensitive one. Stimulation with Heregulin or EGF, or addition of bivalent Herceptin, increased phosphorylation, homodimerization, and internalization of ErbB2 in SKBR3, but not in the resistant line. Furthermore, the ErbB1–ErbB2 heteroassociation and transactivation via ErbB1 using EGF also failed to evoke a response in the resistant line, in spite of its relatively high expression of ErbB1 compared to SKBR3. Interestingly, when a higher local density of Herceptin was applied using Herceptin-derivatized microbeads (see Sect. 2.6.3), specific but localized tyrosine phosphorylation and concomitant structural changes, likely corresponding to an effort to internalize the Herceptin-coated bead, could be detected also in the resistant line. These together hint that (1) ErbB2 dimers or oligomers could be functional in resting, unstimulated cells and may have a relation to focal adhesion points, (2) the therapeutic antibody Herceptin does promote ErbB2 homoassociation, activation and internalization in ErbB2-overexpressing Herceptin-sensitive cells, and (3) Herceptin resistance can occur in ErbB2-overexpressing cells even though these ErbB2 molecules may be functional, possibly because of steric hindrance effects precluding ErbB2 homoassociation and Herceptin binding, and/or the shift of emphasis in signaling from ErbB2 homodimers and ErbB1–ErbB2 heterodimers to other active ErbB dimers, such as the highly mitogenic ErbB2–ErbB3 pair.

2.6.5 Following the Internalization of Receptors Using Fluorescence Approaches

Activation of ErbB receptors is usually followed by their internalization. This mechanism is thought to restrict the duration of receptor signaling (Yarden and Sliwkowski 2001), although claims abound about the signaling competence of endocytosed receptors (Skarpen et al. 1998; Wiley 2003). Ligands or antibodies tagged with conventional fluorescent dyes and most VFPs suffer photobleaching under long-term illumination which is often necessary to follow intracellular trafficking, making quantitative analysis difficult. Therefore, we used fluorescent QDs that show negligible photobleaching. We coupled biotin-EGF to QDs covalently coated with streptavidin (qdEGF). Due to the extreme tight interaction between biotin and streptavidin, this complex can be used as a fluorescently labeled ligand. We observed specific binding and internalization of qdEGF in EGFR-expressing cells followed by directed motions and fusions of vesicles (Lidke et al. 2004). In addition, we described a novel directed motion of ligand-bound EGFR in filopodia transporting EGF/EGFR complexes towards the cell body. Next, we used a combination of flow and image cytometric approaches to study the effect of ErbB2 or ErbB3 expression on the internalization rate of EGFR. Flow cytometry is unable to discriminate between intracellular and cell-surface-associated fluorescence; therefore we used a brief exposure to acidic pH (2.5–3) which removes noncovalently bound material from the cell surface. We showed that EGF induced the formation of EGFR/ErbB2 dimers. Interestingly, the internalization of these heterodimers was retarded as compared to EGFR homodimers. EGFR/ErbB3 heterodimers did not form in response to EGF stimulation.

An alternative, but equally spectacular way to follow the internalization of receptors is to use immunoliposomes loaded with a fluorescent dye, and coated

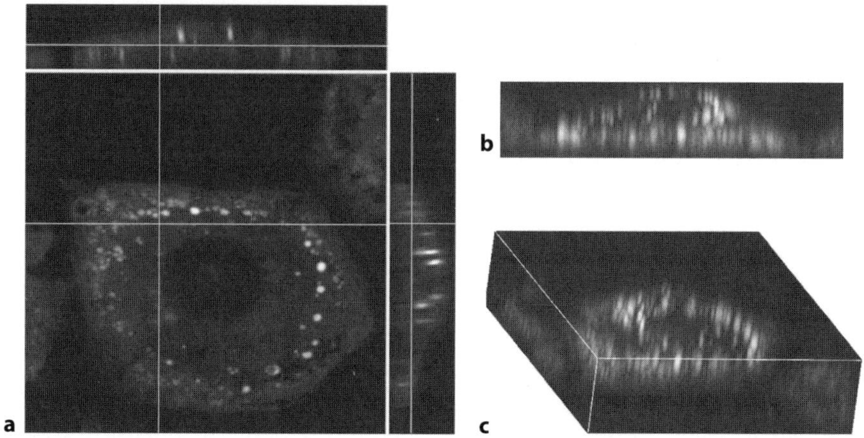

Fig. 2.5. Internalization of 4D5 Fab-coated immunoliposomes in SKBR-3 breast tumor cells. SKBR-3 breast tumor cells were incubated with rhodamine-loaded 4D5 Fab-coated immunoliposomes for 30 min at 37 °C. A stack of images was recorded using confocal microscopy. **a** A single Z slice is shown in the *middle*. An X–Z and a Y–Z vertical section taken at the *horizontal* and *vertical lines* are shown at the *top* and on the *right*, respectively. A side view (**b**) and a 3D projection (**c**) of the same cell are also displayed. **a** was made with Zeiss Image Browser, **b** and **c** were prepared with VoxBlast Light (Vaytek)

with a ligand, antibody or Fab fragment. Such "alternative ligands" have clinical relevance, since immunoliposomes can be loaded with cytostatic drugs and used as targeted drug delivery systems in cancer treatment (Park et al. 1995; Kirpotin et al. 1997). We used anti-ErbB2 Fab or scFv-coated immunoliposomes to show colocalization of liposome binding with lipid rafts and subsequent internalization of ErbB2 (Fig. 2.5; Vereb et al. 2002).

References

Adams SR, Campbell RE, Gross LA, Martin BR, Walkup GK, Yao Y, Llopis J, Tsien RY (2002) New biarsenical ligands and tetracysteine motifs for protein labeling in vitro and in vivo: synthesis and biological applications. J Am Chem Soc 124:6063–6076

Alivisatos AP (1996) Perspectives on the physical chemistry of semiconductor nanocrystals. J Phys Chem 100:13226–13239

Ando R, Hama H, Yamamoto-Hino M, Mizuno H, Miyawaki A (2002) An optical marker based on the UV-induced green-to-red photoconversion of a fluorescent protein. Proc Natl Acad Sci USA 99:12651–12656

Balla T, Varnai P (2002) Visualizing cellular phosphoinositide pools with GFP-fused protein-modules. Sci Signal Transduction Knowledge Environment (STKE), 26 March(125):PL3

Baselga J, Tripathy D, Mendelsohn J, Baughman S, Benz CC, Dantis L, Sklarin NT, Seidman AD, Hudis CA, Moore J, Rosen PP, Twaddell T, Henderson IC, Norton L (1996) Phase II study of weekly intravenous recombinant humanized anti-p185HER2 monoclonal antibody in patients with HER2/neu- overexpressing metastatic breast cancer. J Clin Oncol 14: 737–744

Bastiaens PI, Squire A (1999) Fluorescence lifetime imaging microscopy: spatial resolution of biochemical processes in the cell. Trends Cell Biol 9:48–52

Bastiaens PI, Majoul IV, Verveer PJ, Soling HD, Jovin TM (1996) Imaging the intracellular trafficking and state of the AB5 quaternary structure of cholera toxin. EMBO J 15:4246–4253

Batard P, Szollosi J, Luescher I, Cerottini JC, MacDonald R, Romero P (2002) Use of phycoerythrin and allophycocyanin for fluorescence resonance energy transfer analyzed by flow cytometry: advantages and limitations. Cytometry 48:97–105

Berney C, Danuser G (2003) FRET or no FRET: a quantitative comparison. Biophys J 84: 3992–4010

Betzig E (1995) Proposed method for molecular optical imaging. Opt Lett 20:237–239

Betzig E, Chichester RJ (1993) Single molecules observed by near-field scanning optical microscopy. Science 262:1422–1427

Blackman SM, Piston DW, Beth AH (1998) Oligomeric state of human erythrocyte band 3 measured by fluorescence resonance energy homotransfer. Biophys J 75:1117–1130

Brock R, Jovin TM (2001) Heterogeneity of signal transduction at the subcellular level: microsphere-based focal EGF receptor activation and stimulation of Shc translocation. J Cell Sci 114:2437–2447

Brock R, Jovin TM (2003) Quantitative image analysis of cellular protein translocation induced by magnetic microspheres: application to the EGF receptor. Cytometry 52A:1–11

Brown EB, Campbell RB, Tsuzuki Y, Xu L, Carmeliet P, Fukumura D, Jain RK (2001) In vivo measurement of gene expression, angiogenesis and physiological function in tumors using multiphoton laser scanning microscopy. Nat Med 7:864–868

Bruchez M Jr., Moronne M, Gin P, Weiss S, Alivisatos AP (1998) Semiconductor nanocrystals as fluorescent biological labels. Science 281:2013–2016

Campbell RE, Tour O, Palmer AE, Steinbach PA, Baird GS, Zacharias DA, Tsien RY (2002) A monomeric red fluorescent protein. Proc Natl Acad Sci USA 99:7877–7882

Cardarelli PM, Quinn M, Buckman D, Fang Y, Colcher D, King DJ, Bebbington C, Yarranton G (2002) Binding to CD20 by anti-B1 antibody or $F(ab')^2$ is sufficient for induction of apoptosis in B-cell lines. Cancer Immunol Immunother 51:15–24

Chalfie M (1995) Green fluorescent protein. Photochem Photobiol 62:651–656

Chan WC, Nie S (1998) Quantum dot bioconjugates for ultrasensitive nonisotopic detection. Science 281:2016–2018

Chudakov DM, Belousov VV, Zaraisky AG, Novoselov VV, Staroverov DB, Zorov DB, Lukyanov S, Lukyanov KA (2003) Kindling fluorescent proteins for precise in vivo photolabeling. Nat Biotechnol 21:191–194

Clayton AH, Hanley QS, Arndt-Jovin DJ, Subramaniam V, Jovin TM (2002) Dynamic fluorescence anisotropy imaging microscopy in the frequency domain (rFLIM). Biophys J 83: 1631–1649

Clegg RM (1995) Fluorescence resonance energy transfer. Curr Opin Biotechnol 6:103–110

Clontech (2001) Living colors. User manual. www.bdbiosciences.com/clontech/techinfo/manuals

Dahan M, Levi S, Luccardini C, Rostaing P, Riveau B, Triller A (2003) Diffusion dynamics of glycine receptors revealed by single-quantum dot tracking. Science 302:442–445

Dale RE, Eisinger J, Blumberg WE (1979) The orientational freedom of molecular probes. The orientation factor in intramolecular energy transfer. Biophys J 26:161–193

Damjanovich S, Vereb G, Schaper A, Jenei A, Matkó J, Starink JP, Fox GQ, Arndt Jovin DJ, Jovin TM (1995) Structural hierarchy in the clustering of HLA class I molecules in the plasma membrane of human lymphoblastoid cells. Proc Natl Acad Sci USA 92:1122–1126

Denk W, Strickler JH, Webb WW (1990) Two-photon laser scanning fluorescence microscopy. Science 248:73–76

Desmyter A, Decanniere K, Muyldermans S, Wyns L (2001) Antigen specificity and high affinity binding provided by one single loop of a camel single-domain antibody. J Biol Chem 276:26285–26290

Dexter DL (1953) A theory of sensitized luminescence in solids. J Chem Phys 21:836–850

Dornan S, Sebestyén Z, Gamble J, Nagy P, Bodnár A, Alldridge L, Doe S, Holmes N, Goff LK, Beverley P, Szöllősi J, Alexander DR (2002) Differential association of CD45 isoforms with CD4 and CD8 regulates the actions of specific pools of p56lck tyrosine kinase in T cell antigen receptor signal transduction. J Biol Chem 277:1912–1918

Dyba M, Hell SW (2002) Focal spots of size lambda/23 open up far-field fluorescence microscopy at 33 nm axial resolution. Phys Rev Lett 88:163901–163901-4

Dyba M, Jakobs S, Hell SW (2003) Immunofluorescence stimulated emission depletion microscopy. Nat Biotechnol 21:1303–1304

Edidin M (2001) Shrinking patches and slippery rafts: scales of domains in the plasma membrane. Trends Cell Biol 11:492–496

Egner A, Jakobs S, Hell SW (2002) Fast 100-nm resolution three-dimensional microscope reveals structural plasticity of mitochondria in live yeast. Proc Natl Acad Sci USA 99: 3370–3375

Eisinger J (1976) Energy transfer and dynamic structure. Q Rev Biophys 9:21–33

Elangovan M, Wallrabe H, Chen Y, Day RN, Barroso M, Periasamy A (2003) Characterization of one- and two-photon excitation fluorescence resonance energy transfer microscopy. Methods 29:58–73

Förster T (1946) Energiewanderung und Fluoreszenz. Naturwissenschaften 6:166–175

Gadella TW Jr., Jovin TM (1995) Oligomerization of epidermal growth factor receptors on A431 cells studied by time-resolved fluorescence imaging microscopy. A stereochemical model for tyrosine kinase receptor activation. J Cell Biol 129:1543–1558

Gaietta G, Deerinck TJ, Adams SR, Bouwer J, Tour O, Laird DW, Sosinsky GE, Tsien RY, Ellisman MH (2002) Multicolor and electron microscopic imaging of connexin trafficking. Science 296:503–507

Gautier I, Tramier M, Durieux C, Coppey J, Pansu RB, Nicolas JC, Kemnitz K, Coppey-Moisan M (2001) Homo-FRET microscopy in living cells to measure monomer-dimer transition of GFP-tagged proteins. Biophys J 80:3000–3008

Giordano L, Jovin TM, Irie M, Jares-Erijman EA (2002) Diheteroarylethenes as thermally stable photoswitchable acceptors in photochromic fluorescence resonance energy transfer (pcFRET). J Am Chem Soc 124:7481–7489

Gordon GW, Berry G, Liang XH, Levine B, Herman B (1998) Quantitative fluorescence resonance energy transfer measurements using fluorescence microscopy. Biophys J 74:2702–2713

Griesbeck O, Baird GS, Campbell RE, Zacharias DA, Tsien RY (2001) Reducing the environmental sensitivity of yellow fluorescent protein. Mechanism and applications. J Biol Chem 276: 29188–29194

Griffin BA, Adams SR, Tsien RY (1998) Specific covalent labeling of recombinant protein molecules inside live cells. Science 281:269–272

Griffin BA, Adams SR, Jones J, Tsien RY (2000) Fluorescent labeling of recombinant proteins in living cells with FlAsH. Methods Enzymol 327:565–578

Gustafsson MGL, Agard DA, Sedat JW (1999) I5 M: 3D widefield light microscopy with better than 100 nm axial resolution. J Microsc 195:10–16
Haj FG, Verveer PJ, Squire A, Neel BG, Bastiaens PI (2002) Imaging sites of receptor dephosphorylation by PTP1B on the surface of the endoplasmic reticulum. Science 295:1708–1711
Hanley QS, Verveer PJ, Gemkow MJ, Arndt-Jovin D, Jovin TM (1999) An optical sectioning programmable array microscope implemented with a digital micromirror device. J Microsc 196(3):317–331
Hanley QS, Verveer PJ, Arndt-Jovin DJ, Jovin TM (2000) Three-dimensional spectral imaging by hadamard transform spectroscopy in a programmable array microscope. J Microsc 197(1): 5–14
Harder T, Scheiffele P, Verkade P, Simons K (1998) Lipid domain structure of the plasma membrane revealed by patching of membrane components. J Cell Biol 141:929–942
Harris RC, Chung E, Coffey RJ (2003) EGF receptor ligands. Exp Cell Res 284:2–13
Heim R, Prasher DC, Tsien RY (1994) Wavelength mutations and posttranslational autoxidation of green fluorescent protein. Proc Natl Acad Sci USA 91:12501–12504
Heim R, Cubitt AB, Tsien RY (1995) Improved green fluorescence. Nature 373:663–664
Heim R, Tsien RY (1996) Engineering green fluorescent protein for improved brightness, longer wavelengths and fluorescence resonance energy transfer. Curr Biol 6:178–182
Heintzmann R, Cremer C (2002) Axial tomographic confocal fluorescence microscopy. J Microsc 206:7–23
Heintzmann R, Hanley QS, Arndt-Jovin D, Jovin TM (2001) A dual path programmable array microscope (PAM): simultaneous acquisition of conjugate and non-conjugate images. J Microsc 204:119–135
Hell S, Stelzer EHK (1992) Fundamental improvement of resolution with a 4Pi-confocal fluorescence microscope using two-photon excitation. Opt Commun 93:277–282
Hell SW (2003) Toward fluorescence nanoscopy. Nat Biotechnol 21:1347–1355
Holbro T, Civenni G, Hynes NE (2003) The ErbB receptors and their role in cancer progression. Exp Cell Res 284:99–110
Hollinshead M, Sanderson J, Vaux DJ (1997) Anti-biotin antibodies offer superior organelle-specific labeling of mitochondria over avidin or streptavidin. J Histochem Cytochem 45:1053–1057
Hu CD, Kerppola TK (2003) Simultaneous visualization of multiple protein interactions in living cells using multicolor fluorescence complementation analysis. Nat Biotechnol 21:539–545
Hu CD, Chinenov Y, Kerppola TK (2002) Visualization of interactions among bZIP and Rel family proteins in living cells using bimolecular fluorescence complementation. Mol Cell 9:789–798
Hudson PJ, Souriau C (2003) Engineered antibodies. Nat Med 9:129–134
Hwang J, Tamm LK, Böhm C, Ramalingam TS, Betzig E, Edidin M (1995) Nanoscale complexity of phospholipid monolayers investigated by near-field scanning optical microscopy. Science 270:610–614
Hwang J, Gheber LA, Margolis L, Edidin M (1998) Domains in cell plasma membranes investigated by near-field scanning optical microscopy. Biophys J 74:2184–2190
Jaiswal JK, Mattoussi H, Mauro JM, Simon SM (2003) Long-term multiple color imaging of live cells using quantum dot bioconjugates. Nat Biotechnol 21:47–51
Jares-Erijman EA, Jovin TM (2003) FRET imaging. Nat Biotechnol 21:1387–1395
Jayasena SD (1999) Aptamers: an emerging class of molecules that rival antibodies in diagnostics. Clin Chem 45:1628–1650
Jenei A, Kirsch AK, Subramaniam V, Arndt-Jovin DJ, Jovin TM (1999) Picosecond multiphoton scanning near-field optical microscopy. Biophys J 76:1092–1100

Jovin TM (2003) Quantum dots finally come of age. Nat Biotechnol 21:32–33

Jovin TM, Arndt Jovin DJ (1989) FRET microscopy: digital imaging of fluorescence resonance energy transfer. Application in cell biology. In: Kohen E, Hirschberg JG (eds) Cell structure and function by microspectrofluorometry. Academic Press, San Diego, pp 99–117

Jovin TM, Morris SJ, Striker G, Schultens HA, Digweed M, Arndt-Jovin DJ (1976) Automatic sizing and separation of particles by ratios of light scattering intensities. J Histochem Cytochem 24:269–283

Ju W, Morishita W, Tsui J, Gaietta G, Deerinck TJ, Adams SR, Garner CC, Tsien RY, Ellisman MH, Malenka RC (2004) Activity-dependent regulation of dendritic synthesis and trafficking of AMPA receptors. Nat Neurosci 7:244–253

Kim S, Lim YT, Soltesz EG, De Grand AM, Lee J, Nakayama A, Parker JA, Mihaljevic T, Laurence RG, Dor DM, Cohn LH, Bawendi MG, Frangioni JV (2003) Near-infrared fluorescent type II quantum dots for sentinel lymph node mapping. Nat Biotechnol 22:93–97

Kirpotin D, Park JW, Hong K, Zalipsky S, Li WL, Carter P, Benz CC, Papahadjopoulos D (1997) Sterically stabilized anti-HER2 immunoliposomes: design and targeting to human breast cancer cells in vitro. Biochemistry 36:66–75

Kirsch AK, Subramaniam V, Striker G, Schnetter C, Arndt Jovin D, Jovin TM (1998) Continuous wave two-photon scanning near-field optical microscopy. Biophys J 75:1513–1521

Koyoma-Honda I, Ritchie K, Fujiwara T, Iino R, Murakoshi H, Kasai RS, Kusumi A (2005) Fluorescence imaging for monitoring the colocalization of two single molecules in living cells. Biophys J 88:2126–2136

Kurokawa K, Mochizuki N, Ohba Y, Mizuno H, Miyawaki A, Matsuda M (2001) A pair of fluorescent resonance energy transfer-based probes for tyrosine phosphorylation of the CrkII adaptor protein in vivo. J Biol Chem 276:31305–31310

Labas YA, Gurskaya NG, Yanushevich YG, Fradkov AF, Lukyanov KA, Lukyanov SA, Matz MV (2002) Diversity and evolution of the green fluorescent protein family. Proc Natl Acad Sci USA 99:4256–4261

Le XF, Vadlamudi R, McWatters A, Bae DS, Mills GB, Kumar R, Bast RC Jr (2000) Differential signaling by an anti-p185(HER2) antibody and heregulin. Cancer Res 60:3522–3531

Lewis A (1991) The optical near-field and cell biology. Semin Cell Biol 2:187–192

Lewis A, Isaacson M, Harootunian A, Muray A (1984) Development of a 500 Å spatial resolution light microscope. I. Light is efficiently transmitted through lambda/16 diameter apertures. Ultramicroscopy 13:227–231

Lewis A, Taha H, Strinkovski A, Manevitch A, Khatchatouriants A, Dekhter R, Ammann E (2003) Near-field optics: from subwavelength illumination to nanometric shadowing. Nat Biotechnol 21:1378–1386

Lidke DS, Nagy P, Barisas BG, Heintzmann R, Post JN, Lidke KA, Clayton AH, Arndt-Jovin DJ, Jovin TM (2003) Imaging molecular interactions in cells by dynamic and static fluorescence anisotropy (rFLIM and emFRET). Biochem Soc Trans 31:1020–1027

Lidke DS, Nagy P, Heintzmann R, Arndt-Jovin DJ, Post JN, Grecco HE, Jares-Erijman EA, Jovin TM (2004) Quantum dot ligands provide new insights into erbB/HER receptor-mediated signal transduction. Nat Biotechnol 22:198–203

Matkó J, Szöllősi J, Trón L, Damjanovich S (1988) Luminescence spectroscopic approaches in studying cell surface dynamics. Q Rev Biophys 21:479–544

Matz MV, Fradkov AF, Labas YA, Savitsky AP, Zaraisky AG, Markelov ML, Lukyanov SA (1999) Fluorescent proteins from nonbioluminescent *Anthozoa* species. Nat Biotechnol 17: 969–973

McDonald DM, Choyke PL (2003) Imaging of angiogenesis: from microscope to clinic. Nat Med 9:713–725

Mekler VM (1994) A photochemical technique to enhance sensitivity of detection of fluorescence resonance energy transfer. Photochem Photobiol 59:615–620

Mekler VM, Averbakh AZ, Sudarikov AB, Kharitonova OV (1997) Fluorescence energy transfer-sensitized photobleaching of a fluorescent label as a tool to study donor–acceptor distance distributions and dynamics in protein assemblies: studies of a complex of biotinylated IgM with streptavidin and aggregates of concanavalin A. J Photochem Photobiol B 40:278–287

Meyer D, Birchmeier C (1995) Multiple essential functions of neuregulin in development. Nature 378:386–390

Michalet X, Pinaud F, Lacoste TD, Dahan M, Bruchez MP, Alivisatos AP, Weiss S (2001) Properties of fluorescent semiconductor nanocrystals and their application to biological labeling. Single Mol 2:261–276

Miyawaki A, Llopis J, Heim R, McCaffery JM, Adams JA, Ikura M, Tsien RY (1997) Fluorescent indicators for Ca^{2+} based on green fluorescent proteins and calmodulin. Nature 388:882–887

Miyawaki A, Griesbeck O, Heim R, Tsien RY (1999) Dynamic and quantitative Ca^{2+} measurements using improved cameleons. Proc Natl Acad Sci USA 96:2135–2140

Nagai T, Ibata K, Park ES, Kubota M, Mikoshiba K, Miyawaki A (2002) A variant of yellow fluorescent protein with fast and efficient maturation for cell-biological applications. Nat Biotechnol 20:87–90

Nagy P, Bene L, Balázs M, Hyun WC, Lockett SJ, Chiang NY, Waldman FM, Feuerstein BG, Damjanovich S, Szöllősi J (1998a) EGF-induced redistribution of erbB2 on breast tumor cells: flow and image cytometric energy transfer measurements. Cytometry 32:120–131

Nagy P, Vámosi G, Bodnár A, Lockett SJ, Szöllősi J (1998b) Intensity-based energy transfer measurements in digital imaging microscopy. Eur Biophys J 27:377–389

Nagy P, Jenei A, Damjanovich S, Jovin TM, Szöllősi J (1999a) Complexity of signal transduction mediated by erbB2: clues to the potential of receptor-targeted cancer therapy. Pathol Oncol Res 5:255–271

Nagy P, Jenei A, Kirsch AK, Szöllősi J, Damjanovich S, Jovin TM (1999b) Activation-dependent clustering of the erbB2 receptor tyrosine kinase detected by scanning near-field optical microscopy. J Cell Sci 112:1733–1741

Nagy P, Vereb G, Sebestyen Z, Horvath G, Lockett SJ, Damjanovich S, Park JW, Jovin TM, Szöllősi J (2002) Lipid rafts and the local density of ErbB proteins influence the biological role of homo- and heteroassociations of ErbB2. J Cell Sci 115:4251–4262

Neve RM, Nielsen UB, Kirpotin D, Poul MA, Marks JD, Benz CC (2001) Biological effects of anti-ErbB2 single chain antibodies selected for internalizing function. Biochem Biophys Res Commun 280:274–279

Niswender KD, Blackman SM, Rohde L, Magnuson MA, Piston DW (1995) Quantitative imaging of green fluorescent protein in cultured cells: comparison of microscopic techniques, use in fusion proteins and detection limits. J Microsc 180(2):109–116

Oancea E, Teruel MN, Quest AF, Meyer T (1998) Green fluorescent protein (GFP)-tagged cysteine-rich domains from protein kinase C as fluorescent indicators for diacylglycerol signaling in living cells. J Cell Biol 140:485–498

Oertner TG (2002) Functional imaging of single synapses in brain slices. Exp Physiol 87:733–736

Ormerod MG (1994) Flow cytometry. A practical approach. Oxford University Press, Oxford

Ormo M, Cubitt AB, Kallio K, Gross LA, Tsien RY, Remington SJ (1996) Crystal structure of the *Aequorea victoria* green fluorescent protein. Science 273:1392–1395

Park JW, Hong K, Carter P, Asgari H, Guo LY, Keller GA, Wirth C, Shalaby R, Kotts C, Wood WI, Papahadjopoulos D, Benz CC (1995) Development of anti-p185HER2 immunoliposomes for cancer therapy. Proc Natl Acad Sci USA 92:1327–1331

Patterson GH, Lippincott-Schwartz J (2002) A photoactivatable GFP for selective photolabeling of proteins and cells. Science 297:1873–1877

Patterson GH, Knobel SM, Sharif WD, Kain SR, Piston DW (1997) Use of the green fluorescent protein and its mutants in quantitative fluorescence microscopy. Biophys J 73:2782–2790

Patterson GH, Piston DW, Barisas BG (2000) Forster distances between green fluorescent protein pairs. Anal Biochem 284:438–440

Pawley J (1995) Handbook of biological confocal microscopy. Plenum Press, New York

Peng X, Schlamp MC, Kadavanich AV, Alivisatos AP (1997) Epitaxial growth of highly luminescent CdSe/CdS core/shell nanocrystals with photostability and electronic accessibility. J Am Chem Soc 119:7019–7029

Perrin F (1932) Théorie quantique des transferts d'activation entre molécules de méme espéce. Cas des solutions fluorescentes. Ann Phys (Paris) 17:283–314

Post JN, Lidke KA, Rieger B, Arndt-Jovin DJ (2005) One- and two-photon photoactivation of a paGFP-fusion protein in live *Drosophila* embryos. 579:325–330

Rehm M, Dussmann H, Janicke RU, Tavare JM, Kogel D, Prehn JH (2002) Single-cell fluorescence resonance energy transfer analysis demonstrates that caspase activation during apoptosis is a rapid process. Role of caspase-3. J Biol Chem 277:24506–24514

Rekas A, Alattia JR, Nagai T, Miyawaki A, Ikura M (2002) Crystal structure of venus, a yellow fluorescent protein with improved maturation and reduced environmental sensitivity. J Biol Chem 277:50573–50578

Reynolds AR, Tischer C, Verveer PJ, Rocks O, Bastiaens PI (2003) EGFR activation coupled to inhibition of tyrosine phosphatases causes lateral signal propagation. Nat Cell Biol 5:447–453

Rizzo MA, Shome K, Watkins SC, Romero G (2000) The recruitment of Raf-1 to membranes is mediated by direct interaction with phosphatidic acid and is independent of association with Ras. J Biol Chem 275:23911–23918

Rizzo MA, Springer GH, Granada B, Piston DW (2004) An improved cyan fluorescent protein variant useful for FRET. Nat Biotechnol 22:445–449

Rocheleau JV, Edidin M, Piston DW (2003) Intrasequence GFP in class I MHC molecules, a rigid probe for fluorescence anisotropy measurements of the membrane environment. Biophys J 84:4078–4086

Runnels LW, Scarlata SF (1995) Theory and application of fluorescence homotransfer to melittin oligomerization. Biophys J 69:1569–1583

Sacchetti A, Subramaniam V, Jovin TM, Alberti S (2002) Oligomerization of DsRed is required for the generation of a functional red fluorescent chromophore. FEBS Lett 525:13–19

Sato M, Ozawa T, Inukai K, Asano T, Umezawa Y (2002) Fluorescent indicators for imaging protein phosphorylation in single living cells. Nat Biotechnol 20:287–294

Sausville EA, Elsayed Y, Monga M, Kim G (2003) Signal transduction-directed cancer treatments. Annu Rev Pharmacol Toxicol 43:199–231

Sawano A, Takayama S, Matsuda M, Miyawaki A (2002) Lateral propagation of EGF signaling after local stimulation is dependent on receptor density. Dev Cell 3:245–257

Sebestyén Z, Nagy P, Horváth G, Vámosi G, Debets R, Gratama JW, Alexander DR, Szöllősi J (2002) Long wavelength fluorophores and cell-by-cell correction for autofluorescence significantly improves the accuracy of flow cytometric energy transfer measurements on a dual-laser benchtop flow cytometer. Cytometry 48:124–135

Shapiro HM (2002) Practical flow cytometry. Wiley, New York

Sheppard CJR, Gu M (1990) Image formation in 2-photon fluorescence microscopy. Optik 86:104–106

Shilo BZ (2003) Signaling by the *Drosophila* epidermal growth factor receptor pathway during development. Exp Cell Res 284:140–149

Shimomura O, Johnson FH, Saiga Y (1962) Extraction, purification and properties of aequorin, a bioluminescent protein from the luminous hydromedusan, *Aequorea*. J Cell Comp Physiol 59:223–239

Simons K, Toomre D (2000) Lipid rafts and signal transduction. Nat Rev Mol Cell Biol 1:31–39

Skarpen E, Johannessen LE, Bjerk K, Fasteng H, Guren TK, Lindeman B, Thoresen GH, Christoffersen T, Stang E, Huitfeldt HS, Madshus IH (1998) Endocytosed epidermal growth factor (EGF) receptors contribute to the EGF-mediated growth arrest in A431 cells by inducing a sustained increase in p21/CIP1. Exp Cell Res 243:161–172

Sliwkowski MX, Schaefer G, Akita RW, Lofgren JA, Fitzpatrick VD, Nuijens A, Fendly BM, Cerione RA, Vandlen RL, Carraway KL (1994) Coexpression of erbB2 and erbB3 proteins reconstitutes a high affinity receptor for heregulin. J Biol Chem 269:14661–14665

Soling A, Simm A, Rainov N (2002) Intracellular localization of *Herpes simplex* virus type 1 thymidine kinase fused to different fluorescent proteins depends on choice of fluorescent tag. FEBS Lett 527:153–158

Song L, Hennink EJ, Young IT, Tanke HJ (1995) Photobleaching kinetics of fluorescein in quantitative fluorescence microscopy. Biophys J 68:2588–2600

Song L, Varma CA, Verhoeven JW, Tanke HJ (1996) Influence of the triplet excited state on the photobleaching kinetics of fluorescein in microscopy. Biophys J 70:2959–2968

Song L, van Gijlswijk RPM, Young IT, Tanke HJ (1997) Influence of fluorochrome labeling density on the photobleaching kinetics of fluorescein in microscopy. Cytometry 27:213–223

Sorkin A, McClure M, Huang F, Carter R (2000) Interaction of EGF receptor and grb2 in living cells visualized by fluorescence resonance energy transfer (FRET) microscopy. Curr Biol 10:1395–1398

Squire A, Verveer PJ, Rocks O, Bastiaens PI (2004) Red-edge anisotropy microscopy enables dynamic imaging of homo-FRET between green fluorescent proteins in cells. J Struct Biol 147:62–69

Stryer L (1978) Fluorescence energy transfer as a spectroscopic ruler. Annu Rev Biochem 47:819–846

Szöllősi J, Trón L, Damjanovich S, Helliwell SH, Arndt Jovin D, Jovin TM (1984) Fluorescence energy transfer measurements on cell surfaces: a critical comparison of steady-state fluorimetric and flow cytometric methods. Cytometry 5:210–216

Szöllősi J, Mátyus L, Trón L, Balázs M, Ember I, Fulwyler MJ, Damjanovich S (1987) Flow cytometric measurements of fluorescence energy transfer using single laser excitation. Cytometry 8:120–128

Szöllősi J, Damjanovich S, Mátyus L (1998) Application of fluorescence resonance energy transfer in the clinical laboratory: routine and research. Cytometry 34:159–179

Tarnok A, Gerstner AO (2002) Clinical applications of laser scanning cytometry. Cytometry 50:133–143

Teruel MN, Meyer T (2002) Parallel single-cell monitoring of receptor-triggered membrane translocation of a calcium-sensing protein module. Science 295:1910–1912

Tour O, Meijer RM, Zacharias DA, Adams SR, Tsien RY (2003) Genetically targeted chromophore-assisted light inactivation. Nat Biotechnol 21:1505–1508

Trón L, Szöllősi J, Damjanovich S, Helliwell SH, Arndt Jovin DJ, Jovin TM (1984) Flow cytometric measurement of fluorescence resonance energy transfer on cell surfaces. Quantitative evaluation of the transfer efficiency on a cell-by-cell basis. Biophys J 45:939–946

Tsien RY (1998) The green fluorescent protein. Annu Rev Biochem 67:509–544

Tzahar E, Waterman H, Chen X, Levkowitz G, Karunagaran D, Lavi S, Ratzkin BJ, Yarden Y (1996) A hierarchical network of interreceptor interactions determines signal transduction by Neu differentiation factor/neuregulin and epidermal growth factor. Mol Cell Biol 16:5276–5287

Uster PS, Pagano RE (1986) Resonance energy transfer microscopy: observations of membrane-bound fluorescent probes in model membranes and in living cells. J Cell Biol 103: 1221–1234

van der Linden RH, Frenken LG, de Geus B, Harmsen MM, Ruuls RC, Stok W, de Ron L, Wilson S, Davis P, Verrips CT (1999) Comparison of physical chemical properties of llama VHH antibody fragments and mouse monoclonal antibodies. Biochim Biophys Acta 1431:37–46

Vereb G, Meyer CK, Jovin TM (1997) Novel microscope-based approaches for the investigation of protein–protein interactions in signal transduction. NATO ASI Ser H102:49–52

Vereb G, Matkó J, Vámosi G, Ibrahim SM, Magyar E, Varga S, Szöllősi J, Jenei A, Gáspár R Jr, Waldmann TA, Damjanovich S (2000) Cholesterol-dependent clustering of IL-2Ralpha and its colocalization with HLA and CD48 on T lymphoma cells suggest their functional association with lipid rafts. Proc Natl Acad Sci USA 97:6013–6018

Vereb G, Nagy P, Park JW, Szöllősi J (2002) Signaling revealed by mapping molecular interactions: implications for ErbB-targeted cancer immunotherapies. Clin Appl Immunol Rev 2:169–186

Volkmer A, Subramaniam V, Birch DJ, Jovin TM (2000) One- and two-photon excited fluorescence lifetimes and anisotropy decays of green fluorescent proteins. Biophys J 78: 1589–1598

Weber G, Shinitzky M (1970) Failure of energy transfer between identical aromatic molecules on excitation at the long wave edge of the absorption spectrum. Proc Natl Acad Sci USA 65:823–830

Wiley HS (2003) Trafficking of the ErbB receptors and its influence on signaling. Exp Cell Res 284:78–88

Wu X, Liu H, Liu J, Haley KN, Treadway JA, Larson JP, Ge N, Peale F, Bruchez MP (2003) Immunofluorescent labeling of cancer marker Her2 and other cellular targets with semiconductor quantum dots. Nat Biotechnol 21:41–46

Xu C, Zipfel W, Shear JB, Williams RM, Webb WW (1996) Multiphoton fluorescence excitation: new spectral windows for biological nonlinear microscopy. Proc Natl Acad Sci USA 93: 10763–10768

Yang F, Moss LG, Phillips GN Jr (1996) The molecular structure of green fluorescent protein. Nat Biotechnol 14:1246–1251

Yarden Y, Sliwkowski MX (2001) Untangling the ErbB signalling network. Nat Rev Mol Cell Biol 2:127–137

Young RM, Arnette JK, Roess DA, Barisas BG (1994) Quantitation of fluorescence energy transfer between cell surface proteins via fluorescence donor photobleaching kinetics. Biophys J 67:881–888

Youvan DC, Michel-Beyerle ME (1996) Structure and fluorescence mechanism of GFP. Nat Biotechnol 14:1219–1220

Zacharias DA, Violin JD, Newton AC, Tsien RY (2002) Partitioning of lipid-modified monomeric GFPs into membrane microdomains of live cells. Science 296:913–916

Zhang J, Campbell RE, Ting AY, Tsien RY (2002) Creating new fluorescent probes for cell biology. Nat Rev Mol Cell Biol 3:906–918

Zipfel WR, Williams RM, Webb WW (2003) Nonlinear magic: multiphoton microscopy in the biosciences. Nat Biotechnol 21:1369–1377

Non-Random Patterns of Membrane Proteins and Their Roles in Transmembrane Signaling

Andrea Bodnár, György Vámosi, Katalin Tóth, Attila Jenei, László Mátyus, Sándor Damjanovich

3.1 Introduction

The plasma membrane is the first line of encounter with any "own" or "foreign" molecule or hostile "visitor". Our knowledge of the structure and function of the plasma membrane of mammalian cells was greatly advanced by the famous experiment of Frye and Edidin in 1970, which showed that distinct molecular elements of the surface of mouse and human lymphocytes labeled by green and red fluorescent dyes, respectively, were mixed upon fusing the two cell types. This finding was instrumental in the construction of the Singer–Nicolson fluid mosaic membrane model in 1972, which suggested the free mobility of proteins and other macromolecules in the plasma membrane (Singer and Nicolson 1972).

Experimental observations over past decades have led to the "membrane microdomain" concept describing compartmentalization/organization of membrane components into non-random, well-defined patterns. According to recent knowledge, the lateral order of membrane lipids and proteins is a general phenomenon that may have fundamental importance in the function of individual molecules as well as in that of the whole plasma membrane (Edidin 1993, 2001; Vereb et al. 2003).

The primary target of external stimuli is the plasma membrane, and non-random co-distribution of membrane proteins plays an important role in signal transduction across the cell membrane. The generalized occurrence of such cell-surface patterns of membrane proteins was suggested at the beginning of the 1980s (Damjanovich et al. 1981). Signal transduction processes are often accompanied by dynamic rearrangement of the two-dimensional patterns of the macromolecular constituents at the cell surface (Damjanovich et al. 1992). The structured, and at the same time dynamic, nature of the plasma membrane, i.e., the organization of its components into structures existing at different time and size scales, allows accumulation of the relevant molecules (and exclusion of others) needed for the appropriate response (Vereb et al. 1995, 2003; Edidin 2001).

Protein clusters generated by the physical association/molecular proximity of the molecules (nanometer or small-scale clusters) define the basic organization level of membrane proteins (Damjanovich et al. 1998). Different types of such protein clusters can be present in the plasma membrane (Damjanovich et al. 1997b): (1) in many cases, a given "functional unit" is formed by several components/subunits (e.g., the TCR/CD3 complex; Klausner et al. 1990); (2) rearrange-

Springer Series in Biophysics
S. Damjanovich (Ed.)
Biophysical Aspects of Transmembrane Signaling

ment of proteins in the plane of the plasma membrane caused by an external stimulus may also lead to the formation of protein patterns (e.g., aggregation of the epidermal growth factor receptor upon ligand binding (Zidovetzki et al. 1981)); and (3) apart from the aforesaid examples, where the individual components of protein patterns must be together before or upon ligand binding/external stimuli, seemingly unrelated proteins are also associated in many cases and held together in a dynamic cluster (e.g., Kv1.3 potassium channels and the TCR/CD3 complex (Panyi et al. 2003, 2004a)). Revealing the interactions between such proteins may call our attention to their potential functional relationship (Panyi et al. 2004b, c).

On the basis of presently available data, there are at least two well-defined hierarchical levels of non-random protein patterns in the cell membrane: in addition to the small (nanometer)-scale protein clusters, accumulation of these clusters in larger islands at the submicrometer/micrometer scale can be observed in many cases (Damjanovich et al. 1997b, 2002; Vereb et al. 2003).

The inhomogeneous cell-surface distribution (lateral order) of membrane proteins is the consequence of their restricted lateral diffusion, for which the following factors may be responsible (reviewed in Edidin 1993; Kusumi and Sako 1996; Damjanovich et al. 1997b; Vereb et al. 2003):

1. *Forces acting in the plane of the membrane*: lipids with similar physicochemical character (saturation, length, etc.) tend to form distinct microdomains in model membranes as well as in biological membranes (Edidin 1997). The lipid domain structure of the plasma membrane may cause selective accumulation of membrane proteins or their exclusion from distinct membrane areas through preferential interaction of the transmembrane region of a given protein with a select class of lipids. In addition to lipid–lipid interactions (Welti and Glaser 1994), protein–lipid and protein–protein interactions (e.g., between transmembrane α-helices) may also contribute to the stability of protein clusters and membrane microdomains (Lemmon and Engelman 1994; Tocanne et al. 1994). Microviscosity/orderedness of membrane regions may also physically affect the stability of protein clusters. A lipid raft is a specific type of membrane microdomain characterized by high cholesterol and (glyco)sphingolipid content, glycosyl-phoshatidyl-inositol (GPI)-anchored proteins, and high microviscosity ("liquid ordered phase") (Simons and Ikonen 1997; Edidin 2003; Simons and Vaz 2004). Since rafts also accumulate certain membrane proteins involved in signaling (receptors, ion channels, etc.) and intracellular signaling molecules (e.g., kinases, phosphatases, G-proteins), their role in controlling signaling processes is one of the hottest topics in current membrane biology (Hoessli et al. 2000; Horejsi 2002, 2003; Matkó and Szöllősi 2002; Lucero and Robbins 2004; Yang et al. 2004).
2. *Intracellular factors*: the plasma membrane-associated part of the cytoskeleton (membrane skeleton) can directly (e.g., through interaction with adapter proteins) or indirectly (steric hindrance) restrict the motion of proteins by trapping them in a given membrane area (Ritchie and Kusumi 2004). Free diffusion of membrane proteins can also be hindered by interaction with different cytosolic signaling elements (e.g., G-proteins, kinases, etc.).

3. *Vesicular transport*: vesicular transport mechanisms can also contribute to the selective accumulation of membrane proteins by means of "directed" transport of vesicular components to a given membrane region (Gheber and Edidin 1999; Tang and Edidin 2001). In polarized cells, the localization of lipids and proteins in the apical, basal or lateral surfaces is genetically determined. For quasi-symmetrical cells such as lymphocytes in the blood circulation, such determination seems to be less likely. However, receptor subunits, or proteins destined to cooperate closely in a larger supramolecular assembly, are likely to be synthesized simultaneously in a synchronized way upon the initiation of the same signals. After their synthesis they might be accommodated by the same lipid vesicles, and transported together to the plasma membrane to form a functional membrane microdomain (Damjanovich et al. 2002).
4. *Extracellular factors*: cross-linking ligands and the extracellular matrix may also promote the formation of protein complexes.
5. *Other factors*: changes in the membrane potential or other perturbing factors acting indirectly either through inducing conformational changes or via modification of the factors listed above.

There are several ways to detect molecular associations in the plasma membrane (Damjanovich et al. 1997b, 1999). The most commonly applied biochemical and immunological approaches (chemical cross-linking, co-immunoprecipitation, etc.) provide valuable information on the interaction of membrane proteins, but have several disadvantages. In the first place, with these methods proteins cannot be studied in their native environment. Second, the application of usual extraction and isolation procedures inherent in these techniques may also disrupt protein–protein interactions or, instead, may induce the formation of artificial protein aggregates. Fluorescence resonance energy transfer (FRET) gives us a suitable and convenient tool by which the distribution of membrane proteins can be mapped on intact cells in situ without a major interference with the physiological condition of the cells. Several techniques have been developed to measure FRET on cell surfaces (Damjanovich et al. 1977, 1997b, 1999; Szöllősi et al. 1998). In Chapter 2 (this Vol.), different microscopy-based FRET methodologies are described; in this chapter we give a detailed, protocol-like description of flow cytometric FRET measurements.

Most FRET methods have the limitation that they record a static picture, a "snapshot" of molecular proximities of the studied proteins, and give no information on the stability of protein–protein interactions. Co-localization at the nanometer scale revealed by FRET does not necessarily mean that the studied proteins form stable complexes with one another. Demonstrating the joint diffusive motion of proteins, i.e., their co-mobility, provides evidence of their stable interaction, which can be studied by a dynamic method: fluorescence cross-correlation spectroscopy. This method analyzes the temporal fluctuations of fluorescence signals arising from molecules diffusing across a subfemtoliter volume element, and yields the mobility and co-mobility parameters of molecules.

In this chapter, we will mostly focus on results gained by FRET and fluorescence cross-correlation spectroscopic (FCCS) experiments, but supporting data obtained by other biophysical or biochemical methods will also be mentioned. Functional implications of the presented protein patterns will be discussed as well.

3.2 Fluorescence Resonance Energy Transfer (FRET)

Fluorescence resonance energy transfer (FRET) is a non-radiative process taking place between two dye molecules, whereby the excited donor transfers its excitation energy to a nearby acceptor via dipole–dipole coupling (Förster 1948). In order for FRET to occur it is necessary that the dipole moments of the dyes have a proper relative orientation and the emission spectrum of the donor and the absorption spectrum of the acceptor molecule overlap with each other (Clegg 1996). The rate of energy transfer is inversely proportional to the sixth power of the donor–acceptor distance. In combination with specific fluorescence labeling techniques, the method is a sensitive tool for the determination of inter- or intramolecular distances in the 1- to 10-nm range (Stryer and Haugland 1967).

Measurement of FRET by microscopy (see Chap. 2, this Vol.) provides subcellular mapping of protein–protein interactions, allowing the visual identification of compartments/organelles where the interactions of interest take place. However, besides this undoubted advantage, FRET microscopy has the drawback that the evaluation of a sufficiently high number of cells required for obtaining reasonable statistics is rather time-consuming. Both the acquisition and analysis of microscopic data are labor-intensive. To determine small quantitative changes in protein–protein interactions, usually at least several hundreds or thousands of cells must be evaluated. This can be most easily done by means of flow cytometric FRET measurements, as described later in this chapter (Szöllősi et al. 1998). The high number of cells that can be analyzed by flow cytometry provides for excellent statistics, high accuracy and reproducibility.

3.2.1 Measurement of FRET by Flow Cytometry

The procedure to measure FRET on a cell-by-cell basis by flow cytometry was introduced in 1983 (Damjanovich et al. 1983). The original method (Szöllősi et al. 1984; Trón et al. 1984) was devised for a single-laser setup using two laser lines (the 488- and 514-nm lines of an Ar ion laser), which is the minimum technical requirement of the technique. The sensitivity of the method has been improved by using longer excitation wavelengths along with long wavelength dyes to reduce the contribution of cellular autofluorescence, and by introducing cell-by-cell autofluorescence correction by applying three-color excitation (Sebestyén et al. 2002). In this chapter, the currently most advanced version of the method will be described.

The method, referred to as flow cytometric energy transfer (FCET), is based on the cell-by-cell measurement of three (in the case of cell-by-cell autofluorescence correction, four) fluorescence intensities, which are combinations of autofluorescence, donor fluorescence, sensitized acceptor fluorescence and directly excited acceptor fluorescence signals. The donor and acceptor dyes should be selected to match the wavelengths of the laser lines used for excitation, and should fulfill the requirement of spectral overlap between donor emission and acceptor absorption. In this chapter, the equations will be given as appropriate for the Becton Dickinson FACSDIVA instrument equipped with three lasers: an Ar ion laser (emitting at 488 nm), a solid state laser (532 nm) and a HeNe laser (633 nm). Adequate donor dyes for this setup include Alexa Fluor 546, Alexa Fluor 568 and Cy3, whereas Alexa Fluor 633, Alexa Fluor 647 and Cy5 can be used as acceptor. The experimental and numerical procedures can be used with minor modifications for other instrumental setups as well.

3.2.1.1 Sample Preparation

Cell-surface proteins can be labeled by (preferably monoclonal) antibodies or their Fab fragments that had been tagged with fluorescent dye molecules (Szöllősi et al. 1996; Matkó and Edidin 1997). Polyclonal antibodies may have the capacity to induce aggregation of the labeled proteins, resulting in a rearrangement of protein–protein interactions and the structure of membrane microdomains (patching, capping). For the measurement of FCET, four types of samples are needed: unlabeled, donor-only, acceptor-only and double-labeled ones.

3.2.1.2 Signals and Calculations

The determination of FRET by flow cytometry is based on the measurement of a combination of four fluorescence signals measured from double-labeled cells. The equations given below involve cellular autofluorescence and specific donor and acceptor fluorescence, and account for the decrease in donor intensity ("donor quenching") and the increase in acceptor intensity ("sensitized emission") due to FRET. The measured signals are as follows:

$$FL1(488,510\pm 20) = AF \tag{3.1}$$

$$FL2(532,575\pm 13) = AF \cdot B_2 + I_D(1-E) \tag{3.2}$$

$$FL3(532,>650) = AF \cdot B_3 + I_D(1-E)\cdot S_1 + I_A \cdot S_2 + I_D \cdot E \cdot \alpha \tag{3.3}$$

$$FL4(633,660\pm 10) = AF \cdot B_4 + I_A \tag{3.4}$$

AF is the background or autofluorescence intensity in channel 1, I_D is the unquenched donor fluorescence intensity that would be measured in channel 2 in the absence of FRET, I_A is the directly excited acceptor fluorescence intensity in channel 4 and E is the mean energy transfer efficiency for the given cell. S and B factors characterize the "spectral spillover" between the detection channels (the ratios of the signals from, e. g., the donor-labeled, acceptor-labeled or unlabeled samples detected in the different channels; see Eqs. 3.5–3.9), which depend on hardware parameters [laser intensity, filter transmission, photomultiplier tube (PMT) sensitivity] and instrument settings (amplifier voltages, optical adjustment, etc.) and on the spectral properties of the dyes. S_1 can be determined by using samples labeled only with donor according to the following equation:

$$S_1 = FL3/FL2 \tag{3.5}$$

S_2 can be determined from acceptor-labeled cells according to:

$$S_2 = FL3/FL4 \tag{3.6}$$

B_2, B_3 and B_4 are determined from unlabeled cells as follows:

$$B_2 = FL2/FL1 \tag{3.7}$$

$$B_3 = FL3/FL1 \tag{3.8}$$

$$B_4 = FL4/FL1 \tag{3.9}$$

The factor α defines the relative detection sensitivity of fluorescence from an excited acceptor molecule with respect to that of a donor molecule, and can be determined from measurements with a donor-labeled and an acceptor-labeled sample with the following formula:

$$\alpha = \frac{FL3_A}{FL2_D} \cdot \frac{N_D}{N_A} \cdot \frac{L_D}{L_A} \cdot \frac{\varepsilon_D}{\varepsilon_A} \tag{3.10}$$

where N denotes the mean number of receptors per cell labeled by the corresponding antibody, L is the labeling ratio (the mean number of dye molecules attached to an antibody determined by spectrophotometry) and ε is the molar extinction coefficient of the dyes at 532 nm; subscripts A and D refer to acceptor and donor. The ratios of N, L and ε values are correct for the different numbers of photons absorbed by the acceptor- and donor-only labeled samples.

Solving Eqs. (3.1)–(3.4) yields the mean FRET efficiency value E for each individual cell:

$$A = \frac{E}{1-E} = \frac{1}{\alpha} \cdot \frac{FL1 \cdot (B_2 S_1 + B_4 S_2 - B_3) - FL2 \cdot S_1 + FL3 - FL4 \cdot S_2}{FL2 - FL1 \cdot B_2} \tag{3.11}$$

$$E = A/(1+A) \tag{3.12}$$

Similar expressions of the measured intensities give the cell-by-cell values of I_D, I_A and AF (Sebestyen et al. 2002). The above calculations can be most easily performed by the software FLEX (Szentesi et al. 2004), which is freely downloadable from http://www.biophys.dote.hu/research.htm. In addition to the usual analysis of standard flow cytometric list mode files, this software can generate frequency distribution histograms of cellular FRET efficiencies on selected cell

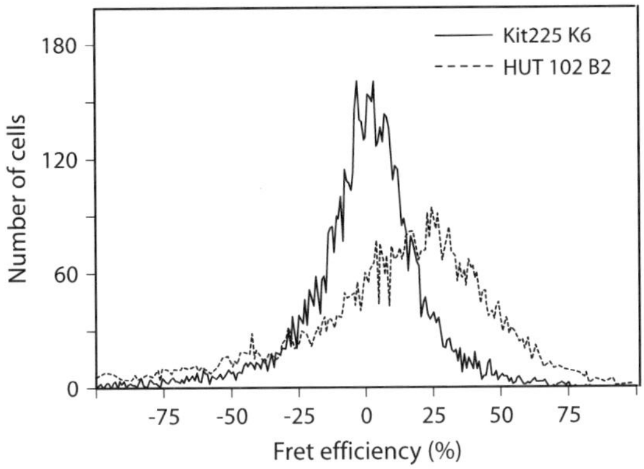

Fig. 3.1. Measurement of interleukin-2 receptor α submit homoassociation by flow cytometric cell-by-cell energy transfer. The FRET efficiency histogram for Kit 225 K6 cells has a mean value of 0, implying that the receptor subunits are in a monomeric form, whereas that for HUT102 B2 cells is ~25% owing to significant homoassociation. Receptor subunits were labeled by donor- and acceptor-tagged anti-Tac mAbs

populations. Figure 3.1 presents a representative example of two FCET measurements with positive and zero mean FRET efficiency.

As a check for the accuracy of the spectral correction factors, and also as a positive control, an intramolecular FRET efficiency can be measured. In our practice, we use the β2-microglobulin (β2m) and the MHC I heavy chain labeled with donor- and acceptor-tagged monoclonal antibodies, since these proteins are expressed in sufficient quantities by practically any human cell type. As a negative control, a donor-only labeled sample can be used, or two proteins that are known to be segregated in distinct membrane microdomains.

3.3 Mapping the Distribution of Cell-Surface Proteins

3.3.1 Self-Association of MHC I Molecules

Using the FRET technique, clustering (self-association) of class I major histocompatibility complex (MHC I; see Chap. 1, this Vol.) glycoproteins was observed on the surface of various human cell types (Bene et al. 1994, 2004; Matkó et al. 1994, 2002). A recent detergent-solubility analysis of their self-association properties has also confirmed that homotypic association is an inherent property of MHC I (and MHC II) molecules (Triantafilou et al. 2000), in accordance with earlier observations of their spontaneous clustering after reconstitution into liposome model systems (Chakrabarti et al. 1992). The degree of MHC I homoassociation highly depends on the culturing conditions, whether the cells are in log phase or in plateau phase (Matkó et al. 1994).

It was shown that expression of free MHC I heavy chains (lacking β2m; "FHC"; Chap. 1, this Vol.) correlated reasonably well with the degree of MHC I oligomerization on numerous cell types: both of them were significant on cells with accelerated cellular metabolism (i.e., activated or transformed/tumor cells) (Matkó et al. 1994). Our FRET and scanning near-field optical microscopic (SNOM) data demonstrated the participation of FHCs and intact MHC I heterodimers in the same small- and large-scale protein clusters at the surface of JY human B lymphoblasts. Furthermore, culturing of JY cells with β2m remarkably decreased the homotypic association of intact MHC I molecules and reduced their co-clustering with free heavy chains, as assessed by FRET experiments. SNOM images indicated a moderate dispersing effect of β2m treatment on the large-scale MHC I clusters, as well (Bodnár et al. 2003). On the basis of these data, FHCs likely make an important contribution to MHC I clustering: their involvement seems to stabilize MHC I clusters and vice versa; their functionally active conformation, which is still capable of rebinding β2m, may also be stabilized by participation in these clusters (Matkó et al. 1994; Bodnár et al. 2003). Otherwise FHCs would undergo irreversible denaturation and become functionally inactive and/or would be either internalized or released in a soluble form from the cell surface (Demaria et al. 1994; Pickl et al. 1996).

Electron and scanning force microscopic experiments disclosed the non-random (clusterized) organization of MHC I molecules at a higher hierarchical level, as well: immunogold-labeled MHC I molecules were observed to form several hundred-nanometer-sized domains (Damjanovich et al. 1995). Both small- and large-scale clusters of MHC I molecules were found to be dynamic, as revealed by SNOM and FRET measurements: dissociation and re-association of small-scale protein complexes and reformation of large-scale associations took place following fusion of cells separately labeled with fluoresceinated and rhodaminated Fab fragments against MHC I heavy chains (Nagy et al. 2001).

On the basis of recent studies with soluble MHC:peptide multimers (dimers, trimers or tetramers), aggregation of MHC molecules may significantly increase the efficiency of activation/immune response of T cells (Boniface et al. 1998; Cochran et al. 2000; Daniels and Jameson 2000). Since clustering of MHC molecules could be observed in "supramolecular activation clusters" within the immunological synapse (IS) (Monks et al. 1998; Grakoui et al. 1999), it is reasonable to assume that in vivo formation of MHC I clusters may promote IS formation, and the high local concentration of MHC I can significantly increase the avidity of APC-TC interaction (Bromley et al. 2001). Indeed, we showed that decreasing the extent of MHC I oligomerization by treatment with β2m on JY target cells significantly reduced the efficiency of activation as well as the effector function of allospecific cytotoxic T lymphocytes (Bodnár et al. 2003).

Modulation of the composition (microdomain structure) of the plasma membrane also influenced MHC I clustering at the surface of JY cells: the addition of cholesterol decreased the membrane fluidity and increased the degree of MHC I oligomerization (Bodnár et al. 1996). It is conceivable that the cholesterol content and fluidity of the plasma membrane also play an important role in the regulation of the degree of MHC I clustering and thus in controlling the efficiency of antigen presentation. This hypothesis is supported by earlier functional data showing that the lysis of target cells by CTLs or NK cells was affected by the cholesterol content of the plasma membrane (Roozemond and Bonavida 1985; Szekeres-Bartho et al. 1989).

3.3.2 Heteroassociation of MHC I Glycoproteins

Co-localization of MHC I with several other membrane proteins has also been detected by biophysical and/or biochemical methods. Table 3.1 shows some examples of the heteroassociations of MHC I molecules, mainly – but not exclusively –

Table 3.1. Some examples of MHC I hetero-associations mainly – but not exclusively – revealed by FCET measurements

Interacting partner of MHC I	Cell type	References
MHC II	JY human B lymphoblasts HUT102 B2 human T lymphoma cells OCM-1 and -3 human uveal melanoma cells	Szöllősi et al. (1989); Bene et al. (1994, 2004)
IL-2Rα	Activated human peripheral lymphocytes Human T lymphoma cells	Harel-Bellan et al. (1990); Bene et al. (1994); Matkó et al. (2002)
IL-2/IL-15Rβ and γ	Human T lymphoma cells	Matkó et al. (2002)
IL-15Rα	Kit225 FT7.10 human T lymphoma cells	Vámosi et al. (2004)
Intercellular adhesion molecule-1 (ICAM-1)	JY human B and HUT102 B2 human T lymphoma cells LS 174T colon carcinoma cells OCM-1 and -3 human uveal melanoma cells	Bene et al. (1994, 2004); Bacsó et al. (2002);
Tetraspan molecules (CD53, CD81, CD82)	Human B lymphoblastoid cell lines (JY, LCL-721)	Szöllősi et al. (1996)
Transferrin receptor	HUT102 B2	Mátyus et al. (1995)
Insulin receptor	Lymphocytes, adipocytes, hepatoma cells	Edidin and Reiland (1990); Liegler et al. (1991); Reiland and Edidin (1993); Ramalingam et al. (1997)

revealed by FCET techniques. Other examples of heteroassociation of MHC I molecules with receptors involved in cell growth and differentiation, along with the potential functional importance of these "*cis* interactions", are summarized in Chapter 1 (this Vol.). According to the examples listed in Table 3.1, MHC I glycoproteins apparently take part in several different association patterns in the plasma membrane, which in some cases may have functional relevance/importance as well. The cell-surface organization of the "partner" molecules also takes place at different hierarchical levels (Jenei et al. 1997; Vereb et al. 2000; Vámosi et al. 2004).

Existence of homo- and heteroassociates formed with the participation of MHC I molecules has also been corroborated by rotational diffusion and single particle tracking (SPT) techniques (Matkó et al. 1994; Smith et al. 1999). The results of the latter experiments indicated the anomalous diffusion of MHC I proteins on HeLa cells (Smith et al. 1999), which can be the consequence of the presence of obstacles (protein associates) or barriers (cytoskeletal elements, lipid domain barriers) trapping the molecules and thus hindering their free diffusion.

3.3.3 Association Patterns of MHC I, MHC II and ICAM-1 Molecules

Our FRET studies revealed the homoassociation of MHC class II molecules as well as their heteroassociation with class I MHC glycoproteins in the plasma membrane of numerous cell types, mainly of lymphoid origin (Table 3.1; Szöllősi et al. 1989; Jenei et al. 1997; Matkó et al. 2002). Atomic force and electron microscopic data showed that, like MHC I, MHC II molecules form homoclusters not only on the nanometer scale attainable by FRET, but also at a higher hierarchical level, in the micrometer distance range. Electron microscopy also revealed that a fraction of MHC II molecules was heteroclustered with MHC I at the same hierarchical level (Jenei et al. 1997). Similarly to MHC I, the large-scale MHC II clusters were also dynamic. Small- and large-scale heteroclusters of MHC I and MHC II also showed dynamic behavior in cell fusion experiments. At the same time, intermixing of components did not take place between small-scale clusters of MHC II (Nagy et al. 2001).

We could detect molecular associations between the ICAM-1 adhesion molecules and both types of MHC glycoproteins at the surface of human T and B lymphoma cell lines (Bene et al. 1994). In addition, a high degree of ICAM-1 self-association was found on HUT102 B2 human T lymphoma cells (Bene et al. 1994).

The above-mentioned association motifs of ICAM-1 and MHC glycoproteins were observed on non-lymphoid cells also. FCET and confocal laser scanning microscope experiments revealed that in OCM-1 and OCM-3 human uveal melanoma cells, MHC I, MHC II and ICAM-1 molecules are mostly confined to the same membrane regions, where they form similar protein patterns (homo- and heteroassociates) to those described above (Bene et al. 2004). Confocal microscopic co-localization experiments with GM1 gangliosides and the GPI-anchored CD59 molecules showed enrichment of MHC I, MHC II and ICAM-1 molecules in lipid rafts of these cells (Bene et al. 2004). Co-distribution of MHC I and ICAM-1 in lipid rafts was also reported for a colon carcinoma cell line, LS-174-T (Bacsó et al. 2002).

IFN-γ changed the expression levels of MHC and ICAM-1 molecules as well as induced rearrangement of their spatial distribution/association patterns on uveal melanoma and on colon carcinoma cells (Bacsó et al. 2002; Bene et al. 2004).

Effective T cell immune response requires well-organized, but dynamic arrangement of molecules present in the target cell–T cell contact region (immunological synapse) (Friedl and Storim 2004). It can be hypothesized that, along with MHC I oligomerization, the association patterns of MHC I, MHC II and ICAM-1 molecules on target (antigen-presenting) cells may play an important role in the regulation of the efficiency of antigen presentation. It is conceivable that by keeping the molecules forming the IS together, these "conservative" association patterns accommodated by lipid rafts may promote the rapid formation of the appropriate intercellular interactions upon target cell–T cell encounter (Bromley et al. 2001). This is corroborated by our observations demonstrating that diminishing the extent of MHC I oligomerization on JY B lymphoblasts was accompanied by their reduced susceptibility to specific lysis by allospecific CTL (Bodnár et al. 2003). Moreover, lipid raft-assisted compartmentation of MHC II was shown to enhance the efficiency of antigen presentation to $CD4^+$ T cells (Anderson et al. 2000). Lipid raft association of proteins may also be important in organizing molecular interactions in CTL-s as well as during the formation of the immunological synapse. Recently, we have demonstrated that Kv1.3, the dominant voltage-gated potassium channel of T lymphocytes (Panyi et al. 1995; Panyi and Deutsch 1996), is recruited into the immunological synapse, which may be mediated by its association to lipid rafts (Hajdu et al. 2003; Panyi et al. 2004a).

3.3.4 Supramolecular Complexes of IL-2/IL-15R and MHC Glycoproteins on Human T Lymphoma Cells

IL-2 and IL-15 receptors share two subunits (IL-2/15Rβ and γ_c) responsible for one set of signaling events evoked by cytokine binding, but both of them have their own, "private" α-chain (for details see Chap. 4, this Vol.). As a result of combining various subunits, several forms of receptor complexes with different affinities may exist at the cell surface. We have shown by using FRET that the three subunits of the high-affinity IL-2R complex (IL-2Rαβγ_c,) are in the molecular vicinity of each other at the surface of human T lymphoma cells even in the absence of IL-2. The addition of IL-2 or IL-15 rearranged the conformation of this preformed receptor complex: addition of IL-2 made it more compact, while IL-15 loosened the interaction of the three subunits (Damjanovich et al. 1997a). A similar preassembly of the heterotrimeric IL-15R (IL-15Rαβγ_c), as well as the molecular proximity of IL-15Rα and IL-2Rα, was also demonstrated on human T cells expressing all the elements of the IL-2/IL-15 receptor system. While the interaction of the two α-chains was unaffected by the relevant cytokines, interaction between IL-2/15Rβ and IL-15Rα became tighter upon IL-15 treatment, as assessed by FRET measurements (Vámosi et al. 2004). Based on these data, a tetrameric IL-2/IL-15 receptor complex can be envisaged, where binding of the appropriate cytokine rearranges the receptor subunits to form a high-affinity receptor trimer ($\alpha\beta\gamma_c$) with the participation of the cytokine-specific α-chain, whereas the "unused" α-chain rotates or shifts away from the site of cytokine–receptor interaction (Vámosi et al. 2004).

Our FRET experiments also indicated the homodimeric/oligomeric molecular association of IL-15Rα (Vámosi et al. 2004). Homoassociation of IL-2Rα may also occur in T lymphoma cells, although in a cell-type-dependent manner (see Fig. 3.1; Eicher et al. 2002; Vámosi et al. 2004).

It was shown by using biophysical (FRET and CLSM) and biochemical approaches that IL-2/IL-15R subunits are mainly partitioned into lipid rafts in the plasma membrane of human T lymphoma cells (Vereb et al. 2000; Matkó et al. 2002; Vámosi et al. 2004). It was also demonstrated that the integrity of lipid rafts has a crucial role in organizing the lateral distribution of IL-2R and it is also essential for IL-2-mediated signaling (Vereb et al. 2000; Matkó et al. 2002). Disruption of the native structure of lipid rafts by cholesterol extraction led to a reduction in tyrosine phosphorylation related to IL-2 signaling (Matkó et al. 2002). Upon cholesterol depletion the size of lipid rafts increased and the boundaries of the microdomains became fuzzier, i.e., the compactness and cohesion within the microdomain declined (Vereb et al. 2000). This implies that the lipid microenvironment, in particular cholesterol, might be an important factor in maintaining the integrity of signaling complexes. Lipid rafts may promote the formation and cytokine-specific modulation of IL-2R/IL-15R complexes and β- and γ_c-subunit "switching" between IL-2 and IL-15 receptors as well.

FCET experiments revealed the co-localization of the IL-2/IL-15R system and MHC I molecules in the plasma membrane of cells of human T lymphoma/leukemia origin (Matkó et al. 2002; Vámosi et al. 2004). Although the exact role of this co-localization has not been elucidated yet, a regulatory tyrosine phosphorylation cross-talk, as suggested earlier for the class I MHC–insulin receptor interaction (Ramalingam et al. 1997), cannot be excluded (Vereb et al. 2000; Matkó et al. 2002). Association of the IL-2/IL-15R with MHC II glycoproteins was also demonstrated (see Fig. 3.2; Matkó et al. 2002; Vámosi et al. 2004). Co-localization of the elements of the IL-2/IL-15R system with MHC glycoproteins also takes place in lipid rafts, as revealed by confocal microscopy (Vereb et al. 2000; Vámosi et al. 2004).

3.4 Fluorescence Correlation Spectroscopy (FCS)

FCS is a fluctuation correlation method that can measure the dynamics of molecular processes by observing spontaneous microscopic fluctuations of local concentrations of molecular species (Fig. 3.3). The driving force for these stochastic fluctuations is the thermal energy rather than a gradient or the non-equilibrium value of a macroscopic parameter. Thus FCS is capable of determining kinetic parameters in an equilibrium system, in contrast to conventional relaxation techniques. In a typical FCS experiment, a subfemtoliter volume element is illuminated by a focused laser beam, which excites fluorophores in the sensitive volume, and the time course of the arising fluorescence signal is detected. Since the rate of signal fluctuations depends on the rate of molecules diffusing in and

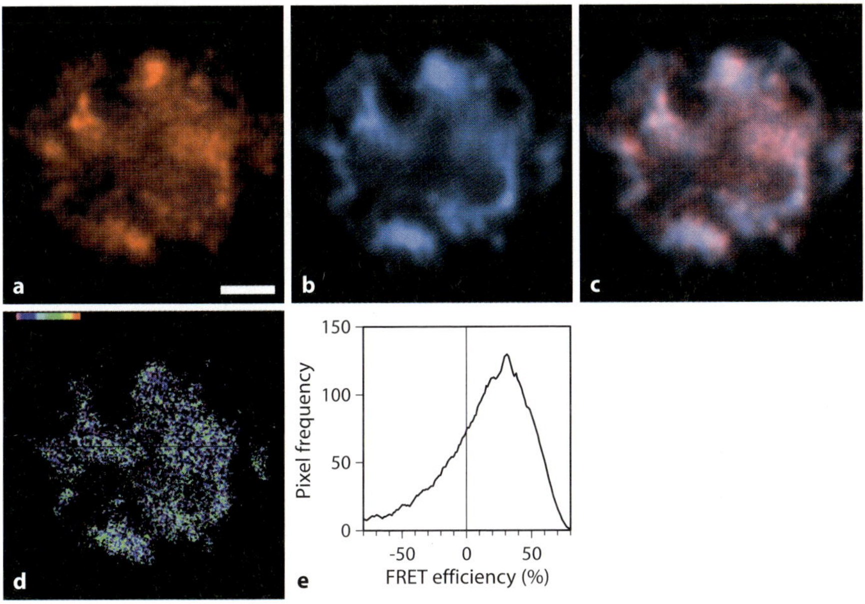

Fig. 3.2. Co-localization and pixel-by-pixel FRET experiment between IL-15Rα and MHC II molecules on Kit225 FT7.10 cells. Confocal microscopic images of the distribution of IL-15Rα subunits (**a**) and MHC II glycoproteins (**b**). Epitopes were labeled with Cy3–7A4 24 and Cy5-L243 mAbs, respectively. The overlay image (**c**) is *pink* where overlap is significant. **d** is a pixel-by-pixel FRET efficiency map between IL-15Rα and MHC II as determined by the acceptor photobleaching technique (description of the method can be found in Chap. 2, this Vol.); the color code ranges between 0% (*black*) and 100% (*red*). **e** is the frequency distribution histogram of FRET efficiencies in the individual pixels. The negative tail of the distribution is due to presence of noise in the original images and consequential error propagation in the calculation of FRET efficiency. *Scale bar* 2 μm

out of the sensitive volume, analysis of the fluctuations can yield the diffusion coefficient of the molecular species (Elson and Magde 1974; Elson et al. 1976). In fact, not only diffusion, but also the kinetics of any other molecular process causing a change in fluorescence intensity can be studied by FCS (e. g., molecular rate constants of chemical reactions and photophysical processes, conformational changes of biological macromolecules, etc.). For this analysis, the autocorrelation function of the fluorescence intensity is calculated, which contains information about the rates and amplitudes (and ideally about the mechanism) of the molecular processes occurring in the system. Fluorescence labeling provides for the specificity of the method. Since the ratio of the size of fluctuations relative to the total signal increases as the concentration of the molecules decreases, the technique has an exceptional signal-to-noise ratio and sensitivity compared to conventional fluorescence spectroscopy down to the nanomolar/picomolar range, allowing the detection of single molecules.

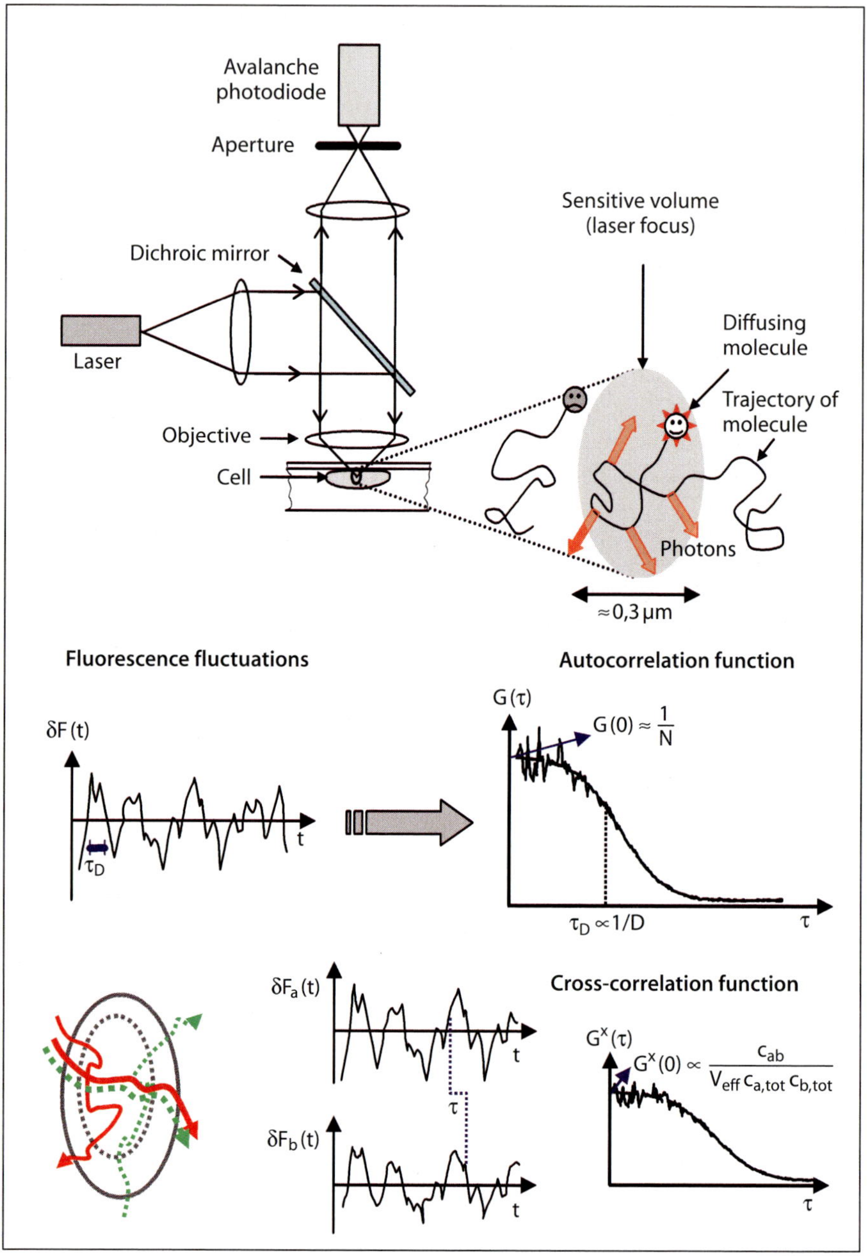

Avalanche photodiode
Aperture
Sensitive volume (laser focus)
Dichroic mirror
Laser
Diffusing molecule
Trajectory of molecule
Objective
Cell
Photons
≈0,3 µm
Fluorescence fluctuations
Autocorrelation function
δF (t)
t
τD
G (τ)
G (0) ≈ 1/N
τD ∝ 1/D
τ
Cross-correlation function
δFa (t)
δFb (t)
Gx (τ)
Gx (0) ∝ cab / Veff ca,tot cb,tot

A recent compilation on the theory and applications of FCS can be found in the book edited by two of the pioneers of the technique, E. Elson and R. Rigler (2001). There has been a renaissance of interest in FCS due to recent developments in the instrumentation, and a growing recognition of the broad spectrum of problems to which it can be applied successfully. Cells are rapidly changing inhomogeneous systems; therefore there is a basic need for selection of the target with a good spatial resolution, and there is a limit to measurement duration over which the cell can be considered relatively invariable. The coupling of FCS with confocal microscopy allowed 3D observation of the sample and accurate selection of the measurement area. It also reduced the sensitive volume to the size of the diffraction-limited spot in the optimal case. This not only improved the spatial resolution, but also significantly shortened the necessary measurement time required to determine diffusion coefficients in the cell. Sensitivity of detection has been greatly improved by the application of avalanche photodetectors in place of conventional photomultiplier tubes. Current galvanometric scanners provide enough stability to allow direct FCS measurements through the confocal scanner at any desired position in the cell with 25-nm positioning accuracy (Wachsmuth et al. 2003).

The dependence of the diffusion coefficient on molecular dimensions makes it possible to detect interactions between molecular species. The binding of a small ligand to a large interacting partner can be readily detected through the decrease in the diffusion coefficient (corresponding to a decrease in the diffusion correlation time) (Pramanik 2004). This is especially the case if a soluble, mobile ligand (or antibody) is bound by its membrane receptor (or antigen), which has low mobility in its highly viscous membrane environment.

However, since the dependence of the diffusion coefficient on the molecular weight of the complex is weak (for spherical particles, $D \propto M^{-1/3}$), the method is not sensitive to smaller changes in the molecular weight of the jointly moving complex. The interaction of molecules of comparable size having a similar mobility can be detected by the dual color version of the technique, fluorescence cross-correlation spectroscopy (FCCS). The method can indicate the joint diffusion of the distinctly labeled and detected molecular species (Rippe 2000; Weidemann et al. 2002; Bacia and Schwille 2003). A non-zero cross-correlation amplitude means that at least a portion of the two molecular species is stably associated for at least the duration of the diffusion time. For detailed explanation of the auto- and cross-correlation functions see Box 3.1.

Fig. 3.3. Fluorescence correlation spectroscopy. Fluorescent molecules diffusing across the laser-illuminated sensitive volume give rise to a fluctuating fluorescence signal. From the signal fluctuations $\delta F(t)$ the autocorrelation function is calculated, which carries information about the diffusion coefficient D and mean number of labeled molecules N in the sensitive volume. Using two-color excitation and two-channel detection, the joint diffusive motion of distinctly labeled molecular species can be analyzed. If the labeled molecules form stable associates, the fluorescence intensities arising from the two types of fluorescent labels will fluctuate in a correlated fashion. Amplitude of the cross-correlation function is proportional to the fraction of molecules in the complex

Box 3.1.
The autocorrelation and cross-correlation functions

The normalized autocorrelation function G(τ) of the fluorescence intensity is calculated from the fluctuations of fluorescence intensity according to the following formula:

$$G(\tau)=\frac{\langle \delta F(t)\cdot \delta F(t+\tau)\rangle}{\langle F\rangle^2}=\frac{\frac{1}{T}\int_0^T \delta F(t)\cdot \delta F(t+\tau)dt}{\langle F\rangle^2} \tag{3.13}$$

The angled brackets in the expression refer to averaging the expression over the duration of the measurement time. The variable τ is the lag time [the time difference between the samples taken from the $F(t)$ curve], $\langle F\rangle$ is the mean fluorescence intensity over the studied time interval and $\delta F(t)=F(t)-\langle F\rangle$ is the deviation of the actual fluorescence intensity from the mean. The rate of stochastic variations of the fluorescence intensity due to the diffusion of emitting particles across the sensitive volume depends on the mobility of the particles and on the geometrical parameters of the sensitive volume. In general, it is assumed that both the distribution of the illumination intensity and that of the detection probability have a 3D Gaussian shape. The product of these distributions is proportional to the probability that a photon will be emitted and detected from a fluorophore residing at the given location. The surface at which this probability drops to e^{-2} times the maximum value at the center of the distribution defines an ellipsoid (the sensitive volume). The autocorrelation function for the diffusion of a single molecular species in 2D (e.g., in the plane of the plasma membrane) undergoing singlet–triplet transition (Widengren et al. 1995) takes on the following form:

$$G(\tau)=\frac{1}{N}\cdot\frac{1-\Theta_{tr}+\Theta_{tr}e^{\tau/\tau_{tr}}}{1-\Theta_{tr}}\cdot\left(1+\frac{\tau}{\tau_D}\right)^{-1}=\frac{1}{N}\cdot\frac{1-\Theta_{tr}+\Theta_{tr}e^{\tau/\tau_{tr}}}{1-\Theta_{tr}}\cdot\left(1+\frac{4D\tau}{\omega_{xy}^2}\right)^{-1} \tag{3.14}$$

where N is the mean number of molecules in the sensitive volume, Θ_{tr} is the fraction of molecules in the triplet state, τ_{tr} is the phosphorescence lifetime and τ_D is the diffusion time (the mean dwell time of a molecule in the sensitive volume), D is the diffusion coefficient and ω_{xy} is the lateral radius of the e^{-2} ellipsoid. The diffusion time is inversely proportional to the diffusion coefficient:

$$D=\omega_{xy}^2/4\tau_D \tag{3.15}$$

If two molecular species are labeled with distinctly excitable and detectable fluorophores, their co-mobility can be studied by means of cross-correlation analysis (Schwille et al. 1997). Usually, distinct laser lines are used for the excitation of the dyes, and the fluorescence emission from the dyes is selectively detected through appropriate optical filters by separate detectors. Analogous to the autocorrelation function, the cross-correlation function of the two fluorescence signals is defined as:

$$G^{\times}(\tau)=\frac{\langle \delta F_a(t)\cdot \delta F_b(t+\tau)\rangle}{\langle F_a\rangle\langle F_b\rangle} \tag{3.16}$$

where the indexes "*a*" and "*b*" refer to the two different molecular species. The form of the cross-correlation function depends on the geometrical parameters of the laser foci, the diffusion properties of the different molecular species, and the concentrations of free and complexed molecules (Bacia et al. 2002; Weidemann et al. 2002). The amplitude of the diffusion cross-correlation function for a mixture of molecules *a*, *b* and their complex *ab* is proportional to the fraction of molecules in the complex:

$$G^{\times}(0)\propto\frac{c_{ab}}{V_{eff}\left(c_{a,tot}\right)\left(c_{b,tot}\right)} \tag{3.17}$$

where c_{ab} is the concentration of the complex, and $c_{a,tot}$ and $c_{b,tot}$ are the total concentrations of molecules *a* and *b*. Non-zero amplitude of the cross-correlation function implies complex formation between the two species.

3.4.1 Instrumentation

The "fluctuation microscope" used for FCS measurements is described in Wachsmuth et al. (2003). The device is a combination of a confocal FCS module (Wachsmuth et al. 2000) and a beam scanning unit attached to the video port of an inverted microscope. The 488- and 568-nm lines of an Ar-Kr laser focused to diffraction-limited spots were used to excite Alexa Fluor 488 and long-wavelength dyes Alexa Fluor 647 or Cy5, respectively. Fluorescence emission of the dyes was split by a 570 DRLP dichroic mirror and detected through 515- to 545-nm and 650- to 690-nm bandpass filters by avalanche photodiodes. Spectral separation practically excluded cross talk between the channels. The instrument can record confocal fluorescence images of the sample and perform FCS measurements at selected points of the image. In FCS measurements, continuous spot illumination by the low-intensity focused laser beam (<0.7 kW/cm^2) resulted in fast photobleaching of immobile molecules (and molecules within slowly moving large macromolecular aggregates) in the first few seconds of the experiment. To separate photobleaching phenomena from diffusion processes, runs of 5×3 or 4×4 s were recorded, in which most of the photobleaching process took place in the first two runs.

3.4.2 Evaluation of FCS Experiments

Measured auto- and cross-correlation curves G(τ) were fitted to a model function assuming a single molecular species diffusing in 2D plus a triplet term (see Eq. 3.14, Box 3.1). The exact formula describing cross-correlation between two distinctly labeled species is more complex (Weidemann et al. 2002) than the autocorrelation function according to Eq. (3.14), but its overall shape is very similar. At the signal-to-noise ratios achievable in the cross-correlation measurements of membrane-bound antibodies, the above function gave a reasonably accurate fit to the experimental curves. The size of the sensitive volume was calibrated by measuring the diffusion correlation time of a standard solution with a known diffusion coefficient. For this purpose, a solution of 50 nM fluorescein in Tris buffer, pH 8, was used ($D=2.6\times10^{-10}$ m^2/s) (Swaminathan et al. 1996).

3.4.3 The Co-mobility of IL-15Rα, IL-2Rα and MHC I Was Studied in FT7.10 Cells

As mentioned above, the molecular proximity between IL-2Rα, IL-15Rα and MHC class I and class II glycoproteins in the plasma membrane of human T lymphoma cells has been demonstrated by flow cytometric and confocal microscopic (acceptor photobleaching) FRET experiments (Matkó et al. 2002; Vámosi et al. 2004). However, significant FRET efficiency values alone only suggest the possibility of stable interactions between the measured donor-/acceptor-labeled pairs, but do not prove the presence of stable complexes or give any information on the temporal stability of protein complexes. Static FRET (co-localization) and dynamic FCCS (co-mobility) are complementary approaches to studying interactions between molecular species. Therefore we probed these interactions by fluorescence cross-correlation spectroscopy as well. In Kit 225 FT7.10 T lymphoma cells, positive cross-correlation amplitude was obtained between IL-15Rα and IL-2Rα in ~20% of cases, whereas IL-15Rα and MHC I yielded positive cross-correlation amplitudes in ~80% of all cases (see Fig. 3.4). In Kit 225 K6 T lymphoma cells, IL-2Rα and MHC I showed non-zero cross-correlation in 80% of cases. These findings are in agreement with the positive FRET results and suggest that the studied molecular complexes are stable on a time scale of at least several tens of milliseconds, the correlation time (see Box 3.1; Vámosi et al. 2004).

As a positive control, MHC I molecules labeled with W6/32 antibodies simultaneously conjugated with Alexa Fluor 488 and Alexa Fluor 647 were used on FT7.10 cells, and resulted in positive cross-correlation. For the negative control experiment, which was performed on K6 cells, we labeled transferrin receptors (known to be accumulated in coated pits and excluded from lipid rafts) and IL-2Rα subunits (localized in lipid rafts). The sensitive volume was placed on the boundary between IL-2Rα and transferrin receptor positive membrane microdomains, but no cross-correlation could be detected.

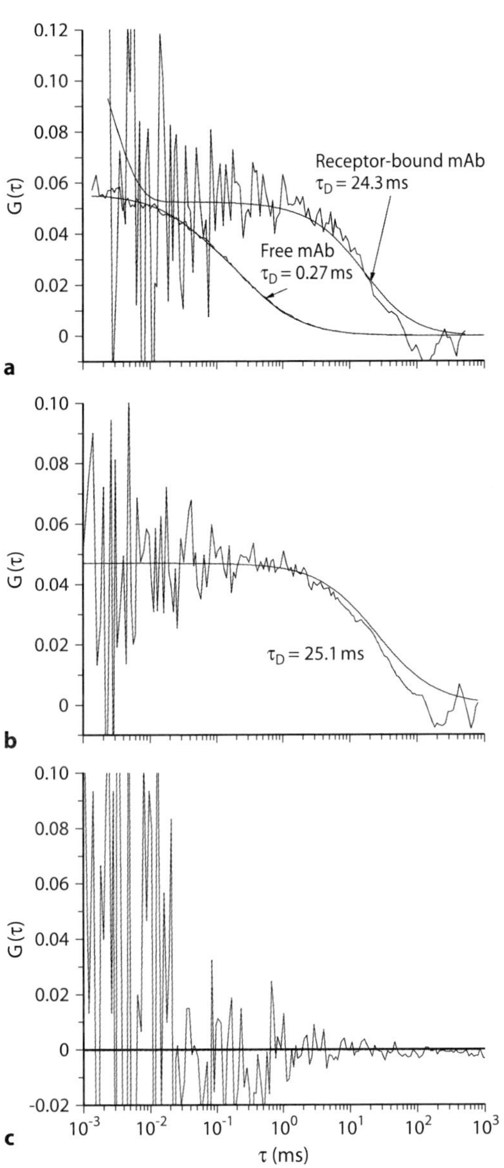

Fig. 3.4. Mobility and co-mobility measurements by FCS on Kit 225 FT7.10 and K6 T lymphoma cells. **a** Autocorrelation curves detected from Alexa 488-tagged anti-FLAG mAbs free in solution and bound to IL-15Rα subunits at the surface of FT7.10 cells. *Smooth curves* represent fits to the experimental data according to Eq. (3.14). Diffusion time τ_D of the antibody increased by two orders of magnitude upon receptor binding. **b** Cross-correlation curve measured between IL-15Rα and MHC I molecules labeled by Alexa 488-anti-FLAG and Cy5-W6/32 mAbs on FT7.10 cells. Non-zero cross-correlation amplitude suggests that at least a certain fraction of the two proteins are hetero-associated and form stable complexes. **c** Cross-correlation curve measured between transferrin receptors and IL-2Rα subunits (tagged with Alexa 488-MEM75 and Cy5-anti-Tac mAbs) on K6 cells. Zero cross-correlation amplitude means that there is no interaction between the labeled proteins

An interesting conclusion to our results follows from the comparison of the expression levels of the studied proteins. The number of IL-15Rα, IL-2Rα and MHC I relates to one another approximately as 1:10:50–100 on the Kit 225 FT7.10 cell line used in our FCCS studies. If complexes of 1:1 stoichiometry were formed between, e. g., IL-15Rα and MHC I, then the out-of-complex fraction of MHC I molecules would suppress the cross-correlation amplitude below the detection level (see Eq. 3.17). This implies that probably higher-order aggregates of MHC class I molecules float together with IL-2 and IL-15 receptors in large supramolecular complexes in the plasma membrane. These results are in line with our previous observations on the homo-association (and hetero-associations) of MHC I (and MHC II) molecules detected by FRET and transmission electron microscopy/atomic force microscopy (see Sect. 3.3).

Acknowledgements. We thank Prof. J. Langowski for useful comments on the manuscript. This work was supported by the following grants: OTKA TS40773, T048745, T42618, T43509, F46497; ETT 602/2003, 603/2003; NATO Life Science and Technology Collaborative Linkage Grant 980200 and Bolyai János Research Fellowships (to G.V. and A.B.). A.B. and G.V. contributed equally to this manuskript.

References

Anderson HA, Hiltbold EM, Roche PA (2000) Concentration of MHC class II molecules in lipid rafts facilitates antigen presentation. Nat Immunol 1:156–162

Bacia K, Schwille P (2003) A dynamic view of cellular processes by in vivo fluorescence auto- and cross-correlation spectroscopy. Methods 29:74–85

Bacia K, Majoul IV, Schwille P (2002) Probing the endocytic pathway in live cells using dual-color fluorescence cross-correlation analysis. Biophys J 83:1184–1193

Bacsó Z, Bene L, Damjanovich L, Damjanovich S (2002) INF-gamma rearranges membrane topography of MHC-I and ICAM-1 in colon carcinoma cells. Biochem Biophys Res Commun 290:635–640

Bene L, Balazs M, Matkó J, Most J, Dierich MP, Szöllősi J, Damjanovich S (1994) Lateral organization of the ICAM-1 molecule at the surface of human lymphoblasts: a possible model for its co-distribution with the IL-2 receptor, class I and class II HLA molecules. Eur J Immunol 24:2115–2123

Bene L, Bodnár A, Damjanovich S, Vámosi G, Bacsó Z, Aradi J, Berta A, Damjanovich J (2004) Membrane topography of HLA I, HLA II, and ICAM-1 is affected by IFN-gamma in lipid rafts of uveal melanomas. Biochem Biophys Res Commun 322:678–683

Bodnár A, Jenei A, Bene L, Damjanovich S, Matkó J (1996) Modification of membrane cholesterol level affects expression and clustering of class I HLA molecules at the surface of JY human lymphoblasts. Immunol Lett 54:221–226

Bodnár A, Bacsó Z, Jenei A, Jovin TM, Edidin M, Damjanovich S, Matkó J (2003) Class I HLA oligomerization at the surface of B cells is controlled by exogenous beta(2)-microglobulin: implications in activation of cytotoxic T lymphocytes. Int Immunol 15:331–339

Boniface J, Rabinowitz J, Wülfing C, Hampl J, Reich Z, Altman J, Kantor R, Beeson C, McConnell H, Davis M (1998) Initiation of signal transduction through the T cell receptor requires the multivalent engagement of peptide/MHC ligands. Immunity 9:459–466

Bromley SK, Burack WR, Johnson KG, Somersalo K, Sims TN, Sumen C, Davis MM, Shaw AS, Allen PM, Dustin ML (2001) The immunological synapse. Annu Rev Immunol 19:375–396

Chakrabarti A, Matkó J, Rahman NA, Barisas BG, Edidin M (1992) Self-association of class I major histocompatibility complex molecules in liposome and cell surface membranes. Biochemistry 31:7182–7189

Clegg RM (1996) Fluorescence resonance energy transfer (FRET). In: Wang XF, Herman B (eds) Fluorescence imaging spectroscopy and microscopy. Wiley, New York, pp 179–252

Cochran J, Cameron T, Stern L (2000) The relationship of MHC-peptide binding and T cell activation probed using chemically defined MHC class II oligomers. Immunity 12:241–250

Damjanovich S, Bahr W, Jovin TM (1977) The functional and fluorescence properties of *Escherichia coli* RNA polymerase reacted with fluorescamine. Eur J Biochem 72:559–569

Damjanovich S, Somogyi B, Trón L (1981) Macromolecular dynamics and information transfer. Adv Physiol Sci 30:9–15

Damjanovich S, Trón L, Szöllősi J, Zidovetzki R, Vaz WL, Regateiro F, Arndt-Jovin DJ, Jovin TM (1983) Distribution and mobility of murine histocompatibility H-2Kk antigen in the cytoplasmic membrane. Proc Natl Acad Sci USA 80:5985–5989

Damjanovich S, Mátyus L, Balazs M, Gáspár R, Krasznai Z, Pieri C, Szöllősi J, Trón L (1992) Dynamic physical interactions of plasma membrane molecules generate cell surface patterns and regulate cell activation processes. Immunobiology 185:337–349

Damjanovich S, Vereb G, Schaper A, Jenei A, Matkó J, Starink JP, Fox GQ, Arndt-Jovin DJ, Jovin TM (1995) Structural hierarchy in the clustering of HLA class I molecules in the plasma membrane of human lymphoblastoid cells. Proc Natl Acad Sci USA 92:1122–1126

Damjanovich S, Bene L, Matkó J, Alileche A, Goldman CK, Sharrow S, Waldmann TA (1997a) Preassembly of interleukin 2 (IL-2) receptor subunits on resting Kit 225 K6 T cells and their modulation by IL-2, IL-7, and IL-15: a fluorescence resonance energy transfer study. Proc Natl Acad Sci USA 94:13134–13139

Damjanovich S, Gáspár RJ, Pieri C (1997b) Dynamic receptor superstructures at the plasma membrane. Q Rev Biophys 30:67–106

Damjanovich S, Matkó J, Mátyus L, Szabo G Jr, Szöllősi J, Pieri JC, Farkas T, Gáspár R Jr (1998) Supramolecular receptor structures in the plasma membrane of lymphocytes revealed by flow cytometric energy transfer, scanning force- and transmission electron-microscopic analyses. Cytometry 33:225–233

Damjanovich S, Bene L, Matkó J, Mátyus L, Krasznai Z, Szabo G, Pieri C, Gáspár RJ, Szöllősi J (1999) Two-dimensional receptor patterns in the plasma membrane of cells. A critical evaluation of their identification, origin and information content. Biophys Chem 82:99–108

Damjanovich S, Mátyus L, Damjanovich L, Bene L, Jenei A, Matkó J, Gáspár R, Szöllősi J (2002) Does mosaicism of the plasma membrane at molecular and higher hierarchical levels in human lymphocytes carry information on the immediate history of cells? Immunol Lett 82:93–99

Daniels MA, Jameson SC (2000) Critical role for CD8 in T cell receptor binding and activation by peptide/major histocompatibility complex multimers. J Exp Med 191:335–345

Demaria S, Schwab R, Gottesman S, Bushkin Y (1994) Soluble β2-microglobulin-free class I heavy chains are released from the surface of activated and leukemia cells by a metalloprotease. J Biol Chem 269:6689–6694

Edidin M (1993) Patches and fences: probing for plasma membrane domains. J Cell Sci Suppl 17:165–169

Edidin M (1997) Lipid microdomains in cell surface membranes. Curr Opin Struct Biol 7:528–532

Edidin M (2001) Shrinking patches and slippery rafts: scales of domains in the plasma membrane. Trends Cell Biol 11:492–496

Edidin M (2003) The state of lipid rafts: from model membranes to cells. Annu Rev Biophys Biomol Struct 32:257–283

Edidin M, Reiland J (1990) Dynamic measurements of the associations between class I MHC antigens and insulin receptors. Mol Immunol 27:1313–1317

Eicher DM, Damjanovich S, Waldmann TA (2002) Oligomerization of IL-2Ralpha. Cytokine 17:82–90

Elson EL, Magde D (1974) Fluorescence correlation spectroscopy. Conceptual basis and theory. Biopolymers 13:1–27

Elson EL, Schlessinger J, Koppel DE, Axelrod D, Webb WW (1976) Measurement of lateral transport on cell surfaces. Prog Clin Biol Res 9:137–147

Förster T (1948) Zwischenmolekulare Energiewanderung und Fluoreszenz. Ann Phys 2:55–75

Friedl P, Storim J (2004) Diversity in immune-cell interactions: states and functions of the immunological synapse. Trends Cell Biol 14:557–567

Frye LD, Edidin M (1970) The rapid intermixing of cell surface antigens after formation of mouse–human heterokaryons. J Cell Sci 7:319–335

Gheber LA, Edidin M (1999) A model for membrane patchiness: lateral diffusion in the presence of barriers and vesicle traffic. Biophys J 77:3163–3175

Grakoui A, Bromley SK, Sumen C, Davis MM, Shaw AS, Allen PM, Dustin ML (1999) The immunological synapse: a molecular machine controlling T cell activation. Science 285:221–227

Hajdu P, Varga Z, Pieri C, Panyi G, Gáspár R Jr (2003) Cholesterol modifies the gating of Kv1.3 in human T lymphocytes. Pflugers Arch 445:674–682

Harel-Bellan A, Krief P, Rimsky L, Farrar WL, Mishal Z (1990) Flow cytometry resonance energy transfer suggests an association between low-affinity interleukin 2 binding sites and HLA class I molecules. Biochem J 268:35–40

Hoessli DC, Ilangumaran S, Soltermann A, Robinson PJ, Borisch B, Nasir UD (2000) Signaling through sphingolipid microdomains of the plasma membrane: the concept of signaling platform. Glycoconj J 17:191–197

Horejsi V (2002) Membrane rafts in immunoreceptor signaling: new doubts, new proofs? Trends Immunol 23:562–564

Horejsi V (2003) The roles of membrane microdomains (rafts) in T cell activation. Immunol Rev 191:148–164

Jenei A, Varga S, Bene L, Mátyus L, Bodnár A, Bacsó Z, Pieri C, Gáspár RJ, Farkas T, Damjanovich S (1997) HLA class I and II antigens are partially co-clustered in the plasma membrane of human lymphoblastoid cells. Proc Natl Acad Sci USA 94:7269–7274

Klausner RD, Lippincott-Schwartz J, Bonifacino JS (1990) The T cell antigen receptor: insights into organelle biology. Annu Rev Cell Biol 6:403–431

Kusumi A, Sako Y (1996) Cell surface organization by the membrane skeleton. Curr Opin Cell Biol 8:566–574

Lemmon MA, Engelman DM (1994) Specificity and promiscuity in membrane helix interactions. Q Rev Biophys 27:157–218

Liegler T, Szöllősi J, Hyun W, Goodenow RS (1991) Proximity measurements between H-2 antigens and the insulin receptor by fluorescence energy transfer: evidence that a close association does not influence insulin binding. Proc Natl Acad Sci USA 88:6755–6759

Lucero HA, Robbins PW (2004) Lipid rafts–protein association and the regulation of protein activity. Arch Biochem Biophys 426:208–224

Matkó J, Edidin M (1997) Energy transfer methods for detecting molecular clusters on cell surfaces. Methods Enzymol 278:444–462

Matkó J, Szöllősi J (2002) Landing of immune receptors and signal proteins on lipid rafts: a safe way to be spatio-temporally coordinated? Immunol Lett 82:3–15

Matkó J, Bushkin Y, Wei T, Edidin M (1994) Clustering of class I HLA molecules on the surfaces of activated and transformed human cells. J Immunol 152:3353–3360

Matkó J, Bodnár A, Vereb G, Bene L, Vámosi G, Szentesi G, Szöllősi J, Gáspár R, Horejsi V, Waldmann TA, Damjanovich S (2002) GPI-microdomains (membrane rafts) and signaling of the multi-chain interleukin-2 receptor in human lymphoma/leukemia T cell lines. Eur J Biochem 269:1199–1208

Mátyus L, Bene L, Heiligen H, Rausch J, Damjanovich S (1995) Distinct association of transferrin receptor with HLA class I molecules on HUT-102B and JY cells. Immunol Lett 44:203–208

Monks C, Freiberg B, Kupfer H, Sciaky N, Kupfer A (1998) Three-dimensional segregation of supramolecular activation clusters in T cells. Nature 395:82–86

Nagy P, Mátyus L, Jenei A, Panyi G, Varga S, Matkó J, Szöllősi J, Gáspár R, Jovin TM, Damjanovich S (2001) Cell fusion experiments reveal distinctly different association characteristics of cell-surface receptors. J Cell Sci 114:4063–4071

Panyi G, Deutsch C (1996) Assembly and suppression of endogenous Kv1.3 channels in human T cells. J Gen Physiol 107:409–420

Panyi G, Sheng Z-F, Tu L-W, Deutsch C (1995) C-type inactivation of a voltage-gated K^+ channel occurs by a cooperative mechanism. Biophys J 69:896–904

Panyi G, Bagdány M, Bodnár A, Vámosi G, Szentesi G, Jenei A, Mátyus L, Varga S, Waldmann TA, Gáspár R, Damjanovich S (2003) Colocalization and nonrandom distribution of Kv1.3 potassium channels and CD3 molecules in the plasma membrane of human T lymphocytes. Proc Natl Acad Sci USA 100:2592–2597

Panyi G, Vámosi G, Bacsó Z, Bagdány M, Bodnár A, Varga Z, Gáspár R, Mátyus L, Damjanovich S (2004a) Kv1.3 potassium channels are localized in the immunological synapse formed between cytotoxic and target cells. Proc Natl Acad Sci USA 101:1285–1290

Panyi G, Vámosi G, Bodnár A, Gáspár R, Damjanovich S (2004b) Looking through ion channels: recharged concepts in T-cell signaling. Trends Immunol 25:565–569

Panyi G, Varga Z, Gáspár R (2004c) Ion channels and lymphocyte activation. Immunol Lett 92:55–66

Pickl W, Holter W, Stöckl J, Majdic O, Knapp W (1996) Expression of LA45 reactive β2-microglobulin free HLA class I α-chains on activated T cells is regulated by internalization, constitutive and protein kinase C inducible release. Tissue Antigens 48:15–21

Pramanik A (2004) Ligand–receptor interactions in live cells by fluorescence correlation spectroscopy. Curr Pharm Biotechnol 5:205–212

Ramalingam TS, Chakrabarti A, Edidin M (1997) Interaction of class I human leukocyte antigen (HLA-I) molecules with insulin receptors and its effect on the insulin-signaling cascade. Mol Biol Cell 8:2463–2474

Reiland J, Edidin M (1993) Chemical cross-linking detects association of insulin receptors with four different class I human leukocyte antigen molecules on cell surfaces. Diabetes 42:619–625

Rigler R, Elson ES (2001) Fluorescence correlation spectroscopy. Theory and applications. Springer, Berlin Heidelberg, New York, 487 pp

Rippe K (2000) Simultaneous binding of two DNA duplexes to the NtrC-enhancer complex studied by two-color fluorescence cross-correlation spectroscopy. Biochemistry 39:2131–2139

Ritchie K, Kusumi A (2004) Role of the membrane skeleton in creation of microdomains. Subcell Biochem 37:233–245

Roozemond RC, Bonavida B (1985) Effect of altered membrane fluidity on NK cell-mediated cytotoxicity. I. Selective inhibition of the recognition or post recognition events in the cytolytic pathway of NK cells. J Immunol 134:2209–2214

Schwille P, Meyer-Almes FJ, Rigler R (1997) Dual-color fluorescence cross-correlation spectroscopy for multicomponent diffusional analysis in solution. Biophys J 72:1878–1886

Sebestyén Z, Nagy P, Horvath G, Vámosi G, Debets R, Gratama JW, Alexander DR, Szöllősi J (2002) Long wavelength fluorophores and cell-by-cell correction for autofluorescence significantly improves the accuracy of flow cytometric energy transfer measurements on a dual-laser benchtop flow cytometer. Cytometry 48:124–135

Simons K, Ikonen E (1997) Functional rafts in cell membranes. Nature 387:569–572

Simons K, Vaz WL (2004) Model systems, lipid rafts, and cell membranes. Annu Rev Biophys Biomol Struct 33:269–295

Singer SJ, Nicolson GL (1972) The fluid mosaic model of the structure of cell membranes. Science 175:720–731

Smith P, Morrison I, Wilson K, Fernandez N, Cherry R (1999) Anomalous diffusion of major histocompatibility complex class I molecules on HeLa cells determined by single particle tracking. Biophys J 76:3331–3344

Stryer L, Haugland RP (1967) Energy transfer: a spectroscopic ruler. Proc Nat Acad Sci USA 58:719–726

Swaminathan R, Bicknese S, Periasamy N, Verkman AS (1996) Cytoplasmic viscosity near the cell plasma membrane: translational diffusion of a small fluorescent solute measured by total internal reflection-fluorescence photobleaching recovery. Biophys J 71:1140–1151

Szekeres-Bartho J, Nemeth A, Varga P, Csernus V, Koszegi T, Paal M (1989) Membrane fluidity of trophoblast cells and susceptibility to natural cytotoxicity. Am J Reprod Immunol 19:92–98

Szentesi G, Horvath G, Bori I, Vámosi G, Szöllősi J, Gáspár R, Damjanovich S, Jenei A, Mátyus L (2004) Computer program for determining fluorescence resonance energy transfer efficiency from flow cytometric data on a cell-by-cell basis. Comput Methods Programs Biomed 75:201–211

Szöllősi J, Trón L, Damjanovich S, Helliwell SH, Arndt-Jovin D, Jovin TM (1984) Fluorescence energy transfer measurements on cell surfaces: a critical comparison of steady-state fluorimetric and flow cytometric methods. Cytometry 5:210–216

Szöllősi J, Damjanovich S, Balazs M, Nagy P, Trón L, Fulwyler MJ, Brodsky FM (1989) Physical association between MHC class I and class II molecules detected on the cell surface by flow cytometric energy transfer. J Immunol 143:208–213

Szöllösi J, Horejsi V, Bene L, Angelisova P, Damjanovich S (1996) Supramolecular complexes of MHC class I, MHC class II, CD20, and tetraspan molecules (CD53, CD81, and CD82) at the surface of a B cell line JY. J Immunol 157:2939–2946

Szöllősi J, Damjanovich S, Mátyus L (1998) Application of fluorescence resonance energy transfer in the clinical laboratory: routine and research. Cytometry 34:159–179

Tang Q, Edidin M (2001) Vesicle trafficking and cell surface membrane patchiness. Biophys J 81:196–203

Tocanne JF, Cezanne L, Lopez A, Piknova B, Schram V, Tournier JF, Welby M (1994) Lipid domains and lipid/protein interactions in biological membranes. Chem Phys Lipids 73:139–158

Triantafilou K, Triantafilou M, Wilson KM, Fernandez N (2000) Human major histocompatibility molecules have the intrinsic ability to form homotypic associations. Hum Immunol 61:585–598

Trón L, Szöllősi J, Damjanovich S, Helliwell SH, Arndt-Jovin DJ, Jovin TM (1984) Flow cytometric measurements of fluorescence resonance energy transfer on cell surfaces. Quantitative evaluation of the transfer efficiency on a cell by cell basis. Biophys J 45:939–946

Vámosi G, Bodnár A, Vereb G, Jenei A, Goldman CK, Langowski J, Tóth K, Mátyus L, Szöllősi J, Waldmann TA, Damjanovich S (2004) IL-2 and IL-15 receptor alpha-subunits are co-expressed in a supramolecular receptor cluster in lipid rafts of T cells. Proc Natl Acad Sci USA 101:11082–11087

Vereb G, Mátyus L, Bene L, Panyi G, Bacsó Z, Balázs M, Matkó J, Szöllősi J, Gáspár R Jr, Damjanovich S (1995) Plasmamembrane bound macromolecules are dynamically aggregated to form nonrandom codistribution patterns of selected functional elements. Do pattern recognition processes govern antigen presentation and intercellular interactions? J Mol Recogn 8:237–246

Vereb G, Matkó J, Vámosi G, Ibrahim SM, Magyar E, Varga S, Szöllősi J, Jenei A, Gáspár RJ, Waldmann TA, Damjanovich S (2000) Cholesterol-dependent clustering of IL-2Ralpha and its colocalization with HLA and CD48 on T lymphoma cells suggest their functional association with lipid rafts. Proc Natl Acad Sci USA 97:6013–6018

Vereb G, Szöllősi J, Matkó J, Nagy P, Farkas T, Vigh L, Mátyus L, Waldmann TA, Damjanovich S (2003) Dynamic, yet structured: the cell membrane three decades after the Singer–Nicolson model. Proc Natl Acad Sci USA 100:8053–8058

Wachsmuth M, Waldeck W, Langowski J (2000) Anomalous diffusion of fluorescent probes inside living cell nuclei investigated by spatially-resolved fluorescence correlation spectroscopy. J Mol Biol 298:677–689

Wachsmuth M, Weidemann T, Muller G, Hoffmann-Rohrer UW, Knoch TA, Waldeck W, Langowski J (2003) Analyzing intracellular binding and diffusion with continuous fluorescence photobleaching. Biophys J 84:3353–3363

Weidemann T, Wachsmuth M, Tewes M, Rippe K, Langowski J (2002) Analysis of ligand binding by two-colour fluorescence cross-correlation spectroscopy. Single Molecules 3:49–61

Welti R, Glaser M (1994) Lipid domains in model and biological membranes. Chem Phys Lipids 73:121–137

Widengren J, Mets Ü, Rigler R (1995) Fluorescence correlation spectroscopy of triplet state in solution: a theoretical and experimental study. J Phys Chem 99:13368–13379

Yang XL, Xiong WC, Mei L (2004) Lipid rafts in neuregulin signaling at synapses. Life Sci 75:2495–2504

Zidovetzki R, Yarden Y, Schlessinger J, Jovin TM (1981) Rotational diffusion of epidermal growth factor complexed to cell surface receptors reflects rapid microaggregation and endocytosis of occupied receptors. Proc Natl Acad Sci USA 78:6981–6985

Transmembrane Signals Mediated by IL-2 and IL-15 Control the Life and Death of Lymphocytes

Thomas A. Waldmann, Sigrid Dubois, Jürgen Müller,
Carolyn Goldman, Sándor Damjanovich

4.1 Introduction

Immune responses are regulated by a series of cytokines that exhibit a high degree of redundancy and pleiotropy controlling a wide range of functions in various cell types. This redundancy is explained in part by the sharing of common receptor subunits among members of the cytokine receptor family. Each cytokine utilizes its own private receptor subunit but may also share public receptor subunits with other cytokines. This sharing of subunits is exemplified by the receptors within the common gamma (γ_c) receptor system. The γ_c is an essential element of the multisubunit receptors for IL-4, IL-7, IL-9, IL-21 as well as for the two cytokines, the focus of this chapter, IL-2 and IL-15 (Takeshita et al. 1992; Kondo et al. 1993; Noguchi et al. 1993; Leonard 2000). The shared actions of IL-2 and IL-15 on T and natural killer (NK) cells also require the expression of IL-2/15Rβ as well as the private alpha chains IL-2Rα and IL-15Rα used by IL-2 and IL-15 respectively (Sharon et al. 1986; Tsudo et al. 1986; Bamford et al. 1994; Giri et al. 1994, 1995; Grabstein et al. 1994). When this receptor constellation is utilized, IL-2 and IL-15 activate similar JAK1 and JAK3 (Janus kinases) as well as STAT5 (signal transducers and transactivators of transcription)-dependent signaling pathways (Witthuhn et al. 1994; Johnston et al. 1995; Lin et al. 1995). As might be anticipated from this sharing of receptor subunits and the use of common elements in one signaling pathway, there is a significant overlap between the immunological functions of IL-2 and IL-15 (Waldmann and Tagaya 1999).

Both cytokines stimulate the proliferation of different subsets of activated T cells involved in the maintenance of NK cells and act as co-stimulatory molecules for B cell immunoglobulin synthesis (Waldmann and Tagaya 1999). In addition to shared functions, IL-2 and IL-15 provide distinct and at times contrasting contributions to T cell-mediated immunity (Waldmann et al. 2001). IL-2 plays a central role in activation-induced cell death (AICD) and thereby is involved in peripheral tolerance to self. In contrast, the transgenic expression of IL-15 in mice inhibits IL-2-mediated AICD (Marks-Konczalik et al. 2000). Furthermore, IL-15 stimulates the persistence of $CD8^+$ memory phenotype cells whereas IL-2 inhibits their survival (Zhang et al. 1998; Marks-Konczalik et al. 2000; Ku et al. 2000; Waldmann et al. 2001). The many biophysical elements involved in the relationships among receptor subunits as well as the transmembrane signals mediated by IL-2 and IL-15 that control the life and death of lymphocytes are the subjects of this chapter.

Springer Series in Biophysics
S. Damjanovich (Ed.)
Biophysical Aspects of Transmembrane Signaling

4.2 The Classical View of the IL-2 and IL-15 Receptor and Signaling Pathways That Underlie the Shared Functions of These Two Cytokines

The IL-2R is made up of at least three distinct membrane components: the 55-kDa alpha chain (IL-2Rα), the 70- to 75-kDa beta chain (IL-2/15Rβ) and the 64-kDa gamma chain (IL-2Rγ or γ_c) (Figs. 4.1, 4.2; Waldmann 1989a,b; Minami et al. 1993; Taniguchi and Minami 1993; Lin et al. 1995; Leonard 2000; Gaffen 2001; Shurin et al. 2003;). The development of the anti-Tac monoclonal antibody led to the identification and cloning of the gene encoding the first IL-2 binding protein now known as IL-2Rα. IL-2Rα, encoded on chromosome 10p14-p15, is composed of 251 amino acids as well as an NH_2-terminal 21 amino acid signal peptide (Uchiyama et al. 1981; Cosman et al. 1984; Leonard et al. 1984; Nikaido et al. 1984). The 219 terminal amino acids make up an extracellular domain that contains two potential N-linked glycosylation sites and multiple possible O-linked carbohydrates. A second 19 amino acid domain that contains a single hydrophobic region near the COOH terminus presumably represents a membrane-spanning region. The third domain is a short (13 amino acid) cytoplasmic domain. IL-2Rα is an inducible receptor that is not expressed by resting cells but is expressed by activated T cells, B cells and NK cells (Waldmann 1989b; Leonard et al. 1984). In particular, IL-2Rα is induced by the addition of anti-CD3 plus anti-CD28 antibodies, mitogens, antigens, the *tax* transactivator protein of HTLV-I and by certain cytokines including IL-15 as well as IL-2 itself (Waldmann 1989a). In addition, soluble forms of the IL-2Rα and IL-15Rα receptor subunits cleaved from the surfaces of receptor-expressing cells circulate in the biological fluids (Rubin and Nelson 1990).

Full-length IL-2/15Rβ encoded on human chromosome 22q11.2–12 is a 551 amino acid receptor that contains a 26 amino acid signal peptide and the mature IL-2/15Rβ subunit composed of 525 amino acids (Hatakeyama et al. 1989b). It contains an extracellular segment of 214 amino acids, a hydrophobic transmembrane stretch of 25 amino acids, followed by a 286 amino acid cytoplasmic domain. IL-2/15Rβ is constitutively expressed by NK cells, monocytes and most T cells (Sharon et al. 1986; Tsudo et al. 1986, 1987). It is further induced by anti-CD3 plus anti-CD28, mitogens, PMA and by cytokines such as IL-4. The cytoplasmic domain of IL-2/15Rβ has been divided into three broad regions: the serine-rich S region (amino acids 267–322), the acidic A region (amino acids 313–382) and the proline-rich H region (amino acids 378–525). The S region is involved in the recruitment of JAK1 at a proline-rich box 1 motif. The H region function relies on the critical tyrosine residues Y392 and Y510 for STAT docking and activation.

The common gamma subunit encoded on chromosome Xq 13.1 is a 269 amino acid residue peptide (Suzuki et al. 1989; Takeshita et al. 1992; Noguchi et al. 1993). This protein contains a 22 amino acid signal peptide, a 255 amino acid extracellular domain, a 29 amino acid hydrophobic transmembrane and an 86 amino acid C-terminal cytoplasmic domain. γ_c is constitutively expressed on lymphocytes

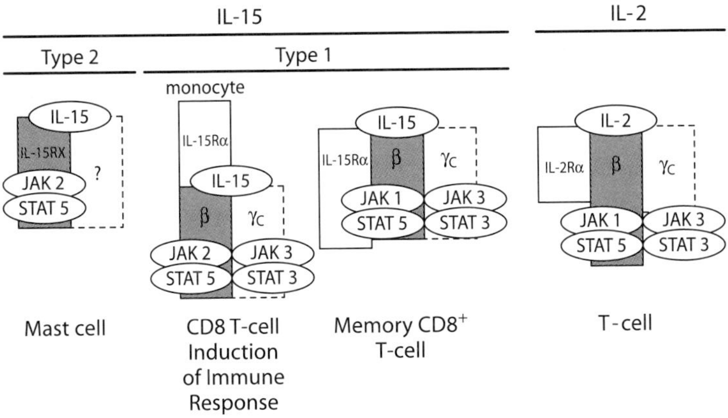

Fig. 4.1. Schematic diagram of the IL-2 and IL-15 receptor systems. IL-2R is made up of three distinct membrane components: IL-2Rα, IL-2/15Rβ and γ_c. There are three forms of the IL-15 receptor. The receptor involved in T cell and NK cell activity utilizes the private IL-15Rα receptor chain in association with IL-2/15Rβ and γ_c. During induction of the immune response, antigen-presenting cells such as monocytes coordinately express IL-15Rα and IL-15. IL-15 is presented in *trans* to the beta and gamma subunits expressed on T and NK cells. Mature memory CD8+ T cells coordinately express IL-15Rα, IL-2/15Rβ and γ_c. Mast cells express a novel IL-15RX receptor. The IL-2 and IL-15 receptors that are expressed in T and NK cells through IL-2/15Rβ and γ_c utilize JAK1/JAK3 and STAT3 and STAT5. The mast cell IL-15RX receptor includes tyrosine phosphorylation of JAK2 and an activation and nuclear translocation process limited to STAT5

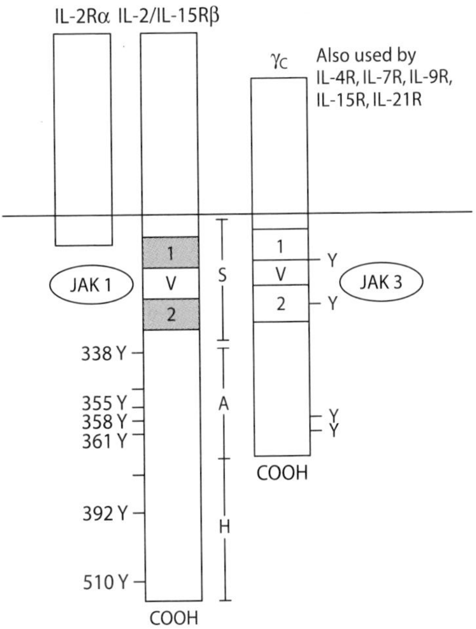

Fig. 4.2. Schematic diagram of the IL-2 receptor complex. IL-2R is composed of three subunits: IL-2Rα, IL-2/IL-15Rβ which is shared with IL-15, and γ_c shared with the IL-4, IL-7, IL-9, IL-15 and IL-21 receptors. *Box 1* and *box 2* domains of IL-2/IL-15Rβ are conserved among members of the cytokine receptor superfamily. A variable spacer region (*V*) intervenes. Tyrosine residues in *S*, *A* and *H* regions are numbered according to Hatakeyama et al. (1989a), *JAK1* associates constitutively with box 1-V and box 2 regions of IL-2/IL-15Rβ and *JAK3* associates with γ_c. Some of the tyrosines on both chains become phosphorylated following receptor activation and serve to recruit STAT molecules that direct downstream signaling events

(Takeshita et al. 1992; Cao et al. 1993). Its expression on monocytes can be up-regulated at the post-transcriptional level by IFN-γ or IL-2.

IL-2/15Rβ and γ_c used by both IL-2 and IL-15 are members of the hematopoietin or cytokine I superfamily of receptors that contain four conserved cystines and a canonical membrane proximal WSXWS (Trp-Ser-X-Trp-Ser) motif (Bazan 1990; Leonard et al. 1999). Mice genetically deficient in γ_c manifest severe combined immunodeficiency since they cannot respond to IL-2, IL-4, IL-7, IL-9, IL-15 and IL-21 (Cao et al. 1995; DiSanto et al. 1995). They have markedly reduced numbers of T cells (related to the deficiency of the response to IL-7) (Puel et al. 1998), NK cells (related to the IL-15-deficient response) (Carson et al. 1994) and an abnormal B cell response related to the failure to respond by this pathway to IL-4 and IL-21.

IL-15 with T and NK cells utilizes IL-2/15Rβ and γ_c as well as a private receptor subunit IL-15Rα (Fig. 4.1; Giri et al. 1995; Kennedy and Park 1996; Tagaya et al. 1996a; Waldmann and Tagaya 1999). IL-15Rα is a type-1 membrane protein with a predicted signal peptide of 32 amino acids, a 173 amino acid extracellular domain, a membrane-spanning region of 21 amino acids and a 37 amino acid cytoplasmic domain. At least eight different IL-15Rα transcripts associated with an exon-splicing mechanism have been reported. IL-15Rα isoforms that include exon 2 are capable of binding IL-15, whereas those without exon 2 are not (Dubois et al. 2002). Functional differences have been ascribed to the two alternative cytoplasmic domains. The 41 amino acid isoform undergoes endosomal cycling, as discussed below, whereas the 21 amino acid isoform does not. IL-15Rα is not a member of the cytokine I superfamily; however, a comparison between IL-2Rα and IL-15Rα revealed the shared presence of a conserved motif known as the "Sushi domain" or a GP-1 motif (Giri et al. 1995). The IL-2Rα and IL-15Rα genes have a similar intron:exon organization and are closely linked in both human (10q 14–15) and murine (chromosome 2) genomes (Anderson et al. 1995). Thus, IL-15Rα and IL-2Rα together define a new cytokine receptor family.

There are three forms of IL-2 receptors based on their affinity for the ligand IL-2. The very high affinity form (K_d 10^{-11} M) is made of the IL-2R alpha, beta and gamma subunits, the intermediate affinity form (10^{-9} M) encompasses the beta and gamma subunits, and the lower affinity form (10^{-8} M) is composed of IL-2Rα alone (Lowenthal and Greene 1987; Wang and Smith 1987). The $\beta\gamma_c$ heterodimeric and the $\alpha\beta\gamma_c$ heterotrimeric forms of the IL-2R can induce a signal following interaction with IL-2. The IL-15Rα receptor subunit binds IL-15 with a K_d of 10^{-11} M, a 1,000-fold higher affinity than that of IL-2Rα for IL-2. The β and γ_c subunits together bind IL-15 at approximately K_d 10^{-9} M, whereas the heterotrimeric receptor involving alpha, beta and gamma has an affinity comparable to that of IL-15Rα alone. There is a widespread cellular distribution but a low-level expression of IL-15Rα (Anderson et al. 1995). IL-15Rα expression is augmented in macrophage cell lines following the treatment with IFN-γ or interferon plus lipopolysaccharide (LPS) (Dubois et al. 2002).

4.3
IL-15 Utilizes a Novel Receptor in Mast Cells

Mast cells lack the required beta chain of the IL-2 receptor and therefore do not respond to IL-2. However, IL-15 is a growth factor for mast cells. where it utilizes a novel receptor system and signal transduction pathway which is distinct from that utilized in T cells. The IL-15 receptor in mast cells not only does not utilize the beta subunit but also does not require the action of γ_c. This latter trait was demonstrated by transfecting these cells with a cytoplasmic-truncated transdominant negative mutant of γ_c (Tagaya et al. 1996a,b). Furthermore, it appears that IL-15Rα is not the critical element of the mast cell IL-15R. Using ^{125}I-IL-15 cross-linked to mast cell lines, these cells were shown to express a novel 60- to 65-kDa receptor, which we have provisionally designated IL-15RX (Fig. 4.1; Tagaya et al. 1996a,b).

4.4
The Shared and Contrasting Functions of IL-2 and IL-15

The roles of IL-2 and IL-15 in the life and death of lymphocytes may be considered in terms of a series of goals of the immune system including: (1) the generation of a rapid and adaptive response to invading pathogens, (2) the maintenance of a specific memory response to these pathogens, and (3) the elimination of autoreactive T cells to yield tolerance to self (Waldmann 1991; Taniguchi and Minami 1993; Waldmann et al. 2001). As a consequence of their sharing the receptor subunits IL-2/15Rβ and γ_c and their use of common JAK/STAT signaling elements, IL-2 and IL-15 also share a number of functions, including the stimulation of the proliferation of activated CD4$^-$, CD8$^-$, CD4$^+$8$^+$, CD4$^+$, CD8$^+$ and γ/δ subsets of T cells (Burton et al. 1994; Grabstein et al. 1994; Edelbaum et al. 1995; Garcia et al. 1998; Zhang et al. 1998; Waldmann and Tagaya 1999). Furthermore, the two cytokines facilitate the induction of cytotoxic effector T cells including CTL cells, and stimulate the proliferation and immunoglobulin synthesis of B cells activated by anti-IgM or CD40 ligand (Armitage et al. 1995). They also induce the generation, proliferation, activation and survival of NK cells (Carson et al. 1994; Mrozek et al. 1996; Mingari et al. 1997; Suzuki et al. 1997a; Ogasawara et al. 1998). In addition to these shared functions, IL-2 and IL-15 manifest contrasting contributions to T cell-mediated immune responses (Zhang et al. 1998; Ku et al. 2000; Ma et al. 2000; Waldmann et al. 2001). Although IL-2 is an important growth and survival factor, it also plays a pivotal role in activation-induced cell death (AICD), a phenomenon involved in peripheral tolerance by acting through the suicidal elimination of self-reactive T cells (Lenardo 1996; Marks-Konczalik et al. 2000; Waldmann et al. 2001). As demonstrated in IL-15 transgenic mice, IL-15 manifests anti-apoptotic actions and inhibits IL-2-mediated AICD (Dooms et al. 1998; Marks-Konczalik et al. 2000). In addition to their actions in AICD, IL-2 and IL-15 play opposing roles in the control of the homeostasis of CD8$^+$ memory phenotype T cells. Zhang and coworkers (Zhang et al. 1998) demonstrated that IL-15, unlike IL-2, provides potent and selective stimulation of memory phenotype (CD44^{hi}) CD8$^+$ T cells in vivo. Furthermore, Ku and coworkers (2000) reported that the division of $CD8^+$ T cells of memory phenotype is stimulated by IL-15 but is inhibited by IL-2 through its actions on $CD4^+$ $CD25^+$ (T-regs) suppressor cells. IL-15 transgenic mice manifest high numbers of $CD8^+$ memory phenotype T cells (Marks-Konczalik et al. 2000; Waldmann et al. 2001). IL-15 was also shown to play a role in the expression of antigen-specific, MHC I restricted memory phenotype $CD8^+$ T cells that participate in the pathogenesis of the neurological disease HAM/TSP caused by infection with the retrovirus HTLV-I (Azimi et al. 1999, 2001).

These conclusions concerning the distinct and competing roles of IL-2 and IL-15 that derive from ex vivo functional studies are supported by an analysis of mice with disrupted cytokine and cytokine-receptor genes. In particular, $IL\text{-}2^{-/-}$ and $IL\text{-}2R\alpha^{-/-}$ mice develop massive enlargement of peripheral lymphoid organs associated with polyclonal T and B cell expansion (Schorle et al. 1991; Sadlack et al. 1994). Furthermore, the IL-2Rα-deficient mice developed autoimmune diseases including hemolytic anemia and inflammatory bowel disease (Willerford et al. 1995). This lymphocyte expansion and associated autoimmune disease represent the loss of the IL-2-mediated checkpoints or brakes on the immune system discussed above. A similar phenotypic pattern is observed in IL-2/15Rβ-deficient mice secondary to the loss of their ability to respond to IL-2. In contrast to the phenotype observed with IL-2-deficient mice, mice genetically deficient in IL-15 did not manifest lymphoid enlargement, high immunoglobulin levels or autoimmune disease (Lodolce et al. 1998; Kennedy et al. 2000). Rather, they displayed a marked reduction in the number of thymic and peripheral blood NK cells, NK T cells and intestinal intraepithelial lymphocytes (IELs). Furthermore, they manifested a marked reduction in memory phenotype $CD8^+$ T cells. Taken together, these studies support the view that in their special adaptive immune functions, IL-2 and IL-15 favor opposing actions that tend to emphasize one or the other of two competing major goals of the immune response: IL-2, through its contribution to AICD for $CD4^+$ cells, its interference with the persistence of memory $CD8^+$ phenotype T cells and its role in the induction of $CD4^+$, $CD25^+$ and T-regs, favors the elimination of select lymphocytes directed toward self-antigens and thus plays a critical role in the maintenance of peripheral self-tolerance; in contrast, IL-15, through its inhibition of IL-2-mediated AICD and its role in the maintenance of $CD8^+$ memory phenotype T cells, favors the maintenance and survival of $CD4^+$ and $CD8^+$ T cells.

4.5 Classical IL-2, IL-15 Signal Transduction Pathways That Underlie the Shared Roles of IL-2 and IL-15

The IL-2 and IL-15 receptor subunits do not manifest intrinsic tyrosine kinase activity. However, in T and NK cells, IL-2 and IL-15, acting through heterotrimeric $\alpha\beta\gamma_c$ or heterodimeric $\beta\gamma_c$ receptors, activate JAK1 and JAK3 of the kinase family (Witthuhn et al. 1994; Johnston et al. 1995; Lin et al. 1995; Ellery and Nicholls 2002; Shurin et al. 2003). Using chimeric receptors composed of the cytoplasmic domains of the beta and gamma subunits and extracellular domains of either the

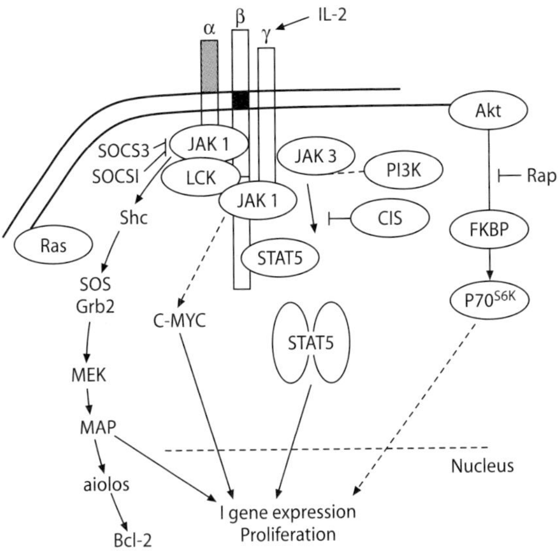

Fig. 4.3. Classical IL-2 receptor signaling pathways. Schematic representation of specific signaling pathways linked to the cytoplasmic domains of the IL-2/IL-15Rβ chain and γ_c

c-kit receptor or IL-2Rα, it was demonstrated that oligomerization of IL-2/15Rβ and γ_c is necessary for signal transduction through the JAK/STAT pathways (Fig. 4.3; Nakamura et al. 1994; Nelson et al. 1994). JAK1 is associated with the S region of IL-2/15Rβ because this region contains a Box 1 and a Box 2 motif required for proliferation, similar to that of other type I cytokine receptors (Fig. 4.2; Miyazaki et al. 1994). JAK3 binds to the cytoplasmic domain of γ_c following cytokine-induced oligomerization of the beta and gamma receptor subunits. JAK3 and JAK1 are activated by transphosphorylation of critical tyrosines of the JAK kinases. The JAK kinases then phosphorylate tyrosines at positions 392 and 510 of IL-2/15Rβ. These phosphorylated tyrosines act as docking sites for the signal transducer and transactivator of transcription (STAT) proteins, which in the case of IL-2 and IL-15 involve STAT3 and STAT5 (Goldsmith et al. 1995; Lin et al. 1995, 1996; Friedmann et al. 1996; Fujii et al. 1998). The docked STAT proteins, when phosphorylated on a critical tyrosine, dimerize and then translocate to the nucleus where they permit transcription of target genes. Since IL-2, IL-4, IL-7, IL-9 and IL-15 activate the same JAK kinases, it is clear that JAK kinases per se do not determine the cytokine specificity of the immune response.

As noted above, in mast cells IL-15 utilizes a novel IL-15RX receptor. In further contrast to T cells, the membrane-proximal events of the mast cell IL-15–IL-15RX signaling pathway include the tyrosine phosphorylation of JAK2 rather than JAK1 and JAK3, as well as an activation and nuclear translocation process limited to STAT5 (Tagaya et al. 1996a,b).

4.6 Suppressors of Cytokine Signaling

Activation of the JAK/STAT pathway is short-lived. Two classes of endogenous cytokine signaling inhibitors (CIS/SOCS/Jab/SSI and PIAS family members) have been identified as cytokine-inducible factors that exert inhibitory activities on JAK and/or STAT proteins (Shurin et al. 2003). CIS (cytokine-inducible SH2-containing protein) is induced by a number of cytokines and growth factors including IL-2 (Yoshimura et al. 1995; Kovanen and Leonard 1999). CIS associates with the A-region of IL-2/15Rβ (Aman et al. 1999). CIS inhibits IL-2-induced STAT5 activation. A second SOCS family gene, SOCS 1, inhibits multiple cytokines including IL-2 by binding to the catalytic loop in JAK kinases, thereby blocking downstream STAT activation (Marine et al. 1999; Sporri et al. 2001). SOCS 3 interacts with JAK1 and can inhibit JAK1 activation. Thus a number of CIS/SOCS/SSI family proteins are involved in the negative regulation of IL-2 signaling in vivo.

4.7 The Distinct Receptor Signaling Pathways That Underlie the Contrasting Roles of IL-2 and IL-15 in the Life and Death of Lymphocytes

A critical issue that remains to be fully resolved is how the two cytokines IL-2 and IL15, which share two signaling receptor subunits and utilize as one of their signaling pathways a common JAK1/JAK3/STAT5 signaling pathway, manifest a number of contrasting functions. Many alternatives have been suggested as explanations for these differences. One distinction between IL-2 and IL-15 is in their binding interactions with IL-2/15Rβ. Antibodies directed toward this receptor subunit such as MiK-Beta-1 block the functions of IL-15 but not those of IL-2 to their respective high affinity heterotrimeric receptors (Tsudo et al. 1989; Guex-Crosier et al. 1997). Furthermore, the response to the cytokines depends not only on the interactions of the subunits with the ligand but also on the interactions of the three subunits with each other. In this regard, the proximity of the IL-2Rα subunit changes the function of the IL-2/IL-15Rβ to alter its IL-2 binding by mechanisms that do not require the binding of IL-2 to the IL-2Rα subunit (Grant et al. 1992; Roessler et al. 1994).

Biophysical techniques have been applied in order to develop insights into the cytokine-dependent subunit assembly, which is indispensable for our understanding of the paradoxical distinct signaling capacities. By using fluorescence resonance energy transfer (FRET) techniques, Damjanovich and coworkers (Damjanovich et al. 1997) demonstrated that the three subunits of the high affinity IL-2R are in molecular vicinity of each other at the surface of T-lymphoma/leukemia cells even in the absence of IL-2 (Fig. 4.4). The addition of IL-2 resulted in the "tightening" of this performed receptor complex. Moreover, IL-15 acts in the opposite direction by opening the triangle, possibly because IL-15Rα becomes more closely associated with the beta and gamma subunits that it shares with the

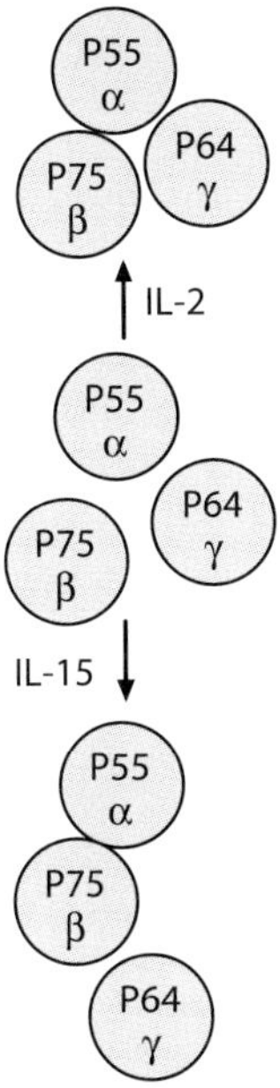

Fig. 4.4. Schematic representation of organization of subunits of the IL-2 receptor complex on Kit 225 K6 T cells. FRET data indicated that IL-2R alpha, beta and gamma subunits are preassembled, forming a heterotrimeric triangle on the surface of resting cells (*middle*). Addition of IL-2 promoted a stronger contact of the alpha subunit with beta and gamma chains (*above*). IL-15 induced a weakening of the contact between the γ_c and IL-2Rα subunits, thereby leading to a somewhat linearized configuration of the IL-2R complex (*below*). (Reprinted with modifications with permission from Sándor Damjanovich et al. 1997)

IL-2R complex. It was theoretically possible that the IL-2 and IL-15 receptor systems might be expressed in different cell-surface compartments, thereby explaining their distinct functions. However, this was not shown to be the case. In particular, by using FRET and confocal microscopy, Vámosi and coworkers (2004) demonstrated that IL-2Rα, IL-15Rα, IL-2/15Rβ and γ_c subunits as well as MHC I and II glycoproteins formed supermolecular receptor complexes in lipid rafts of the T-lymphoma cell line, Kit 225 FT7.10.

Fluorescence cross-correlation microscopy was used to demonstrate the co-mobility of IL-15Rα, IL-2Rα and MHC I. A model was generated for subunit switching between IL-2Rα and IL-15Rα on binding of the appropriate cytokine, resulting in the formation of high affinity heterotrimeric receptors. The FRET data presented were compatible with a tetrameric structure of the IL-2/IL-15 receptor complex. IL-2Rα and IL-15Rα are in the molecular vicinity of each other in the presence and absence of cytokines. Binding of IL-2 or IL-15 leads to an increase in FRET efficiency between IL-2/15Rβ and IL-2Rα or IL-15Rα, respectively. These observations could be explained by the coordinated rotational/translational motion of the IL-2Rα and IL-15Rα subunits with respect to the $\beta\gamma_c$ heterodimer (Vámosi et al. 2004).

Another mechanism that may play a role in the quite distinct outcomes of the biological responses mediated by these two cytokines reflects the mode of action of the cytokines early in the immune response. In general, IL-2 produced by activated CD4$^+$ cells is secreted into the biological fluids where it interacts with the preformed heterotrimeric IL-2 receptor expressed on activated responding T cells, B cells and NK cells. In contrast, we have demonstrated that IL-15 does not

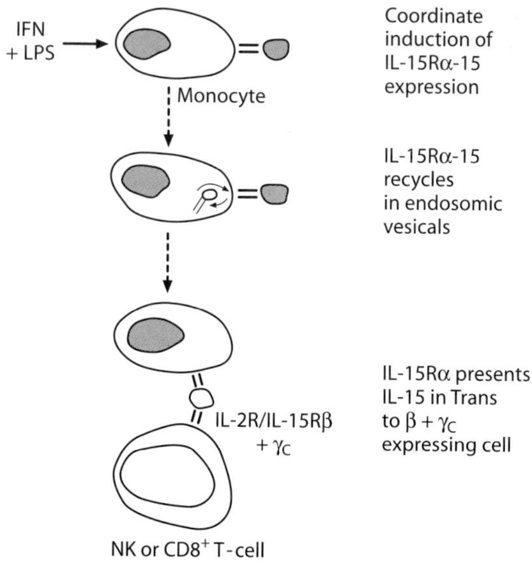

Fig. 4.5. IL-15Rα on antigen-presenting cells presents IL-15 in *trans* to NK and CD8$^+$ memory T cells that express only IL-2/IL-15Rβ and γ_c. During induction of the immune response, IL-15Rα and IL-15 are coordinately expressed on the surface of APCs. On interaction between IL-15Rα and IL-15, the complex recycles in endosomic vesicles, leading to reappearance of IL-15 on the cell surface. IL-15Rα on these APCs then can present the tightly associated IL-15 in *trans* to NK cells or CD8$^+$ memory T cells which express only IL-2/IL-15Rβ and γ_c. Such transpresentation events occur as part of an immunological synapse that provides associated co-stimulatory signals

usually act primarily as a secreted cytokine but rather as a membrane-associated molecule (Dubois et al. 2002). During the induction of the immune response, IL-15Rα and IL-15 are coordinately expressed on the surface of antigen-presenting cells such as monocytes and dendritic cells that have been stimulated by an interferon as well as through toll-like receptor 4 (Fig. 4.5).

The IL-15Rα on these antigen-presenting cells can then present the tightly associated IL-15 in *trans* to NK cells or CD8$^+$ T cells that express only IL-2/15Rβ and γ_c but not IL-15Rα. Such transpresentation events occur as part of an immunological synapse that provides associated co-stimulatory signals such as those mediated by CD40L with CD40 and CD80/86 with CD28 in addition to those initiated by IL-15. It should be noted that during the late phases in the evolution of the immune response, co-expression of all three elements of the IL-15 receptor in the same membrane domain of an individual cell occurs in mature memory CD8$^+$ T cells (Oh et al. 2003). Such co-expression of all three elements of the IL-15 receptor in responding T cells may be involved in the maintenance of long-term CD8$^+$ memory responses by permitting the cells to respond to the exceedingly low physiological concentrations of IL-15 present in the biological fluids.

A final and most important explanation for the distinct actions of IL-2 and IL-15 focuses on the private cytokine-specific receptor elements IL-15Rα and IL-2Rα and on proteins nonrandomly associated with them.

4.8 The Distinct Roles of IL-2 and IL-15 in Signaling for AICD

As noted above, IL-2 plays a pivotal role in AICD, a function that appears to be nonredundant. If IL-2 is neutralized via antibodies or is genetically deleted, T lymphocytes can be repeatedly activated via their TCR without undergoing programmed cell death (Lenardo 1991). No other cytokine, including IL-4, IL-7 or IL-15, can replace IL-2 in this activity. One critical step during AICD proceeds via the death receptors Fas and p75. T cell apoptosis can be achieved by stimulating Fas (Russell and Wang 1993; Mixter et al. 1994; Anderson et al. 1995). T cells that carry mutations in Fas or its ligand (FasL) are resistant to AICD. The presence of IL-2 increases the amounts of the death receptor ligands TNFα and FasL upon TCR re-engagement. In contrast to IL-2, little upregulation of either FasL or TNFα is observed if mature T cells are grown in IL-15 (unpubl. data). This indicates signaling differences between the two cytokines that act upstream of death receptor ligands, resulting in their expression.

A number of transcription factors, including AP-1, SP-1, NF-κB, NFAT, Egr1–3, c-myc and forkhead transcription factor 1, have been described as regulating both basal and inducible as well as tissue-specific transcription of FasL (Green et al. 2003). Our laboratory has focused on the different abilities of IL-2 and IL-15 to induce FasL expression after TCR re-engagement and analyzed the status of transcription factors in these cells. The most striking difference was observed with NFAT1 (unpubl. data). In contrast to previous reports (Lyakh et al. 1997), NFAT1 is not expressed in resting T cells, and the presence of IL-2 is absolutely necessary to induce the expression of NFAT1 in T cells following activation. There is strong evidence for a central role of NFAT1 in AICD as follows: (1) NFAT1-deficient T cells fail to upregulate FasL expression during TCR re-engagement and fail to undergo AICD (Hodge et al. 1996; Xanthoudakis et al. 1996); (2) TCR engagement in resting T cells does not cause apoptosis. While NFAT1 is not expressed at this state, a number of other transcription factors that are implicated in the regulation of FasL are activated during both primary and secondary TCR engagements (unpubl. data); and (3) inhibitors of NFAT activation such as FK506 or CsA reduce AICD if they are present during TCR re-engagement (Shi et al. 1989; unpubl. data). Thus the regulation of FasL via NFAT1 represents one pathway to AICD that is differently regulated by IL-2 and IL-15.

It is presently unclear what mechanism underlies the distinct regulation of NFAT1 by IL-2 and IL-15. Besides effects on the transcription of NFAT1, it is also regulated via degradation processes that could be inhibited by the presence of IL-2. No direct data are available about mediators that affect the presence of NFAT1. Circumstantial evidence points to an involvement of the Tec family kinase Itk (Lucas et al. 2003). Itk-deficient peripheral T lymphocytes are defective in AICD via their inability to upregulate FasL (Miller and Berg 2002). Persistent expression of Itk in activated T cells depends on IL-2 that cannot be replaced by IL-15 (unpubl. data). Another potential mediator of NFAT1 expression could be Lck. Peripheral T cells that do not express Lck fail to undergo AICD (Yu et al. 2004). In contrast to Itk, Lck expression does not appear to be affected by the presence of IL-2. Differences between the effects of IL-2 and IL-15 are observed on the phosphorylation pattern of Lck, which indicates an activation by IL-2 but not by IL-15 in a manner different from that observed with TCR-induced activation (Hatakeyama et al. 1991; unpubl. data).

4.8.1 AICD Signaling Downstream of Death Receptors

Although IL-2 is necessary for the induction of FasL, FasL stimulation via cross-linking with antibodies still proves insufficient to cause apoptosis in IL-2-deficient cells (Kneitz et al. 1995; Nishimura et al. 1995). While many characteristics are shared between IL-$2^{-/-}$ and IL-2/15R$\beta^{-/-}$ cells, T cells with a mutation in this IL-2/15Rβ receptor subunit are reported to respond normally to Fas ligation (Suzuki et al. 1997b). This points to an additional effect of IL-2 on AICD downstream of death receptor ligation that is mediated by either IL-2Rα, γ_c or a molecule associated with them.

Ligation of death receptors results in the recruitment of FADD and caspase-8 (Curtin and Cotter 2003). Caspase-8 is subsequently activated via proteolytic cleavage. An interactor of caspase-8, FLIP, appears to have the potential to both positively and negatively affect caspase-8 activation (Curtin and Cotter 2003). Activated caspase-8 processes the Bcl-2-like protein Bid, setting the mitochondrial cascade in motion. Alternatively, caspase-8 may directly cleave caspase-3 to induce cell death independent of mitochondrial pathways.

IL-2 is implicated in the regulation of several mediators of post-death receptor AICD. Transcription of caspase-3 is induced by IL-2 (unpubl. data). Although caspase-3 expression has been reported to depend on NFAT1, we have not detected differences in caspase-3 expression between wild-type and NFAT$1^{-/-}$ T cells in response to IL-2. IL-2 stimulation also activates the transcription of FLIP, which may represent a positive feedback since increased expression of FLIP in T cells also results in increased production of interleukin-2 (Kataoka et al. 2000).

The effects of IL-2 on post-death receptor pathways are more complex since a number of anti-apoptotic events are also induced by IL-2. These include an increased expression of Bcl-2 and XIAP as well as a downregulation of the pro-apoptotic Bok (Romero et al. 1999; unpubl. data). It appears to be likely that the final choice of an activated T cell between proliferation and death is based on the net-effect of several opposing mechanisms that would allow multiple targets for additional regulations. Some caution has to be employed in interpreting induction patterns, since anti-apoptotic proteins like Bcl-2 can also induce cell death under some circumstances (Cheng et al. 1997).

4.9 Distinct Signals for Proliferation Mediated by IL-2 and IL-15

An in vitro treatment of activated T cells with either IL-2 or IL-15 results in robust proliferation. These proliferation-inducing activities appear to be redundant in vivo since mice that carry a mutation in IL-2/15Rβ show no decrease in the number of most T cells (Suzuki et al. 1995). One exception may be the T regulatory cells that are absent in mice with disruptions in IL-2, IL-2Rα or IL-2/15Rβ (Papiernik et al. 1998). However, controversy exists concerning the effect of IL-2 signaling on the number of T regulatory cells, since IL-2 regulates the expression of IL-2Rα which is commonly used to identify T regulatory cells (Plaetinck et al. 1990). Mice with disruptions in either IL-15 or IL-15Rα have reduced numbers of $CD8^+$ memory cells (Lodolce et al. 1998; Kennedy et al. 2000). This decrease, however, is believed to result from defects in survival rather than expansion of these cells and their precursors (Burkett et al. 2003; Koka et al. 2003).

In attempts to analyze pathways that underlie the mitogenic activities of IL-2 and IL-15, we studied proliferation of T cells with deletions of FKBP12 and FKBP12.6 and in the presence of immunosuppressive drugs (Dubois et al. 2003). To our surprise, the proliferation-inducing activities of IL-2 and IL-15 proved to be quite different. IL-15 appears to employ a single signaling pathway to induce proliferation. This involves signaling through phosphoinositide-3 kinase (PI-3 K) and mammalian target of rappamycin (mTOR) to activate mediators of cell division such as $p70^{S6\,K}$ or 4E-BP. Proliferation was virtually absent if mTOR activity was affected by either the presence of rapamycin or a genetic deletion of FKBP12. Proliferation was equally affected by a pharmacological inhibition of PI-3 K.

In contrast, IL-2 induces at least four distinct signaling pathways leading to proliferation. A reduction in IL-2-induced T cell proliferation requires that at least two of these pathways are inhibited. In contrast to previous reports (Kuo et al. 1992), effective inhibition of mTOR and $p70^{S6\,K}$ activation by low doses of rapamycin only marginally decreases IL-2-induced proliferation, as does a mutation in the FKBP12.6 gene. A combination of both, however, reduces the mitogenic activity of IL-2 by about one third. An inhibition of PI-3 K alone reduces proliferation by about one third, indicating the branching of two signaling pathways downstream of PI-3 K. Proliferation was further reduced if PI-3 K was inhibited on an $FKBP12.6^{-/-}$ background, suggesting independent pathways that are affected by PI-3 K and FKBP12.6 (unpubl. data). Since significant proliferation was still measured, a fourth pathway to proliferation must exist in response to IL-2.

4.9.1 The PI-3 K Pathway

IL-2 stimulation leads to a phosphorylation of the p85 regulatory subunit of PI-3 K via IL-2/15Rβ (Remillard et al. 1991). In the case of both IL-2 and IL-15, PI-3 K activates $p70^{S6\,K}$ via mTOR, resulting in the translation of RNA species that are believed to be critical for cell division. PI-3 K also activates E2F via PKB (Brennan et al. 1997). E2F transcription factors mediate cell cycle entry into S phase. This pathway appears to be independent of mTOR, and only IL-2 but not IL-15 is capable of signaling to E2F in activated T cells (Dubois et al. 2003).

4.9.2 The FKBP12.6 Pathway

The nature of signaling events that depend on FKBP12.6 in the context of IL-2 stimulation is unknown. Both FKBP12 and FKBP12.6 form a complex with the immunosuppressive drug FK506 which affects NFAT activation via calcineurin inhibition (Dumont 2000). We studied NFAT in activated T cells and observed that activation of NFAT2 differs if cells are grown in IL-2 or alternatively IL-15 (unpubl. data). However, effective nuclear translocation of NFAT2 requires TCR engagement, suggesting that IL-2 signaling is unlikely to proceed directly through NFAT2. Alternatively, the presence of FKBP12.6 may be required during TCR-mediated T cell activation, leading to the expression of a signaling molecule that is used by IL-2 but not by IL-15 to stimulate proliferation.

4.9.3 The MAPK Pathway

Activation of MAPK causes proliferation in many cell types. Stimulation by IL-2 results in the JAK3-mediated phosphorylation of IL-2/15Rβ (Gaffen 2001). Phosphorylation of one of the tyrosines (Y338) creates a docking site for the adaptor protein Shc, which activates the MAPK cascade via Grb2, Sos, Ras, Raf (Ravichandran and Burakoff 1994; Evans et al. 1995; Friedmann et al. 1996; Ravichandran et al. 1996). Since cells that express IL-2/15Rβ with mutations of Y338 are still able to divide, it was suggested that MAPK activation is neither necessary nor sufficient for IL-2-stimulated T cells to proliferate (Friedmann et al. 1996; Gaffen et al. 1996; Fujii et al. 1998). However, since the status of MAPK was not determined in these cells, alternative activation mechanisms may exist that provide a proliferative impulse. Along this line, pharmacological inhibition of MAPK p38 affects IL-2-induced proliferation (Crawley et al. 1997). An abrogation of proliferation in response to IL-15 by rapamycin or by a disruption of FKBP12 does not affect MAPK p42/44 (ERK) activation, indicating that ERK activation is not sufficient to cause cell division in response to IL-15 (Dubois et al. 2003).

Other studies have focused on IL-15Rα, the private receptor subunit of IL-15. It has been reported that IL-15 can produce major effects on cells that bear IL-15Rα but do not express IL-2/15Rβ. These IL-15-mediated actions utilize signaling pathways that do not involve JAK1/3 or STAT5. For example, it has been reported but not yet confirmed that following IL-15 addition there is a recruitment of TRAF2 and Syk kinase to the cytoplasmic tail of IL-15Rα (Bulfone-Pau et al. 1999; Bulanova et al. 2001). Furthermore, the existence of two isoforms of IL-15Rα that differ in their cytoplasmic domains and functions supports the concept of signaling through IL-15Rα. Recent data from our laboratory suggest the interactions of a series of cytoplasmic proteins with IL-15Rα.

4.10 Immunotherapy Targeted to the IL-2/IL-2R IL-15/IL-15R System

The IL-2Rα chain has been widely used as a target for immunotherapy (Waldmann 1992, 1993; Waldmann et al. 1993; Morris and Waldmann 2000). The scientific basis for this strategy is that IL-2Rα is not expressed by virtually all of the resting cells of the body, but is constitutively expressed by the T cells from patients with certain lymphoid malignancies, select autoimmune disorders and those involved in allograft rejection. To exploit the difference in IL-2Rα antigen expression between normal resting cells and the abnormal T cells, clinical trials have been performed using unmodified murine anti-IL-2Rα antibodies and humanized antibodies, as well as antibodies armed with toxins and alpha- and beta-emitting radionuclides. Reviews have been published concerning IL-2Rα-directed therapy (Waldmann 1993; Morris and Waldmann 2000). Early human clinical trials involving IL-2Rα-directed therapy focused on HTLV-I-associated adult T cell leukemia. The retrovirus HTLV-I encodes a 42-kDa protein termed *tax* which is responsible for the transactivation of numerous host genes, including those encoding IL-2, IL-2Rα, IL-15 and IL-15Rα which are involved in T cell activation and potentially HTLV-I-mediated leukemogenesis (Mariner et al. 2001). Therapy with the unmodified murine version of the anti-Tac monoclonal antibody (anti-CD25, daclizumab), which blocks the binding of IL-2 to IL-2Rα and which also functions by antibody-dependent cellular cytotoxicity to eliminate target cells, has led to partial or complete remission in select patients with ATL (Waldmann et al. 1993). Increased efficacy of IL-2Rα-directed therapy of leukemia was achieved by humanizing the antibody and by arming it with alpha- and beta-emitting radionuclides (Waldmann et al. 1995). In other studies, following encouraging results in animal models and in phase I/II trials, phase III randomized placebo controlled clinical trials were performed that included 535 evaluated patients receiving renal allografts to determine the value of humanized anti-Tac (daclizumab) in the prevention of renal allograft rejection (Vincenti et al. 1998). On the basis of the success of these phase III trials, Food and Drug Administration (FDA) approval was obtained for the use of daclizumab and in the prevention of renal kidney transplant rejection. In a similar randomized trial it was shown that basiliximab, a mAb, which is also directed toward IL-2Rα, was associated with reduced numbers of acute rejection episodes in renal allograft recipients (Nashan et al. 1997). In a further study, Nussenblatt et al. (1999) showed that daclizumab is of value in the therapy of select T cell-mediated autoimmune disorders. In particular, this humanized monoclonal antibody provided effective treatment for noninfectious intermediate and posterior uveitis in a clinical trial (Nussenblatt et al. 1999). Furthermore, it was demonstrated that there was an approximately 78% reduction in Gadolinium-enhanced MRI lesions in patients with multiple sclerosis who were failing beta interferon therapy when they were treated with daclizumab (Bielekova et al. 2004).

Although the use of monoclonal antibodies directed toward the IL-2Rα subunit has been of great value, such therapy has limitations in certain conditions. In particular, IL-2Rα-directed therapy does not inhibit the actions of IL-15, nor does it act on resting T and NK cells that do not express IL-2Rα. Disorders of IL-15 expression have been observed in a number of inflammatory autoimmune diseases, including rheumatoid arthritis, psoriasis, inflammatory bowel disease, multiple sclerosis and disorders caused by the retrovirus HTLV-I (McInnes et al. 1997; Waldmann and Tagaya 1999; Mention et al. 2003). These disorders are associated with the fact that IL-15 induces the expression of TNFα and IL-1β as well as inflammatory chemokines, inhibits self-tolerance mediated by AICD and facilitates $CD8^+$ memory T cell survival (Waldmann et al. 2001). To circumvent these limitations, IL-15 receptor-directed therapy has been developed for use in autoimmune disorders as well as for the treatment of diseases caused by the retrovirus HTLV-I (Azimi et al. 2001). The soluble IL-15Rα chain prevented the development of collagen-induced arthritis in mice (Ruchatz et al. 1998). An IL-15 antagonist produced by mutating a glutamine residue to aspartic acid within the C-terminus of IL-15 inhibited IL-15-triggered self-proliferation and enhanced pancreatic islet cell allograft survival in mice (Kim et al. 1998). Antibodies to IL-15 have been effectively used in murine models of autoimmune diseases including psoriasis (Villadsen et al. 2003). Furthermore, such an antibody has shown efficacy in a phase I/II trial involving patients with rheumatoid arthritis (Baslund et al. 2003). Our own IL-15-directed therapeutic approach involves an antibody, humanized MiK-Beta-1 (Hu-MiKa-Beta-1), directed toward the IL-2/15Rβ cytokine receptor that is shared by IL-2 and IL-15. This humanized antibody interacts with the IL-2/15Rβ receptor subunit and blocks IL-15-mediated stimulation of NK and T cells ex vivo (Hakimi et al. 1993; Guex-Crosier et al. 1997). This antibody, when used as a single agent, prolonged cardiac allograft survival in cynomolgus monkeys (Tinubu et al. 1994). In an effort to test the hypothesis that IL-15 plays a role in the pathogenesis of select autoimmune disorders, a clinical trial is being initiated with Hu-MiK-Beta-1 in patients with rheumatoid arthritis, multiple sclerosis, T cell LGL associated with granulocytopenia and arthritis, and HTLV-I-associated disorders including the neurological disease HAM/TSP.

As noted above, there is a sharing of receptor subunits and signaling pathway elements among IL-2 and its receptors and those of other cytokines that stimulate T cells. A corollary of this sharing of cytokine signaling pathways among cytokines is that therapy directed toward a shared signal transduction element may yield more profound immunosuppression than can be achieved by an antibody directed toward a private receptor subunit among the redundant cytokine molecules such as IL-2Rα. Many groups have initiated programs directed toward developing and evaluating inhibitors of JAK3 as agents for controlled immunosuppression. JAK3 is involved in the signaling of the cytokines including IL-2 and IL-15 that employ γ_c, but is not essential for signaling by other growth factors (Witthuhn et al. 1994; Johnston et al. 1995). JAK3 expression is largely limited to lymphocytes and hematopoietic cells. JAK3 deficiency in the autosomal form of severe combined immunodeficiency disease (JAK3-deficient SCID) in humans yields immunodeficiency but not disorders in extraimmunological systems (Macchi et al. 1995; Russell et al. 1995). In parallel, mice made JAK3 deficient by gene targeting manifested an absence of NK cells and abnormalities of T and B cells, but, like the JAK3-deficient humans, do not have disorders in nonimmunological systems (Nosaka et al. 1995; Thomis et al. 1995). Taken together, these observations suggest that drugs that inhibit JAK3 action may be of value as therapeutic agents in patients with constitutive JAK3 activation, such as those with T cell leukemia and lymphoma, as well as in patients with autoimmune disorders and those receiving allografts.

In conclusion, our emerging understanding of the IL-2/IL-2R and IL-15/IL-15R systems, including the definition of the actions that these cytokines share as well as those functions where their roles are distinct, provides a better understanding of the mechanisms underlying cytokine-mediated generation, proliferation, survival and apoptosis of lymphocytes. Multiple studies, including those directed toward the definition of new signaling pathways that involve the private receptor subunits and proteins nonrandomly associated with them, are providing insights into the mechanisms underlying the divergent actions of IL-2 and IL-15. Furthermore, our greater understanding of these cytokine/cytokine receptor systems is opening up new possibilities for the development of more rational immune interventions which may be of value in prevention of allograft rejection and in the treatment of lymphocytic leukemia/lymphoma and T cell-mediated autoimmune disorders, as well as diseases associated with the retrovirus HTLV-I.

References

Aman MJ, Migone TS, Sasaki A, Ascherman DP, Zhu M, Soldaini E, Imada K, Miyajima A, Yoshimura A, Leonard WJ (1999) CIS associates with the interleukin-2 receptor beta chain and inhibits interleukin-2-dependent signaling. J Biol Chem 274:30266–30272

Anderson DM, Kumaki S, Ahdieh M, Bertles J, Tometsko M, Loomis A, Giri J, Copeland NG, Gilbert DJ, Jenkins NA, Valentine V, Shapiro DN, Morris SW, Park LS, Cosman D (1995) Functional characterization of the human interleukin-15 receptor alpha chain and close linkage of IL15RA and IL2RA genes. J Biol Chem 270:29862–29869

Armitage RJ, Macduff BM, Eisenman J, Paxton R, Grabstein KH (1995) IL-15 has stimulatory activity for the induction of B cell proliferation and differentiation. J Immunol 154: 483–490

Azimi N, Jacobson S, Leist T, Waldmann TA (1999) Involvement of IL-15 in the pathogenesis of human T lymphotropic virus type I-associated myelopathy/tropical spastic paraparesis: implications for therapy with a monoclonal antibody directed to the IL-2/15R beta receptor. J Immunol 163:4064–4072

Azimi N, Nagai M, Jacobson S, Waldmann TA (2001) IL-15 plays a major role in the persistence of Tax-specific CD8 cells in HAM/TSP patients. Proc Natl Acad Sci USA 98:14559–14564

Bamford RN, Grant AJ, Burton JD, Peters C, Kurys G, Goldman CK, Brennan J, Roessler E, Waldmann TA (1994) The interleukin (IL) 2 receptor beta chain is shared by IL-2 and a cytokine, provisionally designated IL-T, that stimulates T-cell proliferation and the induction of lymphokine-activated killer cells. Proc Natl Acad Sci USA 91:4940–4944

Baslund B, Tvede N, Danneskiold-Samsoe B, Petersen J, Petersen LJ, Schuurman J, de Winkel JV, Larsson P, Panayi G, McInnes LB (2003) A novel human monoclonal antibody against IL-15 (humax-IL15) in patients with active rheumatoid arthritis (RA): results of a double-blind, placebo-controlled phase I/II trial. Arthritis Rheumatism 48(Suppl S):1706

Bazan JF (1990) Structural design and molecular evolution of a cytokine receptor superfamily. Proc Natl Acad Sci USA 87:6934–6938

Bielekova B, Richert N, Howard T, Blevins G, Markovic-Plese S, McCartin J, Wurfel J, Ohayon J, Waldmann TA, McFarland HF, Martin R (2004) Humanized anti-CD25 (daclizumab) inhibits disease activity in multiple sclerosis patients failing to respond to interferon beta. Proc Natl Acad Sci USA 101:8705–8708

Brennan P, Babbage JW, Burgering BM, Groner B, Reif K, Cantrell DA (1997) Phosphatidylinositol 3-kinase couples the interleukin-2 receptor to the cell cycle regulator E2F. Immunity 7:679–689

Bulanova E, Budagian V, Pohl T, Krause H, Durkop H, Paus R, Bulfone-Paus S (2001) The IL-15R alpha chain signals through association with Syk in human B cells. J Immunol 167: 6292–6302

Bulfone-Pau SS, Bulanova E, Pohl T, Budagian V, Durkop H, Ruckert R, Kunzendorf U, Paus R, Krause H (1999) Death deflected: IL-15 inhibits TNF-alpha-mediated apoptosis in fibroblasts by TRAF2 recruitment to the IL-15Ralpha chain. FASEB J 13:1575–1585

Burkett PR, Koka R, Chien M, Chai S, Chan F, Ma A, Boone DL (2003) IL-15R alpha expression on CD8+ T cells is dispensable for T cell memory. Proc Natl Acad Sci USA 100:4724–4729

Burton JD, Bamford RN, Peters C, Grant AJ, Kurys G, Goldman CK, Brennan J, Roessler E, Waldmann TA (1994) A lymphokine, provisionally designated interleukin T and produced by a human adult T-cell leukemia line, stimulates T-cell proliferation and the induction of lymphokine-activated killer cells. Proc Natl Acad Sci USA 91:4935–4939

Cao X, Kozak CA, Liu YJ, Noguchi M, O'Connell E, Leonard WJ (1993) Characterization of cDNAs encoding the murine interleukin 2 receptor (IL-2R) gamma chain: chromosomal mapping and tissue specificity of IL-2R gamma chain expression. Proc Natl Acad Sci USA 90:8464–8468

Cao X, Shores EW, Hu-Li J, Anver MR, Kelsall BL, Russell SM, Drago J, Noguchi M, Grinberg A, Bloom ET, Paul WE, Katz SI, Love PE, Leonard WJ (1995) Defective lymphoid development in mice lacking expression of the common cytokine receptor gamma chain. Immunity 2:223–238

Carson WE, Giri JG, Lindemann MJ, Linett ML, Ahdieh M, Paxton R, Anderson D, Eisenmann J, Grabstein K, Caligiuri MA (1994) Interleukin (IL) 15 is a novel cytokine that activates human natural killer cells via components of the IL-2 receptor. J Exp Med 180:1395–1403

Cheng EH, Kirsch DG, Clem RJ, Ravi R, Kastan MB, Bedi A, Ueno K, Hardwick JM (1997) Conversion of Bcl-2 to a Bax-like death effector by caspases. Science 278:1966–1968

Cosman D, Cerretti DP, Larsen A, Park L, March C, Dower S, Gillis S, Urdal D (1984) Cloning, sequence and expression of human interleukin-2 receptor. Nature 312:768–771

Crawley JB, Rawlinson L, Lali FV, Page TH, Saklatvala J, Foxwell BM (1997) T cell proliferation in response to interleukins 2 and 7 requires p38MAP kinase activation. J Biol Chem 272:15023–15027

Curtin JF, Cotter TG (2003) Live and let die: regulatory mechanisms in Fas-mediated apoptosis. Cell Signal 15:983–992

Damjanovich S, Bene L, Matko J, Alileche A, Goldman CK, Sharrow S, Waldmann TA (1997) Preassembly of interleukin 2 (IL-2) receptor subunits on resting Kit 225 K6 T cells and their modulation by IL-2, IL-7, and IL-15: a fluorescence resonance energy transfer study. Proc Natl Acad Sci USA 94:13134–13139

DiSanto JP, Muller W, Guy-Grand D, Fischer A, Rajewsky K (1995) Lymphoid development in mice with a targeted deletion of the interleukin 2 receptor gamma chain. Proc Natl Acad Sci USA 92:377–381

Dooms H, Desmedt M, Vancaeneghem S, Rottiers P, Goossens V, Fiers W, Grooten J (1998) Quiescence-inducing and antiapoptotic activities of IL-15 enhance secondary CD4+ T cell responsiveness to antigen. J Immunol 161:2141–2150

Dubois S, Mariner J, Waldmann TA, Tagaya Y (2002) IL-15Ralpha recycles and presents IL-15 in *trans* to neighboring cells. Immunity 17:537–547

Dubois S, Shou W, Haneline LS, Fleischer S, Waldmann TA, Muller JR (2003) Distinct pathways involving the FK506-binding proteins 12 and 12.6 underlie IL-2-versus IL-15-mediated proliferation of T cells. Proc Natl Acad Sci USA 100:14169–14174

Dumont FJ (2000) FK506, an immunosuppressant targeting calcineurin function. Curr Med Chem 7:731–748

Edelbaum D, Mohamadzadeh M, Bergstresser PR, Sugamura K, Takashima A (1995) Interleukin (IL)-15 promotes the growth of murine epidermal gamma delta T cells by a mechanism involving the beta- and gamma c-chains of the IL-2 receptor. J Invest Dermatol 105:837–843

Ellery JM, Nicholls PJ (2002) Alternate signalling pathways from the interleukin-2 receptor. Cytokine Growth Factor Rev 13:27–40

Evans GA, Goldsmith MA, Johnston JA, Xu W, Weiler SR, Erwin R, Howard OM, Abraham RT, O'Shea JJ, Greene WC, Farrar WL (1995) Analysis of interleukin-2-dependent signal transduction through the Shc/Grb2 adapter pathway. Interleukin-2-dependent mitogenesis does not require Shc phosphorylation or receptor association. J Biol Chem 270:28858–28863

Friedmann MC, Migone TS, Russell SM, Leonard WJ (1996) Different interleukin 2 receptor beta-chain tyrosines couple to at least two signaling pathways and synergistically mediate interleukin 2-induced proliferation. Proc Natl Acad Sci USA 93:2077–2082

Fujii H, Ogasawara K, Otsuka H, Suzuki M, Yamamura K, Yokochi T, Miyazaki T, Suzuki H, Mak TW, Taki S, Taniguchi T (1998) Functional dissection of the cytoplasmic subregions of the IL-2 receptor beta chain in primary lymphocyte populations. Embo J 17:6551–6557

Gaffen SL (2001) Signaling domains of the interleukin 2 receptor. Cytokine 14:63–77

Gaffen SL, Lai SY, Ha M, Liu X, Hennighausen L, Greene WC, Goldsmith MA (1996) Distinct tyrosine residues within the interleukin-2 receptor beta chain drive signal transduction specificity, redundancy, and diversity. J Biol Chem 271:21381–21390

Garcia VE, Jullien D, Song M, Uyemura K, Shuai K, Morita CT, Modlin RL (1998) IL-15 enhances the response of human gamma delta T cells to nonpeptide [correction of nonpetide] microbial antigens. J Immunol 160:4322–4329

Giri JG, Ahdieh M, Eisenman J, Shanebeck K, Grabstein K, Kumaki S, Namen A, Park LS, Cosman D, Anderson D (1994) Utilization of the beta and gamma chains of the IL-2 receptor by the novel cytokine IL-15. EMBO J 13:2822–2830

Giri JG, Kumaki S, Ahdieh M, Friend DJ, Loomis A, Shanebeck K, DuBose R, Cosman D, Park LS, Anderson DM (1995) Identification and cloning of a novel IL-15 binding protein that is structurally related to the alpha chain of the IL-2 receptor. EMBO J 14:3654–3663

Goldsmith MA, Lai SY, Xu W, Amaral MC, Kuczek ES, Parent LJ, Mills GB, Tarr KL, Longmore GD, Greene WC (1995) Growth signal transduction by the human interleukin-2 receptor requires cytoplasmic tyrosines of the beta chain and non-tyrosine residues of the gamma c chain. J Biol Chem 270:21729–21737

Grabstein KH, Eisenman J, Shanebeck K, Rauch C, Srinivasan S, Fung V, Beers C, Richardson J, Schoenborn MA, Ahdieh M, Johnson L, Alderson MR, Watson JD, Anderson DM, Giri JG (1994) Cloning of a T cell growth factor that interacts with the beta chain of the interleukin-2 receptor. Science 264:965–968

Grant AJ, Roessler E, Ju G, Tsudo M, Sugamura K, Waldmann TA (1992) The interleukin 2 receptor (IL-2R): the IL-2R alpha subunit alters the function of the IL-2R beta subunit to enhance IL-2 binding and signaling by mechanisms that do not require binding of IL-2 to IL-2R alpha subunit. Proc Natl Acad Sci USA 89:2165–2169

Green DR, Droin N, Pinkoski M (2003) Activation-induced cell death in T cells. Immunol Rev 193:70–81

Guex-Crosier Y, Raber J, Chan CC, Kriete MS, Benichou J, Pilson RS, Kerwin JA, Waldmann TA, Hakimi J, Roberge FG (1997) Humanized antibodies against the alpha-chain of the IL-2 receptor and against the beta-chain shared by the IL-2 and IL-15 receptors in a monkey uveitis model of autoimmune diseases. J Immunol 158:452–458

Hakimi J, Ha VC, Lin P, Campbell E, Gately MK, Tsudo M, Payne PW, Waldmann TA, Grant AJ, Tsien WH (1993) Humanized Mik beta 1, a humanized antibody to the IL-2 receptor beta-chain that acts synergistically with humanized anti-TAC. J Immunol 151:1075–1085

Hatakeyama M, Mori H, Doi T, Taniguchi T (1989a) A restricted cytoplasmic region of IL-2 receptor beta chain is essential for growth signal transduction but not for ligand binding and internalization. Cell 59:837–845

Hatakeyama M, Tsudo M, Minamoto S, Kono T, Doi T, Miyata T, Miyasaka M, Taniguchi T (1989b) Interleukin-2 receptor beta chain gene: generation of three receptor forms by cloned human alpha and beta chain cDNAs. Science 244:551–556

Hatakeyama M, Kono T, Kobayashi N, Kawahara A, Levin SD, Perlmutter RM, Taniguchi T (1991) Interaction of the IL-2 receptor with the src-family kinase p56lck: identification of novel intermolecular association. Science 252:1523–1528

Hodge MR, Ranger AM, Charles de la Brousse F, Hoey T, Grusby MJ, Glimcher LH (1996) Hyperproliferation and dysregulation of IL-4 expression in NF-ATP-deficient mice. Immunity 4:397–405

Johnston JA, Bacon CM, Finbloom DS, Rees RC, Kaplan D, Shibuya K, Ortaldo JR, Gupta S, Chen YQ, Giri JD, O'Shea JJ (1995) Tyrosine phosphorylation and activation of STAT5, STAT3, and Janus kinases by interleukins 2 and 15. Proc Natl Acad Sci USA 92:8705–8709

Kataoka T, Budd RC, Holler N, Thome M, Martinon F, Irmler M, Burns K, Hahne M, Kennedy N, Kovacsovics M, Tschopp J (2000) The caspase-8 inhibitor FLIP promotes activation of NF-kappaB and Erk signaling pathways. Curr Biol 10:640–648

Kennedy MK, Park LS (1996) Characterization of interleukin-15 (IL-15) and the IL-15 receptor complex. J Clin Immunol 16:134–143

Kennedy MK, Glaccum M, Brown SN, Butz EA, Viney JL, Embers M, Matsuki N, Charrier K, Sedger L, Willis CR, Brasel K, Morrissey PJ, Stocking K, Schuh JC, Joyce S, Peschon JJ (2000) Reversible defects in natural killer and memory CD8 T cell lineages in interleukin 15-deficient mice. J Exp Med 191:771–780

Kim YS, Maslinski W, Zheng XX, Stevens AC, Li XC, Tesch GH, Kelley VR, Strom TB (1998) Targeting the IL-15 receptor with an antagonist IL-15 mutant/Fc gamma2a protein blocks delayed-type hypersensitivity. J Immunol 160:5742–5748

Kneitz B, Herrmann T, Yonehara S, Schimpl A (1995) Normal clonal expansion but impaired Fas-mediated cell death and anergy induction in interleukin-2-deficient mice. Eur J Immunol 25:2572–2577

Koka R, Burkett PR, Chien M, Chai S, Chan F, Lodolce JP, Boone DL, Ma A (2003) Interleukin (IL)-15R[alpha]-deficient natural killer cells survive in normal but not IL-15R[alpha]-deficient mice. J Exp Med 197:977–984

Kondo M, Takeshita T, Ishii N, Nakamura M, Watanabe S, Arai K, Sugamura K (1993) Sharing of the interleukin-2 (IL-2) receptor gamma chain between receptors for IL-2 and IL-4. Science 262:1874–1877

Kovanen PE, Leonard WJ (1999) Inhibitors keep cytokines in check. Curr Biol 9:R899–R902

Ku CC, Murakami M, Sakamoto A, Kappler J, Marrack P (2000) Control of homeostasis of CD8+ memory T cells by opposing cytokines. Science 288:675–678

Kuo CJ, Chung J, Fiorentino DF, Flanagan WM, Blenis J, Crabtree GR (1992) Rapamycin selectively inhibits interleukin-2 activation of p70 S6 kinase. Nature 358:70–73

Lenardo MJ (1991) Interleukin-2 programs mouse alpha beta T lymphocytes for apoptosis. Nature 353:858–861

Lenardo MJ (1996) Fas and the art of lymphocyte maintenance. J Exp Med 183:721–724

Leonard WJ (2000) IL-2 family cytokines and their receptor. Academic Press, San Diego

Leonard WJ, Depper JM, Crabtree GR, Rudikoff S, Pumphrey J, Robb RJ, Kronke M, Svetlik PB, Peffer NJ, Waldmann TA, Greene WC (1984) Molecular cloning and expression of cDNAs for the human interleukin-2 receptor. Nature 311:626–631

Leonard WJ, Imada K, Nakajima H, Puel A, Soldaini E, John S (1999) Signaling via the IL-2 and IL-7 receptors from the membrane to the nucleus. Cold Spring Harbor Symp Quant Biol 64:417–424

Lin JX, Migone TS, Tsang M, Friedmann M, Weatherbee JA, Zhou L, Yamauchi A, Bloom ET, Mietz J, John S, Leonard WJ (1995) The role of shared receptor motifs and common Stat proteins in the generation of cytokine pleiotropy and redundancy by IL-2, IL-4, IL-7, IL-13, and IL-15. Immunity 2:331–339

Lin JX, Mietz J, Modi WS, John S, Leonard WJ (1996) Cloning of human Stat5B. Reconstitution of interleukin-2-induced Stat5A and Stat5B DNA binding activity in COS-7 cells. J Biol Chem 271:10738–10744

Lodolce JP, Boone DL, Chai S, Swain RE, Dassopoulos T, Trettin S, Ma A (1998) IL-15 receptor maintains lymphoid homeostasis by supporting lymphocyte homing and proliferation. Immunity 9:669–676

Lowenthal JW, Greene WC (1987) Contrasting interleukin 2 binding properties of the alpha (p55) and beta (p70) protein subunits of the human high-affinity interleukin 2 receptor. J Exp Med 166:1156–1161

Lucas JA, Miller AT, Atherly LO, Berg LJ (2003) The role of Tec family kinases in T cell development and function. Immunol Rev 191:119–138

Lyakh L, Ghosh P, Rice NR (1997) Expression of NFAT family proteins in normal human T cells. Mol Cell Biol 17:2475–2484

Ma A, Boone DL, Lodolce JP (2000) The pleiotropic functions of interleukin 15: not so interleukin 2-like after all. J Exp Med 191:753–756

Macchi P, Villa A, Giliani S, Sacco MG, Frattini A, Porta F, Ugazio AG, Johnston JA, Candotti F, O'Shea JJ, Vessoni P, Notarangelo LD (1995) Mutations of Jak-3 gene in patients with autosomal severe combined immune deficiency (SCID). Nature 377:65–68

Marine JC, Topham DJ, McKay C, Wang D, Parganas E, Stravopodis D, Yoshimura A, Ihle JN (1999) SOCS1 deficiency causes a lymphocyte-dependent perinatal lethality. Cell 98:609–616

Mariner JM, Lantz V, Waldmann TA, Azimi N (2001) Human T cell lymphotropic virus type I Tax activates IL-15R alpha gene expression through an NF-kappa B site. J Immunol 166:2602–2609

Marks-Konczalik J, Dubois S, Losi JM, Sabzevari H, Yamada N, Feigenbaum L, Waldmann TA, Tagaya Y (2000) IL-2-induced activation-induced cell death is inhibited in IL-15 transgenic mice. Proc Natl Acad Sci USA 97:11445–11450

McInnes IB, Leung BP, Sturrock RD, Field M, Liew FY (1997) Interleukin-15 mediates T cell-dependent regulation of tumor necrosis factor-alpha production in rheumatoid arthritis. Nat Med 3:189–195

Mention JJ, Ben Ahmed M, Begue B, Barbe U, Verkarre V, Asnafi V, Colombel JF, Cugnenc PH, Ruemmele FM, McIntyre E, Brousse N, Cellier C, Cerf-Bensussan N (2003) Interleukin 15: a key to disrupted intraepithelial lymphocyte homeostasis and lymphomagenesis in celiac disease. Gastroenterology 125:730–745

Miller AT, Berg LJ (2002) Defective Fas ligand expression and activation-induced cell death in the absence of IL-2-inducible T cell kinase. J Immunol 168:2163–2172

Minami Y, Kono T, Miyazaki T, Taniguchi T (1993) The IL-2 receptor complex: its structure, function, and target genes. Annu Rev Immunol 11:245–268

Mingari MC, Vitale C, Cantoni C, Bellomo R, Ponte M, Schiavetti F, Bertone S, Moretta A, Moretta L (1997) Interleukin-15-induced maturation of human natural killer cells from early thymic precursors: selective expression of CD94/NKG2-A as the only HLA class I-specific inhibitory receptor. Eur J Immunol 27:1374–1380

Mixter PF, Russell JQ, Budd RC (1994) Delayed kinetics of T lymphocyte anergy and deletion in lpr mice. J Autoimmun 7:697–710

Miyazaki T, Kawahara A, Fujii H, Nakagawa Y, Minami Y, Liu ZJ, Oishi I, Silvennoinen O, Witthuhn BA, Ihle JN, Taniguchi T (1994) Functional activation of Jak1 and Jak3 by selective association with IL-2 receptor subunits. Science 266:1045–1047

Morris JC, Waldmann TA (2000) Advances in interleukin 2 receptor targeted treatment. Ann Rheum Dis 59(Suppl 1):109–114

Mrozek E, Anderson P, Caligiuri MA (1996) Role of interleukin-15 in the development of human CD56+ natural killer cells from CD34+ hematopoietic progenitor cells. Blood 87:2632–2640

Nakamura Y, Russell SM, Mess SA, Friedmann M, Erdos M, Francois C, Jacques Y, Adelstein S, Leonard WJ (1994) Heterodimerization of the IL-2 receptor beta- and gamma-chain cytoplasmic domains is required for signalling. Nature 369:330–333

Nashan B, Moore R, Amlot P, Schmidt AG, Abeywickrama K, Soulillou JP (1997) Randomised trial of basiliximab versus placebo for control of acute cellular rejection in renal allograft recipients. CHIB 201 International Study Group. Lancet 350:1193–1198

Nelson BH, Lord JD, Greenberg PD (1994) Cytoplasmic domains of the interleukin-2 receptor beta and gamma chains mediate the signal for T-cell proliferation. Nature 369:333–336

Nikaido T, Shimizu A, Ishida N, Sabe H, Teshigawara K, Maeda M, Uchiyama T, Yodoi J, Honjo T (1984) Molecular cloning of cDNA encoding human interleukin-2 receptor. Nature 311: 631–635

Nishimura Y, Ishii A, Kobayashi Y, Yamasaki Y, Yonehara S (1995) Expression and function of mouse Fas antigen on immature and mature T cells. J Immunol 154:4395–4403

Noguchi M, Nakamura Y, Russell SM, Ziegler SF, Tsang M, Cao X, Leonard WJ (1993) Interleukin-2 receptor gamma chain: a functional component of the interleukin-7 receptor. Science 262:1877–1880

Nosaka T, van Deursen JM, Tripp RA, Thierfelder WE, Witthuhn BA, McMickle AP, Doherty PC, Grosveld GC, Ihle JN (1995) Defective lymphoid development in mice lacking Jak3. Science 270:800–802

Nussenblatt RB, Fortin E, Schiffman R, Rizzo L, Smith J, Van Veldhuisen P, Sran P, Yaffe A, Goldman CK, Waldmann TA, Whitcup SM (1999) Treatment of noninfectious intermediate and posterior uveitis with the humanized anti-Tac mAb: a phase I/II clinical trial. Proc Natl Acad Sci USA 96:7462–7466

Ogasawara K, Hida S, Azimi N, Tagaya Y, Sato T, Yokochi-Fukuda T, Waldmann TA, Taniguchi T, Taki S (1998) Requirement for IRF-1 in the microenvironment supporting development of natural killer cells. Nature 391:700–703

Oh S, Berzofsky JA, Burke DS, Waldmann TA, Perera LP (2003) Coadministration of HIV vaccine vectors with vaccinia viruses expressing IL-15 but not IL-2 induces long-lasting cellular immunity. Proc Natl Acad Sci USA 100 3392–3397

Papiernik M, de Moraes ML, Pontoux C, Vasseur F, Penit C (1998) Regulatory CD4 T cells: expression of IL-2R alpha chain, resistance to clonal deletion and IL-2 dependency. Int Immunol 10:371–378

Plaetinck G, Combe MC, Corthesy P, Sperisen P, Kanamori H, Honjo T, Nabholz M (1990) Control of IL-2 receptor-alpha expression by IL-1, tumor necrosis factor, and IL-2. Complex regulation via elements in the 5′ flanking region. J Immunol 145:3340–3347

Puel A, Ziegler SF, Buckley RH, Leonard WJ (1998) Defective IL7R expression in T(–)B(+)NK(+) severe combined immunodeficiency. Nat Genet 20:394–397

Ravichandran KS, Burakoff SJ (1994) The adapter protein Shc interacts with the interleukin-2 (IL-2) receptor upon IL-2 stimulation. J Biol Chem 269:1599–1602

Ravichandran KS, Igras V, Shoelson SE, Fesik SW, Burakoff SJ (1996) Evidence for a role for the phosphotyrosine-binding domain of Shc in interleukin 2 signaling. Proc Natl Acad Sci USA 93:5275–5280

Remillard B, Petrillo R, Maslinski W, Tsudo M, Strom TB, Cantley L, Varticovski L (1991) Interleukin-2 receptor regulates activation of phosphatidylinositol 3-kinase. J Biol Chem 266:14167–14170

Roessler E, Grant A, Ju G, Tsudo M, Sugamura K, Waldmann TA (1994) Cooperative interactions between the interleukin 2 receptor alpha and beta chains alter the interleukin 2-binding affinity of the receptor subunits. Proc Natl Acad Sci USA 91:3344–3347

Romero F, Martinez AC, Camonis J, Rebollo A (1999) Aiolos transcription factor controls cell death in T cells by regulating Bcl-2 expression and its cellular localization. Embo J 18:3419–3430

Rubin LA, Nelson DL (1990) The soluble interleukin-2 receptor: biology, function, and clinical application. Ann Intern Med 113:619–627

Ruchatz H, Leung BP, Wei XQ, McInnes IB, Liew FY (1998) Soluble IL-15 receptor alpha-chain administration prevents murine collagen-induced arthritis: a role for IL-15 in development of antigen-induced immunopathology. J Immunol 160:5654–5660

Russell JH, Wang R (1993) Autoimmune gld mutation uncouples suicide and cytokine/proliferation pathways in activated, mature T cells. Eur J Immunol 23:2379–2382

Russell SM, Tayebi N, Nakajima H, Riedy MC, Roberts JL, Aman MJ, Migone TS, Noguchi M, Markert ML, Buckley RH, O'Shea JJ, Leonard WJ (1995) Mutation of Jak3 in a patient with SCID: essential role of Jak3 in lymphoid development. Science 270:797–800

Sadlack B, Kuhn R, Schorle H, Rajewsky K, Muller W, Horak I (1994) Development and proliferation of lymphocytes in mice deficient for both interleukins-2 and -4. Eur J Immunol 24:281–284

Schorle H, Holtschke T, Hunig T, Schimpl A, Horak I (1991) Development and function of T cells in mice rendered interleukin-2 deficient by gene targeting. Nature 352:621–624

Sharon M, Klausner RD, Cullen BR, Chizzonite R, Leonard WJ (1986) Novel interleukin-2 receptor subunit detected by cross-linking under high-affinity conditions. Science 234:859–863

Shi YF, Sahai BM, Green DR (1989) Cyclosporin A inhibits activation-induced cell death in T-cell hybridomas and thymocytes. Nature 339:625–626

Shurin MR, Tourkova IL, Hackstein H, Shurin GV (2003) Interleukin-15 and 21. In: Thompson AW, Lotze MJ (eds) The cytokine handbook. Academic Press, Amsterdam, pp 431–464

Sporri B, Kovanen PE, Sasaki A, Yoshimura A, Leonard WJ (2001) JAB/SOCS1/SSI-1 is an interleukin-2-induced inhibitor of IL-2 signaling. Blood 97:221–226

Suzuki H, Kundig TM, Furlonger C, Wakeham A, Timms E, Matsuyama T, Schmits R, Simard JJ, Ohashi PS, Griesser H, Taniguchi T, Paige CJ, Mak TW (1995) Deregulated T cell activation and autoimmunity in mice lacking interleukin-2 receptor beta. Science 268:1472–1476

Suzuki H, Duncan GS, Takimoto H, Mak TW (1997a) Abnormal development of intestinal intraepithelial lymphocytes and peripheral natural killer cells in mice lacking the IL-2 receptor beta chain. J Exp Med 185:499–505

Suzuki H, Hayakawa A, Bouchard D, Nakashima I, Mak TW (1997b) Normal thymic selection, superantigen-induced deletion and Fas-mediated apoptosis of T cells in IL-2 receptor beta chain-deficient mice. Int Immunol 9:1367–1374

Suzuki J, Takeshita T, Ohbo K, Asao H, Tada K, Sugamura K (1989) IL-2 receptor subunit, p75: direct demonstration of its IL-2 binding ability by using a novel monoclonal antibody. Int Immunol 1:373–377

Tagaya Y, Bamford RN, DeFilippis AP, Waldmann TA (1996a) IL-15: a pleiotropic cytokine with diverse receptor/signaling pathways whose expression is controlled at multiple levels. Immunity 4:329–336

Tagaya Y, Burton JD, Miyamoto Y, Waldmann TA (1996b) Identification of a novel receptor/signal transduction pathway for IL-15/T in mast cells. Embo J 15:4928–4939

Takeshita T, Asao H, Ohtani K, Ishii N, Kumaki S, Tanaka N, Munakata H, Nakamura M, Sugamura K (1992) Cloning of the gamma chain of the human IL-2 receptor. Science 257:379–382

Taniguchi T, Minami Y (1993) The IL-2/IL-2 receptor system: a current overview. Cell 73:5–8

Thomis DC, Gurniak CB, Tivol E, Sharpe AH, Berg LJ (1995) Defects in B lymphocyte maturation and T lymphocyte activation in mice lacking Jak3. Science 270:794–797

Tinubu SA, Hakimi J, Kondas JA, Bailon P, Familletti PC, Spence C, Crittenden MD, Parenteau GL, Dirbas FM, Tsudo M, Bacher JD, Kasten-sportes C, Martinucci TL, Goldman CK, Clark RE, Waldmann TA (1994) Humanized antibody directed to the IL-2 receptor beta-chain prolongs primate cardiac allograft survival. J Immunol 153:4330–4338

Tsudo M, Kozak RW, Goldman CK, Waldmann TA (1986) Demonstration of a non-Tac peptide that binds interleukin 2: a potential participant in a multichain interleukin 2 receptor complex. Proc Natl Acad Sci USA 83:9694–9698

Tsudo M, Goldman CK, Bongiovanni KF, Chan WC, Winton EF, Yagita M, Grimm EA, Waldmann TA (1987) The p75 peptide is the receptor for interleukin 2 expressed on large granular lymphocytes and is responsible for the interleukin 2 activation of these cells. Proc Natl Acad Sci USA 84:5394–5398

Tsudo M, Kitamura F, Miyasaka M (1989) Characterization of the interleukin 2 receptor beta chain using three distinct monoclonal antibodies. Proc Natl Acad Sci USA 86:1982–1986

Uchiyama T, Broder S, Waldmann TA (1981) A monoclonal antibody (anti-Tac) reactive with activated and functionally mature human T cells. I. Production of anti-Tac monoclonal antibody and distribution of Tac (+) cells. J Immunol 126:1393–1397

Vámosi G, Bodnar A, Vereb G, Jenei A, Goldman CK, Langowski J, Toth K, Matyus L, Szöllősi J, Waldmann T, Damjanovich S (2004) IL-2 and IL-15 receptor α-subunits are coexpressed in a supramolecular cluster in lipid rafts of T cells. Proc Natl Acad Sci USA 101:11082–11084

Villadsen LS, Schuurman J, Beurskens F, Dam TN, Dagnaes-Hansen F, Skov L, Rygaard J, Voorhorst-Ogink MM, Gerritsen AF, van Dijk MA, Parren PW, Baadsgaard O, can de Winkel

JG (2003) Resolution of psoriasis upon blockade of IL-15 biological activity in a xenograft mouse model. J Clin Invest 112:1571–1580

Vincenti F, Kirkman R, Light S, Bumgardner G, Pescovitz M, Halloran P, Neylan J, Wilkinson A, Ekberg H, Gaston R, Backman L, Burdick J (1998) Interleukin-2-receptor blockade with daclizumab to prevent acute rejection in renal transplantation. Daclizumab Triple Therapy Study Group. N Engl J Med 338:161–165

Waldmann TA (1989a) The multi-subunit interleukin-2 receptor. Annu Rev Biochem 58:875–911

Waldmann TA (1989b) Multichain interleukin-2 receptor: a target for immunotherapy in lymphoma. J Natl Cancer Inst 81:914–923

Waldmann TA (1991) The interleukin-2 receptor. J Biol Chem 266:2681–2684

Waldmann TA (1992) Immune receptors: targets for therapy of leukemia/lymphoma, autoimmune diseases and for the prevention of allograft rejection. Annu Rev Immunol 10:675–704

Waldmann TA (1993) The IL-2/IL-2 receptor system: a target for rational immune intervention. Immunol Today 14:264–270

Waldmann TA, Tagaya Y (1999) The multifaceted regulation of interleukin-15 expression and the role of this cytokine in NK cell differentiation and host response to intracellular pathogens. Annu Rev Immunol 17:19–49

Waldmann TA, White JD, Goldman CK, Top L, Grant A, Bamford R, Roessler E, Horak ID, Zaknoen S, Kasten-Sportes C (1993) The interleukin-2 receptor: a target for monoclonal antibody treatment of human T-cell lymphotrophic virus I-induced adult T-cell leukemia. Blood 82:1701–1712

Waldmann TA, White JD, Carrasquillo JA, Reynolds JC, Paik CH, Gansow OA, Brechbiel MW, Jaffe ES, Fleisher TA, Goldman CK (1995) Radioimmunotherapy of interleukin-2R alpha-expressing adult T-cell leukemia with Yttrium-90-labeled anti-Tac. Blood 86:4063–4075

Waldmann TA, Dubois S, Tagaya Y (2001) Contrasting roles of IL-2 and IL-15 in the life and death of lymphocytes: implications for immunotherapy. Immunity 14:105–110

Wang HM, Smith KA (1987) The interleukin 2 receptor. Functional consequences of its bimolecular structure. J Exp Med 166:1055–1069

Willerford DM, Chen J, Ferry JA, Davidson L, Ma A, Alt FW (1995) Interleukin-2 receptor alpha chain regulates the size and content of the peripheral lymphoid compartment. Immunity 3:521–530

Witthuhn BA, Silvennoinen O, Miura O, Lai KS, Cwik C, Liu ET, Ihle JN (1994) Involvement of the Jak-3 Janus kinase in signalling by interleukins 2 and 4 in lymphoid and myeloid cells. Nature 370:153–157

Xanthoudakis S, Viola JP, Shaw KT, Luo C, Wallace JD, Bozza PT, Luk DC, Curran T, Rao A (1996) An enhanced immune response in mice lacking the transcription factor NFAT1. Science 272:892–895

Yoshimura A, Ohkubo T, Kiguchi T, Jenkins NA, Gilbert DJ, Copeland NG, Hara T, Miyajima A (1995) A novel cytokine-inducible gene CIS encodes an SH2-containing protein that binds to tyrosine-phosphorylated interleukin 3 and erythropoietin receptors. Embo J 14:2816–2826

Yu XZ, Levin SD, Madrenas J, Anasetti C (2004) Lck is required for activation-induced T cell death after TCR ligation with partial agonists. J Immunol 172:1437–1443

Zhang X, Sun S, Hwang I, Tough DF, Sprent J (1998) Potent and selective stimulation of memory-phenotype CD8+ T cells in vivo by IL-15. Immunity 8:591–599

Single-Molecule Imaging of Diffusion, Recruitment, and Activation of Signaling Molecules in Living Cells

Akihiro Kusumi, Hideji Murakoshi, Kotono Murase,
Takahiro Fujiwara

5.1 Introduction

Techniques that allow researchers to track single molecules or small groups of molecules in the cell membrane in living cells are becoming important tools in the investigation of the cell membrane. These methods include single fluorescent molecule video imaging (SFVI) using fluorescent probes, and single-particle tracking (SPT) using colloidal gold probes (De Brabander et al. 1985, 1988, 1991; Gelles et al. 1988; Sheetz et al. 1989; Kusumi et al. 1993; Schmidt et al. 1995; Sako et al. 2000; Schütz et al. 2000b; Harms et al. 2001; Iino and Kusumi 2001). They have given researchers an unprecedented ability: directly observing the movement, assembly, and localization of individual single molecules in the plasma membrane of living cells in culture (Kusumi and Sako 1996; Saxton and Jacobson 1997; Schütz et al. 2000a).

This ability has fostered a new fundamental understanding of molecular diffusion in the cell membrane. High-speed SPT and SFVI have revealed that the plasma membranes of virtually all mammalian cells in culture are compartmentalized with regard to the translational diffusion of practically all of the membrane molecules, and that virtually all of these molecules undergo non-Brownian diffusion in the plasma membrane, i.e., short-term confined diffusion in a compartment and long-term hop diffusion over the compartments in the cell membrane (see Figs. 5.1–5.3; details will be given later; Kusumi and Sako 1996; Fujiwara et al. 2002; Murase et al. 2004). This entails a paradigm shift of the structure of the plasma membrane: the plasma membrane in space scales greater than several tens of nanometers should not be considered as a two-dimensional continuum fluid, but rather as being partitioned in closely apposed compartments (*compartmentalized fluid model*; see Fig. 5.2).

An important corollary of these results is that the rate of diffusion cannot be described by a single diffusion coefficient; thus all of the diffusion coefficients obtained by fluorescence redistribution after photobleaching (FRAP) or those estimated by single-molecule techniques at slow rates (like a video rate) must be thought of as "*effective diffusion coefficients*", which may be useful only when the involved time window is specified (Nagle 1992; Ghosh and Webb 1994; Saxton 1994, 1996; Feder et al. 1996; Kusumi and Sako 1996; Fujiwara et al. 2002; Murase et al. 2004). However, the fundamental importance of the fact that membrane molecules do not undergo simple Brownian diffusion (although they may undergo "effective simple Brownian

Springer Series in Biophysics
S. Damjanovich (Ed.)
Biophysical Aspects of Transmembrane Signaling

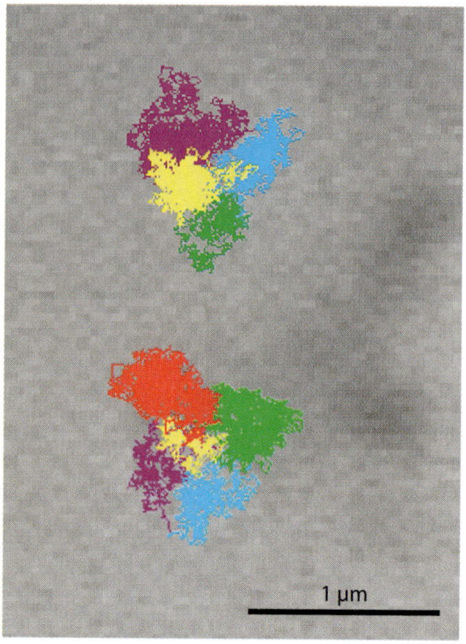

Fig. 5.1. Typical trajectories of 40-nm gold probes bound to a transferrin receptor, recorded at a time resolution of 25 μs for a duration of 250 ms (10,000 frames). The *background* is the actual image of a single frame. Different *colors* represent various plausible compartments, detected by computer software developed in our laboratory (in a time sequence of *purple*, *blue*, *green*, *yellow*, and *red*). These trajectories suggest that transferrin receptor molecules undergo short-term confined diffusion in a compartment and long-term hop diffusion between compartments. This was confirmed by the statistical analysis method developed by Fujiwara et al. (2002)

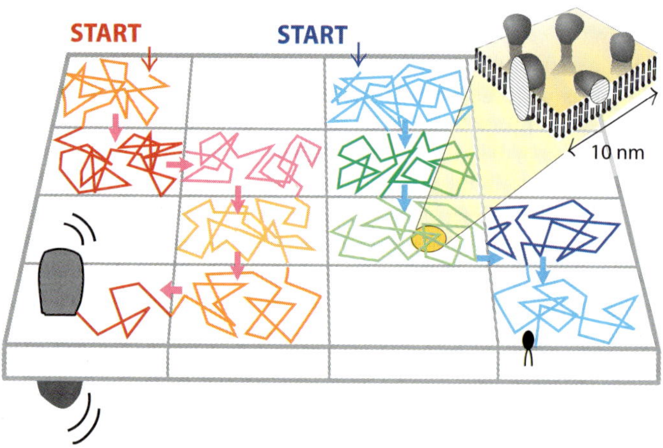

Fig. 5.2. A paradigm shift for the plasma membrane concept is required from the (two-dimensional) continuum model to the compartmentalized fluid model, in which membrane constituent molecules undergo short-term confined diffusion within a compartment and long-term hop diffusion between compartments. The plasma membrane is partitioned into many compartments with regard to translational diffusion of membrane-incorporated molecules, and practically all of the molecules undergo macroscopic diffusion by repeating their confinement within a compartment and hopping to an adjacent one. The two-dimensional fluid-mosaic model of Singer and Nicolson is perfectly adequate if the size is limited within 10 nm (*inset*), as shown in the original cartoon, but it cannot be overextended to a cell membrane structure over several tens of nanometers

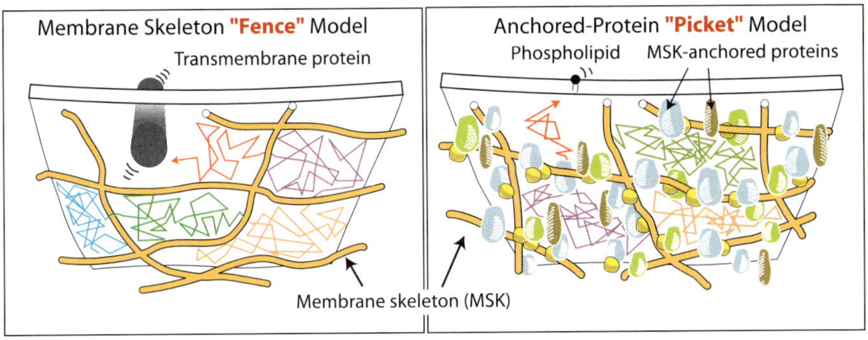

Fig. 5.3. Effects of the membrane-skeleton "fence" (*left*) and anchored-protein "pickets" (*right*) that together partition the entire plasma membrane into small compartments. See text for further details. The hydrodynamic friction-like effect was first described by Bussell et al. (1994, 1995)

diffusion" in limited time windows) has not been widely recognized (Harms et al. 2001; Niv et al. 2002; Vrljic et al. 2002).

Therefore, in this chapter, we begin by briefly describing the plasma membrane compartmentalization, which induces hop diffusion of practically all of the molecules in the cell membrane. Then, we discuss the important consequence of this plasma membrane compartmentalization regarding signal transduction in the plasma membrane: when membrane molecules, including receptor molecules and other signaling molecules in the membrane, form oligomers or molecular complexes, their diffusion rates drop dramatically or they may be temporarily immobilized (*oligomerization-induced trapping* model; Iino et al. 2001; also see Nelson et al. 1999; Hegener et al. 2004). This is completely opposite to the general view of membrane biologists: if one believes in the two-dimensional continuum fluid model for the plasma membrane at the macroscopic scale (over several tens of nanometers), then what diffusion theory teaches us is that oligomerization or the formation of molecular complexes hardly reduces the diffusion rate (Saffman and Delbrück 1975). However, in the compartmentalized fluid model, in which the plasma membrane is partitioned by the actin-based membrane skeleton and various transmembrane proteins are anchored to and lined up along the membrane skeleton, molecular complexes and oligomers will stay longer within a compartment (because they cannot hop across the compartment boundaries as fast as monomers) or become tethered to the actin-based membrane skeleton (due to their avidity effects) (Iino et al. 2001). The experimental results indicated that, upon oligomerization, the diffusion coefficients drop dramatically, consistent with the compartmentalized membrane and oligomerization-induced trapping models (Iino et al. 2001; Murase et al. 2004).

To understand the signal transduction process in the plasma membrane, observing the diffusion of signaling membrane molecules within the plasma membrane and the recruitment of cytoplasmic signaling molecules to the plasma membrane, would be very useful (Hibino et al. 2003; Lommerse et al. 2004;

Mashanov et al. 2004; Murakoshi et al. 2004). However, it would be even more useful and interesting if we could observe the *activation of each individual signaling molecule* in the plasma membrane as we observe their movement and recruitment (Murakoshi et al. 2004). In this way, the activation state of the signaling molecule in the cell membrane and that of the recruited cytoplasmic molecules can be determined, perhaps leading to an understanding of how these molecules become activated in the cell membrane. Furthermore, this would allow us to observe the changes in the diffusion of signaling molecules upon activation, perhaps due to its binding to scaffolding proteins and/or stabilized, enlarged raft domains, and the formation of molecular complexes for signaling. As described in the previous paragraph (and contrary to the general view of membrane biologists), *molecular motion is a particularly good parameter for detecting molecular complexes in and/or on the cell membrane.* Thus, in this chapter, we describe (1) the compartmentalization of the plasma membrane, which induces the hop diffusion of practically all molecules in the cell membrane, (2) the oligomerization-induced trapping and tethering of membrane (signaling) molecules, and (3) the single-molecule imaging of the activation of signaling molecules in living cells; in particular, the activation of the small G protein, Ras.

5.2 The 30-Year-Old Enigma Concerning the Diffusion Rate of Membrane Molecules in the Plasma Membrane

If one considers the statement "*molecular motion is a particularly good parameter for detecting molecular complexes in and/or on the cell membrane*" to be wrong, then it is a good indication of an understanding of the molecular diffusion events occurring in liposomes and reconstituted membranes, but not those taking place in the cell membrane. Based on the fluid-mosaic model of Singer and Nicolson (1972), Saffman and Delbrück (1975) derived an equation that relates the diffusant size to the translational diffusion coefficient in a two-dimensional continuum fluid. For a cylinder (a transmembrane protein) of radius a and height h, floating in a two-dimensional fluid of viscosity μ with a matched thickness (h) immersed in an aqueous medium of viscosity μ', the translational and rotational diffusion coefficients, D_T and D_R, respectively, for the cylinder can be expressed as:

$$D_T = \frac{k_B T}{4\pi\mu h}\left(\log\frac{\mu h}{\mu' a} - \gamma\right) \tag{5.1}$$

$$D_R = \frac{k_B T}{4\pi\mu a^2 h} \tag{5.2}$$

where γ is the Euler constant (~0.5772). This equation predicts that translational diffusion is very *insensitive* to the diffusant size: tetramer formation from monomers (an increase in radius by a factor of 2) will decrease the diffusion rate by only a factor of 1.1, and even 100 mers (an increase in radius by a factor of 10) will have their diffusion rate reduced by only a factor of 1.4 from monomers, assuming a 0.5-nm monomer radius of the membrane-spanning domain. (In contrast, the rotational diffusion rate is quite sensitive to changes in the oligomerization size. It is inversely proportional to a^2, i.e., to the number of proteins in a complex in larger oligomers.)

However, actual observations revealed that the macroscopic (over several 100 nm) diffusion coefficient of the transmembrane protein E-cadherin was decreased by a factor of 40 upon oligomerization (including various sizes of clusters ranging from dimers up to perhaps 10 mers) in the plasma membrane (Iino et al. 2001). Here, "macroscopic" generally refers to space scales greater than several hundred nanometers, corresponding to the typical photobleaching area size in the FRAP experiments (generally 300 nm or greater in diameter) or to the compartment size in the compartmentalized (or partitioned) membrane model (compartment sizes range between 30 and 230 nm, depending on the cell type). Even an unsaturated phospholipid, L-α-dioleoylphosphatidylethanolamine, upon artificial cross-linking, exhibited a fivefold slower macroscopic diffusion rate (over several 100 nm) (Murase et al. 2004). These results clearly indicate that the plasma membrane cannot be considered as a two-dimensional continuum fluid.

There has been another problem regarding the diffusion coefficient for all of the membrane molecules in the plasma membrane. For 30 years, membrane biologists have wondered why the diffusion coefficients for both proteins and lipids in the plasma membrane are lower than those found in artificially reconstituted membranes and liposomes, by factors of 5–50 (Murase et al. 2004). Note that here the diffusion coefficients are those measured at length scales greater than 300 nm, using FRAP, or those observed by SFVI or SPT at the normal video rate (time window of about 100 ms). Further, Peters and Cherry (1982) found that the Saffman–Delbrück theory, which failed in the plasma membrane as described above, worked well in the reconstituted membranes of bacteriorhodopsin.

5.3 Corralling Effects of the Membrane Skeleton for Transmembrane Proteins (the Membrane-Skeleton Fence Model)

Even prior to the single-molecule era, numerous FRAP reports indicated that the reduction in the macroscopic diffusion coefficient in the plasma membrane from that found in artificial membranes may be caused by the actin-based membrane skeleton (Sheetz 1983; Sheetz et al. 1980; Tank et al. 1982; Wu et al. 1982; Tsuji and Ohnishi 1986; Tsuji et al. 1988; Saxton 1989; Paller 1994). This is summarized in Box 5.1. All the results indicated that the membrane skeleton is involved in slowing the diffusion of membrane molecules in the plasma membrane, although the means by which the membrane skeleton causes such a large reduction, by a factor of 5–50, could not be resolved. Among the explanations, the picket-fence model of the membrane skeleton for the reduction in the translational diffusion coefficient was first proposed by Sheetz (1983). Tsuji and Ohnishi (1986) and Tsuji et al. (1988) carried out *both translational and rotational diffu-*

sion measurements for the transmembrane protein band 3 in human red blood cell ghost membranes, providing the first data substantiating the picket-fence model. Furthermore, they found that the passage across the membrane skeleton fence is due to the temporary dissociation of the spectrin tetramers into dimers (spectrin tetramer–dimer equilibrium gate model or SPEQ gate model, see Box 5.1). The percolation threshold idea advanced by Saxton played an important role in these studies (Saxton 1990).

The single-particle tracking (SPT) technique, in which a single or a small number of membrane molecules are conjugated to a colloidal gold particle of 20 or 40 nm in diameter and their movements are imaged by optical transmission microscopy, was developed in the late 1980s (De Brabander et al. 1985, 1988, 1991; Gelles et al. 1988; Schnapp et al. 1988; Kucik et al. 1989; Sheetz et al. 1989; Kusumi

▶ ***Box 5.1.***
Effect of the actin-based membrane skeleton on molecular diffusion in plasma membrane measured by frap, before the single-molecule era

Sheetz et al. (1980) found that the transmembrane protein band 3 (the majority of the ConA receptor observed in this study is known to be band 3) diffuses about ten times faster (Sheetz et al. 1980) in spectrin-deficient mutant mouse erythrocytes than in normal cells. In mammalian red blood cells, the spectrin meshwork, instead of the f-actin network, forms the underlying structure of the plasma membrane, called the membrane skeleton. Furthermore, a number of reports have shown that the lateral diffusion coefficients of transmembrane proteins were increased in blebbed membranes or after partial depolymerization of actin filaments (for example, see Tank et al. 1982; Wu et al. 1982; Paller 1994).

Tsuji and Ohnishi (1986) and Tsuji et al. (1988) showed that the translational diffusion coefficient of band 3 was increased (decreased) when the spectrin network that forms the underlying structure of the human erythrocyte membrane was stabilized [destabilized, i.e., the tetramer–dimer equilibrium of spectrin was shifted toward the tetramer (dimer)], whereas the rotational diffusion coefficient of band 3 was unaffected. These results clearly indicate that (1) the spectrin meshwork partitions the membrane into small compartments, (2) the non-specific collision of band 3 with the spectrin tetramer, which forms the compartment boundary, is responsible for the reduction in the translational diffusion coefficient in the erythrocyte membrane, and (3) the spectrin tetramer is the effective barrier, and when it temporarily dissociates into dimers, band 3 molecules can cross the compartment boundary. Based on these observations, Tsuji et al. (1988) proposed a "spectrin dimer–tetramer equilibrium" gate model (SPEQ gate model). As such, data showing the involvement of the membrane skeleton in the reduction of the translational diffusion rate were abundant in the era before single-molecule observations, but direct observations of molecules undergoing short-term diffusion within a compartment (made of the membrane skeleton) and long-term hop diffusion over the compartments had to wait until the single-molecule technologies became available.

et al. 1993). Using SPT, Sako and Kusumi (1994) were the first to directly observe the "hop diffusion" of membrane molecules: transferrin receptor, a transmembrane protein, is temporarily confined in a compartment of about 700 nm in diameter in the membrane, and then it hops to an adjacent apposed compartment, where it again becomes trapped temporarily. By repeating such confinement and hop movements between the compartments, which was termed hop diffusion, the receptor covers macroscopic areas (Figs. 5.1 and 5.2). Since virtually all of the examined transferrin receptor molecules and α2-macroglobulin receptor molecules were found to undergo hop diffusion, it was proposed that the entire plasma membrane is basically partitioned into small compartments (except for specialized membrane domains, such as clathrin-coated pits, cell–cell and cell–substrate junctions, and microvilli).

The transferrin receptor molecule hop rates for the smaller and greater compartments in NRK cells (NRK cells have a plasma membrane with nested double compartments with sizes of 260 and 710 nm; Fujiwara et al. 2002) have been recently found to be on average every 58 and 530 ms; direct SPT measurements gave 55 and 1,800 ms, respectively, but the latter value needed to be corrected with a macroscopic diffusion coefficient determined by SFVI (a time window of 3 s, giving 0.24 $\mu m^2/s$) and the compartment size determined by SPT (710 nm) (Fujiwara et al. 2002); however, see Box 5.2 for the history of reaching this correct value. Furthermore, all of the transmembrane proteins examined thus far, including E-cadherin (Sako et al. 1998), transferrin receptor (Sako and Kusumi 1994), α2-macroglobulin receptor (Sako and Kusumi 1994), CD44 (Ritchie and Kusumi, unpubl. observ.), band 3 (Tomishige et al. 1998), stem cell factor receptor (Kobayashi, Murakami, and Kusumi, unpubl. observ.), and various G protein-coupled receptors (GPCRs) (Kasai, Prossnitz, Suzuki, and Kusumi, unpubl. observ.), undergo hop diffusion. The macroscopic diffusion coefficients for these molecules, determined by SPT (reflecting hop diffusion rate over many compartments), are basically consistent with the single fluorescent molecule video imaging (SFVI) data and the FRAP data (due to cross-linking effects, the diffusion coefficients may be smaller by a factor of 1–3, but this was corrected using the SFVI data).

What makes the boundaries between these compartments for transmembrane proteins? Based on the single-molecule observation data and the optical trapping results using specimens with modulated cytoskeleton or the modulated cytoplasmic domain of transmembrane proteins, as summarized in Box 5.3, as well as on the FRAP and the rotational diffusion data before the single-molecule era, as summarized in Box 5.1, it was concluded that the part of the actin-based membrane skeleton mesh that is associated with the cytoplasmic surface of the plasma membrane forms the plasma membrane compartment boundaries. Here, transmembrane proteins are temporarily confined in the membrane skeleton mesh, and then undergo a hop movement to an adjacent compartment (Kusumi et al. 1993; Sako and Kusumi 1994, 1995; Sako et al. 1998; Tomishige et al. 1998; Fujiwara et al. 2002). Based on these results, the "membrane-skeleton fence" or "membrane-skeleton corralling" model was proposed (Fig. 5.3, left). Transmembrane proteins protrude into the cytoplasm, and, in this model, their cytoplasmic domains collide with the membrane skeleton, which induces temporal confinement or cor-

▶ *Box 5.2.*
Beginner's luck in single particle tracking, which highlights the pitfalls of these observations

In the early years of single particle tracking, the observation frame rate was only 30 Hz (video rate). In hindsight, the observation of hop diffusion at such a low frame rate was possible at that time due to beginner's luck, in two ways (everybody was a beginner at that time). First, NRK cells were selected for the SPT studies, because at that time we hoped that the previous endocytosis work, by Sako et al. (1990), in which NRK cells were used, would be combined with observations of transferrin receptor movement on the cell surface, in future studies of the endocytosis mechanism for this receptor. It turned out that the plasma membrane of NRK cells has nested double compartmentalization (Fujiwara et al. 2002), and that the greater compartment is unusually large (~ 700 nm), as compared with the compartments found in other cell types (30–230 nm; Murase et al. 2004); this made the 700-nm compartments easy to detect even at the low video-rate time resolution.

The second piece of luck was the choice of transferrin receptor as our target molecule (for the same reason as above). As beginners, it was difficult to prepare good gold particles coated with transferrin that did not cross-link receptor molecules. These gold particles cross-linked transferrin receptor molecules much more readily than the ones we prepare presently. This greatly decreased the hop rate of the gold particle bound to the transferrin receptor across the barrier between the compartments (this in itself is consistent with the oligomerization-induced trapping concept, and is inconsistent with the two-dimensional continuum idea for the plasma membrane. Furthermore, this result is inconsistent with the simple viscoelastic concept of the plasma membrane and the simple anomalous diffusion concept of membrane molecules). Due to the slow intercompartmental hop rate (once every 19 s on average) and the unusually large compartment size of the NRK cell membrane (~700 nm), the hop diffusion of (cross-linked) transferrin receptor became clearly visible, even in video-rate observations (the macroscopic diffusion coefficient obtained was about 0.0014 $\mu m^2/s$, compatible with the FRAP results for various other transmembrane proteins, probably because FRAP could not differentiate the temporary incorporation of the receptor molecules in the clathrin-coated pits). The correct hop rates for the transferrin receptor molecule in the smaller and larger compartments in NRK cells have been recently found to be 58 and 530 ms, respectively (Fujiwara et al. 2002).

ralling of the transmembrane proteins in the membrane skeleton mesh. Transmembrane proteins may hop between the compartments when a space that allows the passage of the cytoplasmic domain of the transmembrane protein is formed between the membrane and the membrane skeleton. This space is formed as a consequence of the thermal fluctuation of these structures, when the actin filament that forms the compartment boundary temporarily dissociates, and when the transmembrane protein incidentally has sufficient kinetic energy to overcome

Box 5.3.
Summary of evidence supporting the membrane skeleton fence and anchored transmembrane protein picket models for compartmentalization of the cell membrane

Other models could explain one or several of the following observations, but not *all* of these observations:

1. Hop diffusion of transmembrane proteins and lipids (determined by statistical analyses) depends on the integrity of the membrane skeleton (Sako and Kusumi 1994; Tomishige et al. 1998; Fujiwara et al. 2002; Murase et al. 2004). Very mild latrunculin and cytochalasin D treatments increased the compartment size. Note that the effect on the macroscopic diffusion coefficient, under these mild conditions, can be very small, between a factor of 1 and 2. This is a consequence of the slightly decreased hop rate due to the increased compartment size (Fujiwara et al. 2002; Nakada et al. 2003; Murase et al. 2004). Hop diffusion cannot be found in liposomes and in membrane blebs, where the membrane skeleton is largely lost (Fujiwara et al. 2002; Murase et al. 2004; Suzuki and Kusumi, unpubl. observ.). In these membranes, the membrane molecules undergo rapid, simple Brownian diffusion that can be characterized by a single diffusion coefficient in the range of 5–10 $\mu m^2/s$ for DOPE or 3 $\mu m^2/s$ for transmembrane proteins.
2. Mild treatments with jasplakinolide, which stabilizes the actin filaments, reduced the macroscopic diffusion coefficient, without strongly affecting the compartment size, by decreasing the hop frequency (Fujiwara et al. 2002; Nakada et al. 2003; Murase et al. 2004).
3. The instances of hops are clearly visible and also detectable with a computer program in the analysis of single-molecule observations with sufficient time resolution (Fujiwara et al. 2002; Murase et al. 2004; Suzuki and Kusumi, unpubl. observ.).
4. The hop rate increases after the partial removal of the cytoplasmic domain of transmembrane proteins (Sako et al. 1998; Tomishige et al. 1998).
5. The hop diffusion is not affected by removing the major fraction of the extracellular domains of transmembrane proteins and the extracellular matrix (Fujiwara et al. 2002; Murase et al. 2004; Suzuki and Kusumi, unpubl. observ.), indicating that these are not the major causes of the induction of hop diffusion.
6. The removal of cholesterol has no major effects on hop diffusion (Fujiwara et al. 2002; Murase et al. 2004; Suzuki and Kusumi, unpubl. observ.), suggesting that lipid rafts are not the major causes of membrane compartmentalization or hop diffusion.
7. The idea that the presence of membrane protrusions and dips over the entire membrane causes apparent confinement within a compartment, reducing the macroscopic diffusion coefficient by a factor of 10, is rejected, because such high protrusions or deep dips (100- to 300-nm structures) that are present throughout in the cell membrane are clearly inconsistent with electron microscope observations (possibly except for the brush-border membranes of epithelial cells) (Rothberg et al. 1992; Morone and Kusumi, unpubl. observ.).

8. Oligomerization of transmembrane proteins or lipids reduces the macroscopic diffusion coefficient and the intercompartmental hop rate without affecting the compartment size (oligomerization-induced trapping; Iino et al. 2001; Murase et al. 2004). This can be easily explained by the models of fences and pickets, but cannot be naturally explained by the two-dimensional continuum fluid model, the viscoelastic model for the suppression of diffusion of membrane molecules, the general anomalous diffusion model, or the model of long membrane protrusions and deep dips throughout in the membrane.
9. The compartment size detected by the free dragging length in optical trap experiments employing very weak trapping forces agrees with the compartment size detected by single-molecule diffusion measurements with high time resolutions (Sako and Kusumi 1995; Sako et al. 1998).
10. The optical trapping experiments revealed that the compartment boundaries are elastic, consistent with the model in which the basis for the compartment barrier is the membrane skeleton meshwork (Sako and Kusumi 1995; Sako et al. 1998; Tomishige et al. 1998). This cannot be explained by models involving raft-induced compartmentalization of the plasma membrane, crowding of the extracellular surface by the extracellular domain of membrane molecules, anomalous diffusion, protrusion and dips throughout in the membrane, or a two-dimensional continuum fluid.
11. Dragging of the membrane skeleton moved transmembrane proteins that are not bound to the membrane skeleton (Tomishige et al. 1998).
12. The compartment sizes detected by transmembrane proteins (transferrin receptor, α_2-macroglobulin receptor) and the phospholipid DOPE are the same in all of the cell types examined thus far (Sako and Kusumi 1994; Fujiwara, Iwasawa, and Kusumi, unpubl. observ.; Murase and Kusumi, unpubl. observ.).
13. The compartment sizes detected from diffusion measurements of transmembrane proteins and lipids are consistent with the mesh size of the membrane skeleton on the cytoplasmic surface of the plasma membrane, determined by atomic force microscopy (AFM) or electron microscope computed tomography (Takeuchi et al. 1998; Tomishige et al. 1998; Morone and Kusumi, unpubl. observ.).
14. Monte Carlo simulations reproduced the experimentally observed residency times when only 20–30% of the compartment boundaries were occupied by the anchored transmembrane protein pickets (Fujiwara et al. 2002; Nakada et al. 2003; Murase et al. 2004). This represents the anchoring of only about 15% of the total transmembrane proteins in the plasma membrane.

the confining potential energy of the compartment barrier when it is in the boundary region.

Several reports employing FRAP or low-speed (video rate or slower) single-molecule tracking have noted the absence of an effect of actin depolymerization on the movement of various membrane molecules (Vrljic et al. 2002; Schmidt and Nichols 2004). Latrunculin or cytochalasin D treatment may slightly increase the compartment size, while at the same time it tends to slightly decrease the hop rate (because the frequency of collisions of membrane molecules with the compartment boundaries decreases due to the increase in the compartment size) (Fujiwara et al. 2002; Murase et al. 2004), leading to minor increases in the *macroscopic* diffusion coefficients, by a factor of 1–2, depending on the cell type (Fujiwara et al. 2002; Murase et al. 2004). A clear example is found in Table 5 in Murase et al. (2004). After cytochalasin treatment, the average compartment size increased from 45 to 87 nm, but the hop rate for gold-tagged dioleoylphosphatidylethanolamine (DOPE) decreased from an average of once every 15 ms to once every 39 ms, thus resulting in practically the same diffusion coefficient before and after the cytochalasin treatment (0.042 or 0.046 $\mu m^2/s$, respectively). In addition, if FRAP, a technique for observing the ensemble-averaged behavior of many molecules, or single-molecule techniques with low time resolutions like a video rate are employed, changes in the motional characteristics of the membrane molecules may easily be missed. Furthermore, the use of high concentrations of actin-depolymerizing drugs and/or the long incubation periods often employed in cell biological studies may complicate the results, as summarized in Box 5.4.

The evidence, obtained by a variety of observational methods and under various conditions, supporting the membrane-skeleton fence model (and anchored transmembrane protein picket model) is summarized in Box 5.3. Furthermore, how hop diffusion could be observed very clearly, even at the low time resolution available in the early era of single-particle tracking, is described in Box 5.2.

5.4 Lipids Also Undergo Hop Diffusion in the Plasma Membrane

The next natural question is "What about lipids?" Surprisingly, Fujiwara et al. (2002) and Murase et al. (2004), using high-speed SPT with a time resolution of 25 μs, found that an unsaturated phospholipid, L-α-DOPE, undergoes hop diffusion (Fig. 5.4). To observe DOPE movement, it was tagged with the fluorescent dye Cy3 or with a 40-nm-diameter colloidal gold particle. In video-rate observations and in NRK cells, both yielded about the same diffusion coefficients, as long as we remained in time windows shorter than 100 ms, which justifies the use of such a large colloidal gold particle as a probe. Over longer time scales, the diffusion coefficient of gold-tagged DOPE was smaller than that of Cy3-tagged DOPE by a factor of 3, due to the cross-linking effect of gold probes (Fujiwara et al. 2002; which, by the way, in itself shows that the plasma membrane cannot be considered as two-dimensional continuum fluid). In the case of FRSK cells (a cell line derived from a fetal rat skin keratinocyte), which have much smaller compartments of ~40 nm, even in the time window of 100 ms, the diffusion of gold-tagged DOPE was slower than that of Cy3-tagged DOPE by a factor of 3. Therefore, to obtain the correct residency time within a compartment, the macroscopic diffusion coefficient obtained from single fluorescent molecule observations at a video rate (SFVI) and the compartment size obtained from the high-speed SPT (this should not be affected by oligomerization) were used ($[\text{compartment size}]^2/4D$). This

Box 5.4.

How can FRAP and low-time-resolution single-molecule tracking miss the large changes in the diffusion characteristics of membrane molecules after actin depolymerization?

The observation of the effect of actin depolymerization on the diffusion of membrane molecules is tricky (Sako and Kusumi 1994; Fujiwara et al. 2002; Nakada et al. 2003). One might think that, based on the membrane skeleton "fence (or corral)" and anchored-transmembrane protein "picket" models, depolymerization of actin filaments, using drugs like cytochalasin D or latrunculin, would restore the fast simple Brownian diffusion to that found in artificial membranes, but this turned out to be an enormous oversimplification.

The behavior of actin filaments upon drug-induced depolymerization is complex because, after drug treatment, they form various types of aggregates, with different levels of interactions with the membrane and membrane molecules, which are compounded by the dependence on the drug concentration and the duration of the drug treatment. We strongly recommend very mild actin depolymerization, in which the drug treatment is carried out with microscopic observation of the specimen at 37°C, at very low concentrations (50–250 nM) of latrunculin (A or B) or cytochalasin D, with preincubation periods of 2–5 min, and with all of the observations completed within 10–15 min after the addition of these drugs (Fujiwara et al. 2002). In some epithelial cell lines, higher concentrations of the drugs may be required: for example, 1 μM for KB cells (Murakoshi et al. 2004) and 13μM for FRSK cells (Murase et al. 2004); however, the preincubation periods must be kept short, normally less than 5 min, and all of the experiments should be finished within 15 min. To induce changes in the neuronal initial segment membrane where actin and its associating proteins are concentrated, 1.7μM of latrunculin A (experiments are finished within 5 min) was used (Nakada et al. 2003). One might think that lower concentrations of the drugs with longer incubation periods would give similar results and simplify the experiments. However, this does not seem to work, probably due to the occurrences of compounding effects over time, resulting in large cell-to-cell variations in the effects and greater tendencies toward side effects, such as major morphological changes and cell death.

Observations under these conditions using high-definition SPT at a frame rate of every 25μs detected a slight increase in the compartment size with a slight decrease in the hop rate (since the collision frequency of the observed molecule with the compartment boundaries was decreased, due to the increased compartment size). These countering effects tend to exert only very minor effects on the macroscopic diffusion rate (for apparent simple Brownian diffusion) over many compartments, by a factor of only 1–2 (Fujiwara et al. 2002; Murase et al. 2004), which may not be detectable with FRAP or low-time-resolution single-molecule tracking techniques (Vrljic et al. 2002; Schmidt and Nichols 2004).

The employment of higher concentrations of actin-depolymerizing drugs and/or longer incubation periods is more common in cell biological studies (such as incubation in 0.3–20 μM of latrunculin cytochalasin for over 20 min). Under these conditions, actin aggregates may form beneath the cell membrane (Wakatsuki et al. 2001), and as membrane molecules become trapped in or interact with these aggregates, they exhibit peculiar behaviors (as observed at the single-molecule level). When membrane molecules are observed at the level of single molecules at video rate under these conditions, they tend to exhibit either an increase or a decrease of (1) the immobile components, (2) the fraction of molecules diffusing faster, and (3) the fraction of molecules diffusing more slowly, perhaps depending on the sizes of the aggregates, their interactions with the remaining membrane skeleton mesh, and the extent to which the membrane molecules under study are trapped in the actin aggregates. Thus, the end results regarding the diffusion of membrane molecules vary greatly, depending on the cell type and molecules under study, and are unpredictable (Paller 1994; Sako and Kusumi 1994; Murase et al. 2004). Under these conditions, changes may or may not be detectable using a low-time-resolution single-molecule tracking technique or FRAP. In the presence of messy actin aggregates, using methods that carry out averaging over time (e.g., due to the long frame time of the camera, in single-molecule observations) and/or over all of the molecules under observation (ensemble averaging, like FRAP) makes the interpretation of the data difficult or impossible. Since there may be many reasons for the same averaged behaviors, it will be impossible to determine exactly how the membrane molecules behave after partial depolymerization of actin filaments from these measurements. There is no question that techniques such as FRAP and single-molecule tracking at video rate can be used for observations of various membrane processes, as described in the text, but it may be unwise to try using these techniques for *examinations of diffusion models in messed-up membranes*. In addition, even when single-molecule observation methods with high time resolutions are employed, if the obtained trajectories are short (which tends to happen when high-speed imaging of single fluorescent molecules is carried out, because, to enhance the signal-to-noise ratio, strong illumination is employed, causing rapid photobleaching), then anomalies in diffusion or hop diffusion can easily be missed. Particular attention should be paid to both the time resolution and the duration of the observation.

was necessary because, with single fluorophore observations, obtaining trajectories with both a sufficient length and sufficient time resolution is difficult, due to poor signal-to-noise ratios and photobleaching.

Figure 5.4 shows representative trajectories of gold-tagged DOPE in the plasma membrane of NRK cells, recorded at time resolutions of 33 ms and 25 μs. At a 33-ms resolution (normal video rate), practically all of the DOPE trajectories were classified into the simple Brownian diffusion mode. However, when the time resolution was enhanced to 25 μs, it became clear that the simple Brownian nature found at the 33-ms resolution is only an apparent one. The hop diffusion is clear even visibly (individual compartments were detected by the computer program we developed), and the statistical analysis method developed by Fujiwara et al. (2002) indeed showed that over 85% of DOPE molecules undergo hop diffusion, rather than simple Brownian diffusion.

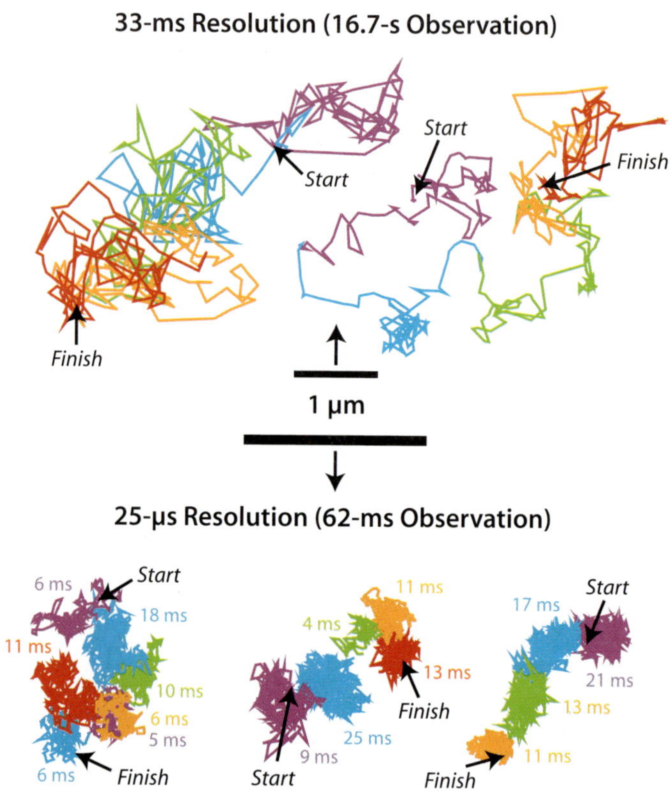

Fig. 5.4. Representative trajectories of single or small groups of DOPE molecules recorded at time resolutions of 33 ms and 25 μs. See text for details. *Above* The different *colors* of trajectories obtained at a 33-ms resolution simply represent a time sequence of every 3.3 s. *Below* The different *colors* of trajectories obtained at a 25-μs resolution represent various plausible compartments, detected by computer software developed in our laboratory (in a time sequence of *purple*, *blue*, *green*, *orange*, and *red* for both time resolutions). These results show that the simple Brownian nature of diffusion observed at a video rate (33 ms/frame) is only superficial, and that it is due to the low time resolution of the observation: the confined + hop movement of each DOPE molecule is totally smeared out at video rate. To resolve such movement, the time resolution must be considerably shorter than average residency time within a compartment

Quantitative analysis showed that the average compartment size was 230 nm. The average residency time within each 230-nm compartment was 11 ms. No wonder we did not see hop movement at video rate, with a time resolution of only 33 ms. In fact, all of the trajectories shown in Fig. 5.4 are 62 ms long, and if we had used video rate observations, there would have been only two or three points in the whole trajectory, and there would have been no way of detecting the hop diffusion of DOPE molecules.

The diffusion rate within the 230-nm compartment, which is 5.4 μm²/s on average, is interesting. It is almost as large as that of DOPE molecules observed in artificial membranes, such as giant liposomes. Therefore, lipid diffusion in the cell membrane is slow, not because the diffusion per se is slow, but because (1) the plasma membrane is compartmentalized with regard to the translational diffusion of phospholipids, (2) the lipid molecules undergo hop diffusion over these compartments, and (3) it takes time to hop from a compartment to an adjacent one (Fig. 5.2). This solved the 30-year-old enigma in the fluid mosaic model: the mechanism underlying the reduction of the diffusion rate in the plasma membrane by a factor of 50–50 from that found in artificial membranes.

What makes the boundaries between these compartments, which even work for phospholipids located in the outer leaflet of the membrane? Fujiwara et al. (2002) and Murase et al. (2004) examined the involvement of the membrane skeleton, as well as the effects of the extracellular matrices, the extracellular domains of membrane proteins, and the cholesterol-rich raft domains, on phospholipid hop diffusion. They found that when they modulated the membrane skeleton, the phospholipid movement was only affected when they modulated the membrane skeleton, consistent with previous FRAP observations, in the sense that the modulation of the membrane skeleton influences the lipid movement (although FRAP did not allow researchers to observe such detailed motion; see Paller 1994). Furthermore, Fujiwara et al. (2002) observed DOPE diffusion in membrane blebs (balloon-like structures of the plasma membranes, where the membrane skeleton is largely lost, and these authors further reduced the actin-based membrane skeleton by treating cells with latrunculin) as well as in liposomes, and found that DOPE molecules undergo rapid, simple Brownian diffusion with a diffusion coefficient of ~9 μm²/s in these membranes. All of these results point to the involvement of the membrane skeleton in both the temporal corralling and hop diffusion of phospholipids.

However, this is very strange. Since the DOPE molecules they observed were located in the extracellular leaflet of the membrane (DOPE may flip, but the large gold particle cannot get into the cytoplasm), but the membrane skeleton is located on the cytoplasmic surface of the membrane, DOPE and the membrane skeleton cannot interact directly. To explain this apparent discrepancy, the "anchored transmembrane-protein picket model" was proposed (Fig. 5.3, right). In this model, various transmembrane proteins anchored to and lined up along the membrane skeleton (fence) effectively act as rows of pickets (these transmembrane proteins act like posts for the fence, and are thus termed pickets) against the free diffusion of phospholipids, due to steric hindrance as well as the hydrodynamic friction-like effects of immobilized anchored membrane protein pickets. The hydrodynamic friction-like effect, first proposed by Hammer's group (Bussell et al. 1994, 1995), is particularly strong when exerted by immobile molecules, and it propagates over about several nanometers (this effect is prominent in the membrane because the membrane viscosity is much greater than that in water, by a factor of ~100). Therefore, when these transmembrane proteins are lined up along the membrane skeleton at a density over a certain threshold (a series of Monte Carlo simulations by Fujiwara et al. indicated that 20–30% coverage of the intercompartmental boundary by these anchored transmembrane protein pickets is sufficient to reproduce the experimentally observed residency time of 11 ms in a 230-nm compartment in NRK cells), the rows of pickets on the membrane-skeleton fences become effective diffusion barriers that confine phospholipids for some time. Note that these transmembrane picket proteins do not have to be stably bound to the membrane skeleton for a long time. Assuming that the boundary region between the compartments is 10 nm wide, it takes about 10 μs for a molecule to traverse this region. Therefore, the zeroth approximation is that if a transmembrane protein is bound to the membrane skeleton for at least 10 μs, then it becomes an effective picket to participate in the formation of the diffusion barrier. Note that, in this model, the transmembrane proteins anchored to the membrane skeleton are coupling the membrane skeleton, which is located on the cytoplasmic surface of the membrane, with the phospholipids that are located in the outer leaflet of the membrane. The evidence, obtained by a variety of observational methods and under various conditions, supporting the anchored transmembrane protein picket model (and membrane-skeleton fence model) is summarized in Box 5.3.

How universal is this plasma membrane compartmentalization? Using the unsaturated phospholipid DOPE, Murase et al. (2004, and unpubl. observ.) found such plasma membrane compartmentalization in all of the nine mammalian cells examined thus far. However, the compartment size varies greatly, from 30 nm up to 230 nm, and also the residency time of DOPE varies between 1 and 17 ms.

The anchored transmembrane protein pickets would be operative on any molecules incorporated in the membrane, including transmembrane proteins. Therefore, the diffusion of transmembrane proteins will be doubly suppressed in the membrane. Both the fence and picket will act on transmembrane proteins.

5.5 Oligomerization-Induced Trapping

The partitioning of the plasma membrane into many small compartments could explain why the diffusion in the plasma membrane is very sensitive to oligomerization or the formation of molecular complexes (Fig. 5.5, left), in contrast to the prediction from the two-dimensional continuum fluid model (Fig. 5.5, right). Monomers of membrane molecules may hop across the picket line with relative ease, but upon oligomerization or molecular complex formation, the oligomers or the complexes as a whole, rather than single molecules, have to hop across the picket-fence line all at once; therefore, these complexes are likely to have a much slower rate of hopping between the compartments, as found with E-cadherin-GFP (Iino et al. 2001) and oligomers of DOPE (Murase et al. 2004). In addition, molecular complexes are more likely to be bound or tethered to the membrane skeleton, perhaps temporarily, which also induces (temporary) immobilization or trapping of molecular complexes. Such enhanced confinement and binding effects induced by oligomerization or molecular complex formation were collectively termed "*oligomerization-induced trapping*" (Iino et al. 2001).

Such oligomerization-induced trapping might play very important roles in the temporary spatial confinement of a cytoplasmic signal at the very early stages of

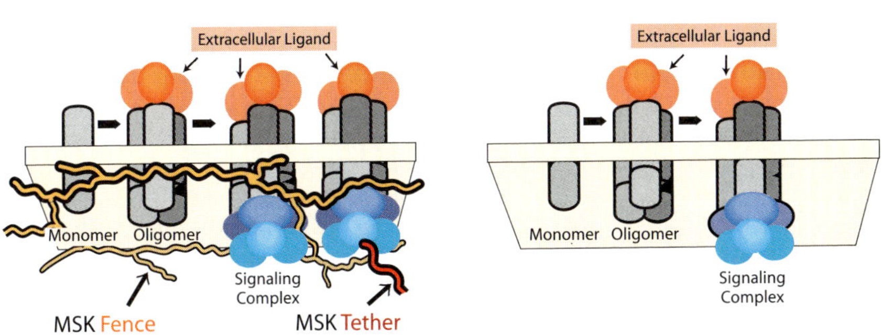

Fig. 5.5. Oligomerization-induced trapping model, indicating how slowing or immobilization of membrane molecules may be induced upon oligomerization or formation of greater molecular complexes. Upon oligomerization or molecular complex formation (*left*), the hop rate across the intercompartmental barrier would be reduced greatly, because, in contrast to monomers, in the case of molecular complexes, all of the molecules that form the complex have to hop across the picket-fence line simultaneously. In addition, due to the avidity effect, molecular complexes are more likely to be tethered to the membrane skeleton, perhaps temporarily, which also reduces their overall diffusion rate. This enhanced confinement and the binding effects induced by oligomerization or molecular complex formation are collectively termed "oligomerization-induced trapping" (Iino et al. 2001). This would not occur in the absence of membrane skeleton fences and pickets (*right*): the diffusion theory by Saffman and Delbrück (1975) predicts that diffusion rates of oligomeric complexes would be almost the same as single receptor molecules

signal transduction. When an extracellular signal is received by a receptor molecule, the receptor often forms oligomers and signaling complexes by recruiting cytoplasmic signaling molecules. Due to the "oligomerization-induced trapping", these oligomeric complexes tend to be trapped in the same membrane skeleton compartment as that where the extracellular signal was received. Therefore, the membrane skeleton fence and the anchored transmembrane-protein pickets temporarily help to confine the cytoplasmic signal to the place where the extracellular signal was received. Such spatial confinement is particularly important for signals that induce local or polarized reorganization of the cytoskeleton or chemotactic events.

This would not occur in the absence of membrane skeleton fences and pickets (Fig. 5.5, right). If there were no such structures, then even when the signaling complex is formed, the diffusion rate of such a complex would be almost the same as that of the single receptor molecules, as the diffusion theory teaches us (Saffman and Delbrück 1975).

Therefore, in the plasma membrane, oligomerization or molecular complex formation is tied to immobilization by the membrane skeleton fence and anchored-protein pickets.

5.6 A Paradigm Shift of the Plasma Membrane Structure Concept Is Necessary: From the Simple Two-Dimensional Continuum Fluid Model to the Compartmentalized Fluid Model

As described above, the membrane-skeleton "fence" and the anchored-protein "picket" together solved the two long-standing problems of molecular diffusion in the plasma membrane: (1) the oligomerization-induced slowing of diffusion and (2) the reduced diffusion coefficients of membrane molecules in the plasma membrane, compared with that found in artificial membranes, by a factor of 5–50. These results could not be explained by the two-dimensional continuum fluid model. The two-dimensional continuum fluid model is fine, as long as the spatial scale is limited to the size of the original cartoon depicted by Singer and Nicolson (1972) (although at the smaller limit of the molecular scale, the continuum model would also fail), which is about 10×10 nm, based on the number of lipid molecules in the cartoon (Fig. 5.2, inset). However, in spatial scales over several tens of nanometers in the plasma membrane, simple extensions of the fluid-mosaic model of Singer and Nicolson and the theory by Saffman–Delbrück fail. The cell appears to have developed (during evolution) a means to control the long-range diffusion of membrane molecules, and to make it sensitive to the diffusant size. The long-range control of diffusion appears to be carried out by the actin-based membrane skeleton, as indicated by the partitioning (corralling) effect of the membrane skeleton and the rows of anchored-protein pickets.

Furthermore, when the cell needs to build a macroscopic diffusion barrier that blocks the diffusion of even phospholipids over the barrier region, like that found in the initial segment region of the neuronal cell membrane, the cell achieves this task by forming very dense picket-fence lines in the barrier region, effectively blocking the diffusion of membrane molecules across this region (Kobayashi et al. 1992; Winckler et al. 1999; Nakada et al. 2003).

As such, a paradigm shift in the concept of plasma membrane structures in the space scales greater than several tens of nanometers is required from the two-dimensional continuum fluid to the compartmentalized fluid, in which its constituent molecules are undergoing hop diffusion over the compartments.

5.7 FRET Strategy for Detecting the Activation of Single Molecules of Ras

In studies of the temporal-spatial regulation of signal transduction, it would be quite useful if we could study, at the level of individual molecules, the activation states of signaling molecules on the plasma membrane, as well as their movement in the plasma membrane (Lommerse et al. 2004) and their recruitment to specific sites in the plasma membrane (Prior et al. 2001). Murakoshi et al. (2004) were the first to achieve such observations using the small GTP-binding proteins, H- and K-Ras.

Ras activation is initiated by the binding of a GDP-GTP exchange factor (GEF) such as Sos, which promotes the release of GDP from Ras, thereby allowing free GTP in the cytoplasm to bind to Ras [it is generally believed that the GTP concentration is an order of magnitude higher than the GDP concentration in the cytoplasm (Gamberucci et al. 1994)]. GTP binding activates the Ras molecule; i.e., the GTP-bound Ras molecules are in the activated state. These Ras molecules are turned off by the binding of GTPase-activating protein (GAP), which stimulates the weak intrinsic GTPase activity of Ras, thereby inducing the hydrolysis of GTP to GDP by Ras itself and returning Ras to its inactive conformation (Satoh et al. 1992). The key function of Ras in the signaling cascade for cell proliferation has been established (Satoh et al. 1992; Rebollo and Martinez 1999).

Murakoshi et al. (2004) have developed a method to *visualize the activation of individual molecules* of Ras and the behavior of activated Ras molecules at video rate. Such a single-molecule method would allow direct investigations of the interactions of activated Ras, separate from inactive Ras, with its effector and scaffolding proteins and its localization in specialized domains. These studies would provide valuable information in the understanding of the signal transduction mechanism after Ras becomes activated.

In order to visualize the instances of activation of single individual Ras molecules, Murakoshi et al. observed GTP binding to Ras using a single-molecule FRET technique, i.e., single-molecule FRET between YFP-Ras and BodipyTR-GTP (Fig. 5.6). First, KB (human epidermoid mouth carcinoma) cells stably expressing YFP-H-Ras were microinjected with GTP conjugated with a fluorescent tag, BodipyTR (BodipyTR-GTP) (Draganescu et al. 2000; McEwen et al. 2001). We then stimulated these cells with 20 nM epidermal growth factor (EGF), which should lead to the release of pre-bound GDP from Ras and the binding of BodipyTR-GTP to the YFP-Ras molecule. The Förster distance for this dye pair was estimated to be 5.5 nm, assuming that the orientation factor is 2/3. Since the distance between the YFP chromophore and the BodipyTR on Ras was estimated to be about

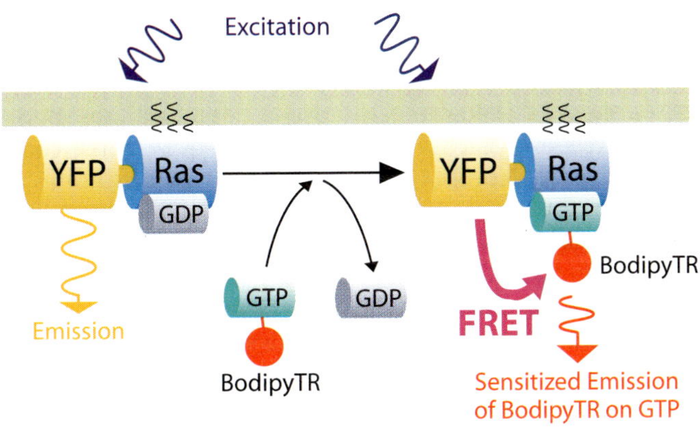

Fig. 5.6. Schematic design of the single-molecule FRET experiment for detecting activation of individual single Ras molecules by monitoring GTP binding. See text for details

3–5 nm, based on crystallographic data (Krengel et al. 1990; Ormo et al. 1996; Yang et al. 1996), the sensitized emission of BodipyTR conjugated to GTP, due to energy transfer from YFP on Ras, should be observed. In fact, Murakoshi et al. succeeded in observing FRET in vitro. Thus Ras activation, i.e., GTP binding, may be detected as the appearance of a sensitized emission spot of BodipyTR-GTP at the place superimposable with the YFP-H-Ras spot, which would become dimmer upon the appearance of the BodipyTR spot.

To image single-molecule FRET, an objective-lens-type total internal reflection fluorescence microscope (TIRFM) was employed (Fig. 5.7). The YFP and BodipyTR signals were chromatically separated and simultaneously observed on two detection arms, using two video cameras working synchronously (at video rate, 33 ms/frame).

5.8 Visualizing the Activation of Single Molecules of H-Ras in Living Cells

The occurrence of FRET from YFP-H-Ras to BodipyTR-GTP in KB cells was examined by exciting YFP, using a 488-nm laser line. YFP-H-Ras expressed in KB cells exhibited a fluorescence intensity distribution with two peaks; 75% of the spots exhibited a distribution similar to that of single molecules of purified YFP (YFP was expressed in *E. coli* and then purified) non-specifically adsorbed on the coverslip, whereas the remaining 25% of the YFP-H-Ras spots had higher intensities (Murakoshi et al. 2004). The latter spots may represent YFP-H-Ras in clusters, microdomains, or an incidental proximity. The majority of the YFP-H-Ras spots with single YFP intensities were photobleached in a single step. This single-step photobleaching suggests that the fluorescent spots we observed were single molecules of YFP-H-Ras.

Approximately 30 s after stimulation with 20 nM EGF, BodipyTR-GTP spots undergoing sensitized emission started appearing on the plasma membrane, exactly where the YFP-H-Ras spots had been observed, and they remained superimposable as long as the BodipyTR signal was visible (Fig. 5.8A). Figure 5.8B, C shows the time-dependent changes in the fluorescence images and intensities, respectively, of the YFP-H-Ras (energy donor) spot and the BodipyTR-GTP (acceptor) spot excited by FRET from YFP-H-Ras. In the time code shown in Fig. 5.8B, C, at 0.6 s, a BodipyTR-GTP spot started appearing at the place superimposable with the pre-existing YFP-H-Ras spot, and as the YFP-H-Ras spot dimmed, the sensitized fluorescence spot of BodipyTR-GTP became brighter. At ~0.9 s, the BodipyTR-GTP molecule was photobleached or released from the YFP-H-Ras with the concomitant recovery of the fluorescent intensity of the YFP-H-Ras donor molecule, until it also was photobleached at ~1.3 s. The single-molecule FRET results for YFP-K-Ras were very similar to those for YFP-H-Ras. The 488-nm beam could directly excite the BodipyTR-GTP acceptor, but the efficiency of direct excitation was so low that the signals from individual molecules did not form identifiable spots in the presence of the background noise.

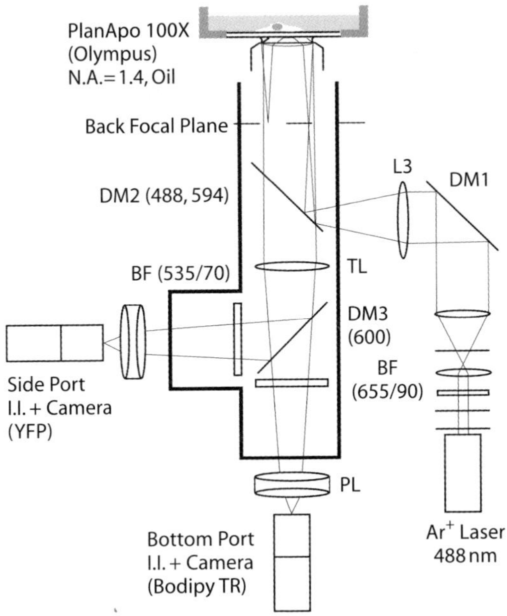

Fig. 5.7. Microscope setup for single-molecule FRET. An Olympus IX-70 microscope was used as a base. The mirror turret was modified to allow side entry of excitation laser beams into the microscope. YFP is excited with a 488-nm Ar^+ laser beam. Excitation was performed through a laser beam steered into the microscope by a dual color dichroic mirror (Chroma), so as to focus at the back focal plane of the objective. The excitation arm consists of the following optical components: *S* electronic shutter; *ND* neutral density filter; *l/4* quarter-wave plate; *L1* and *L2* working as a 10× beam expander (L1, F=150 mm; L2, F=15 mm); *FD* field diaphragm; *M* mirror, *DM* dichroic mirror; *L3* focusing lens (F=35 mm); *BP* band-pass filter. When direct observation of the acceptor is needed, a 594-nm beam from a He-Ne laser incident into DM 1 (passing) was used. An Olympus PlanApo 100× oil immersion with a numerical aperture of 1.4 was used. The fluorescence images of YFP (FRET donor) and BodipyTR (FRET acceptor) were separated by a dichroic mirror (600 nm, DM 3), and were projected into two detection arms with band-pass filters (500–570 nm for YFP, and 605–700 nm for BodipyTR, on the side and bottom ports, respectively). A microchannel plate intensifier (VS4–1845, VideoScope, Sterling, Virginia) and a silicon-intensified target tube (SIT) camera (Hamamatsu Photonics) were used in each arm, enabling us to observe single-molecule FRET from both the donor and acceptor sides simultaneously and synchronously at video rate (33 ms/frame). *TL* Tube lens (1× or 2×); *BF* barrier filter; *PL* projection lens (2×); *I.I.* image intensifier

5.9 Ras Diffusion Slows Upon Activation

Figure 5.9A shows typical trajectories of YFP-H-Ras on the cell membrane. Before EGF stimulation, the majority of the H-Ras molecules diffuse rapidly (Fig. 5.9A, left; Niv et al. 1999, 2002), as fast as phospholipids. In contrast, the activated

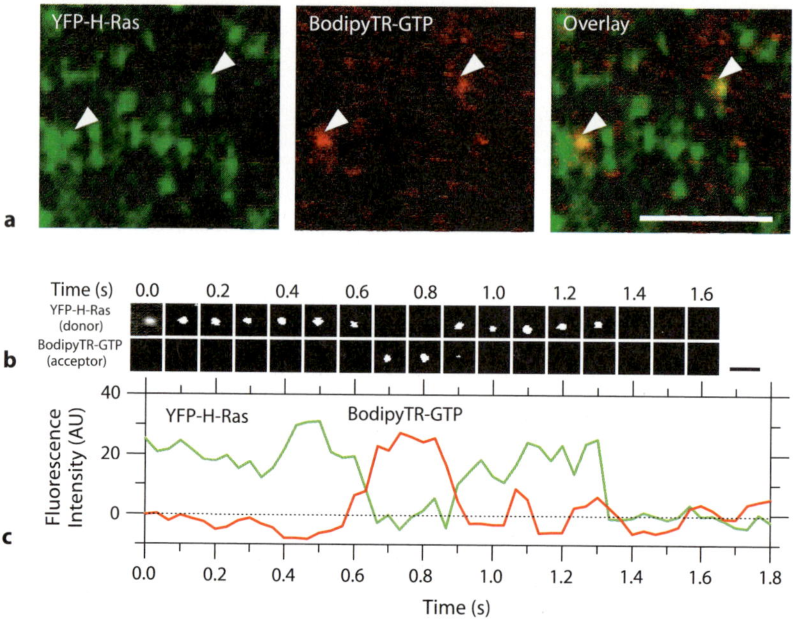

Fig. 5.8. Activation of single H-Ras molecules, monitored by single-molecule FRET from YFP-H-Ras to BodipyTR-GTP upon their binding. **a** Single-molecule FRET observations with simultaneous imaging in the YFP (*left*) and BodipyTR (*middle*) channels. Single frames (33 ms) in the video-rate recordings are shown. Excitation was at 488 nm with an Ar^+ laser beam. Fluorescent spots shown in the BodipyTR channel appeared as a result of FRET from YFP on Ras to BodipyTR on GTP (*arrows*). *Scale bar* 5 μm. **b** Images of single molecules of YFP-H-Ras (donor) and BodipyTR-GTP (acceptor) undergoing FRET. Note that time scales for **b** and **c** are the same. *Scale bar* 0.5 μm. **c** Fluorescence intensities of the spots displayed in B plotted as a function of time: a representative example of anticorrelation between donor and acceptor fluorescence intensities. (Adapted from Fig. 2 of Murakoshi et al. 2004, with permission)

Ras molecules, which could be tracked by following both the donor and the FRET signals, exhibited trajectories indicating that their diffusion is substantially slowed or blocked (Fig. 5.9A, right). These results are consistent with the data reported by the Parton–Hancock group (Roy et al. 1999; Prior et al. 2001, 2003; Parton and Hancock 2004; Rotblat et al. 2004) in the sense that, upon activation, the H- and K-Ras molecules change their distributions in different domains in the plasma membrane and/or their binding to different scaffolding molecules or structures, in a very dynamic manner.

In the cells at the resting (steady) state before EGF stimulation, the trajectories indicate that only about 9% of the YFP-H-Ras (16% of the K-Ras) molecules are immobile. Among the mobile YFP-H-Ras molecules, over 90% of them exhibited rapid diffusion similar to that of the non-raft phospholipid DOPE (Fujiwara et al. 2002). This result suggests that the H-Ras molecules may make only temporary (perhaps shorter than 100 ms, which is our time window here) interactions with

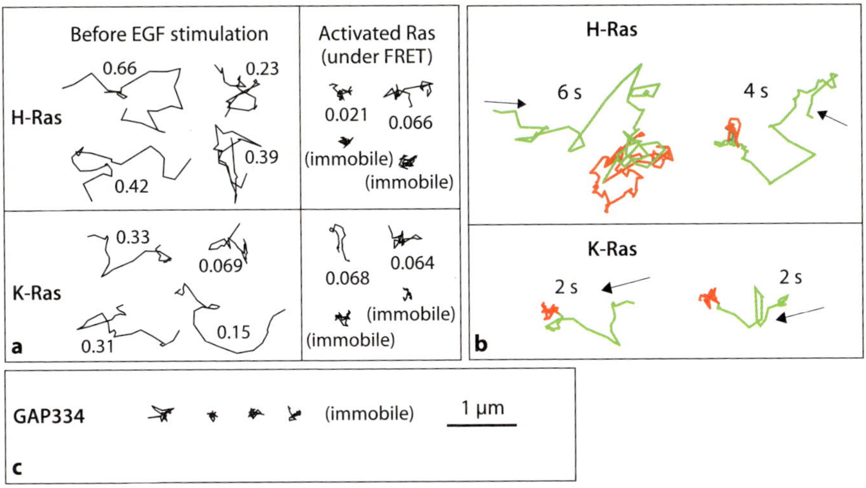

Fig. 5.9. Representative trajectories of YFP-Ras and GAP334-GFP on the cytoplasmic surface of plasma membrane, indicating that Ras molecules become largely immobile upon activation. **a** (*left*) Typical 1-s trajectories of single YFP (H and K)-Ras molecules on the cell membrane, recorded at a video rate. *Numbers* represent $D_{100\,ms}$ (*left column*) before EGF stimulation (inactive YFP-Ras). **a** (*right*) Typical trajectories of *activated* Ras molecules, i.e., those of single-molecule pairs of YFP (H or K)-Ras and BodipyTR-GTP undergoing FRET. **b** Representative trajectories of single YFP-Ras (donor, *green*) molecules, which later became activated by binding of BodipyTR-GTP, as detected by the FRET signal (*red*). During the occurrence of FRET, the donor signal became very weak, and therefore the YFP (H or K)-Ras trajectory shown in *green* is connected with the *red* trajectory of the sensitized emission spot of BodipyTR-GTP. YFP-Ras diffusion is substantially blocked or slowed during FRET periods. **c** Typical 1-s trajectories of single GAP334-GFP molecules recruited to the cell membrane, recorded at a video rate. In a time course similar to that of H- and K-Ras activation, GAP334-GFP molecules were recruited to the cell membrane, where they hardly exhibited diffusion

caveolae or large and stable rafts (Mineo et al. 1996; Wu et al. 1997; Roy et al. 1999; Prior et al. 2003), if they interact with these structures at all in the steady-state resting cells. This is consistent with the discussion advanced by Prior et al. (2001, 2003), who argued that the partitioning of Ras molecules into rafts may be very dynamic, with Ras entering and exiting the rafts rapidly.

The activated H- and K-Ras molecules undergoing FRET after the EGF application exhibited substantial slowing/immobilization (Fig. 5.9A, right). About half of the activated Ras molecules (observed between 1 and 3 min after EGF stimulation) became immobilized. The remaining half were classified into the mobile mode, but their diffusion rates were decreased by a factor of 3–4 from those before EGF application. Typical trajectories exhibiting the slowing of diffusion upon the activation of H-Ras molecules are shown in Fig. 5.9B.

5.10 A Model for the Formation of the Active Ras Signal Transduction Complex

Based on these observations, we propose a model in which activated Ras molecules may be bound by activated Ras-specific scaffolding proteins, like SUR-8 (Sieburth et al. 1998; Li et al. 2000), spred (Wakioka et al. 2001), and galectin-1 (Paz et al. 2001; Prior et al. 2003), which might initiate the cooperative formation of transient signaling complexes including effector molecules, such as Raf-1, and deactivating proteins for Ras, such as RasGAP (Fig. 5.10). The formation of such

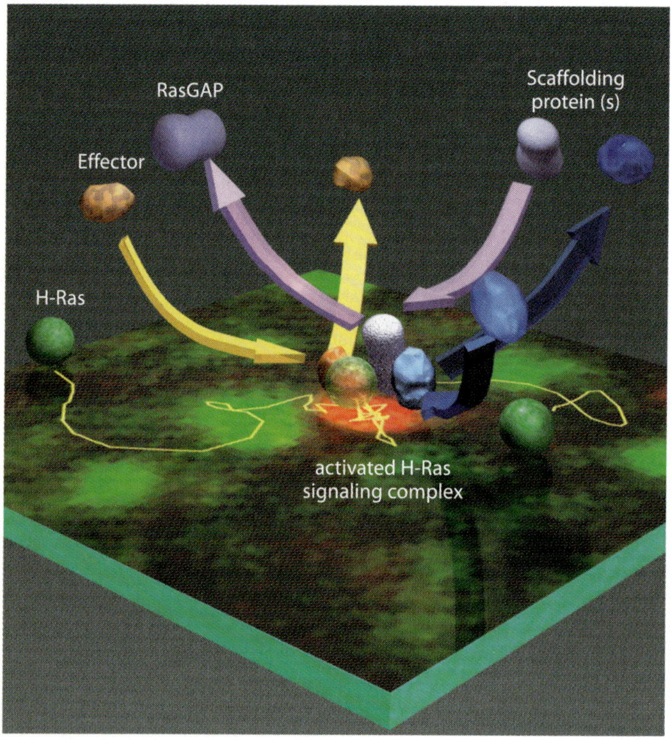

Fig. 5.10. Model for formation of the activated-Ras signaling complex with the aid of scaffolding proteins like SUR-8, spred, and/or galectin 1. On top of the actual FRET image of an activated H-Ras molecule (see *red spot in center*) superimposed by the actual trajectory of the H-Ras molecule (*yellow line*), the view of an artist (Junko Kondo) for the transient assembly of the activated-Ras-dependent signaling complex is further superimposed. When an H-Ras molecule becomes activated (*red* FRET signal in *center*), diffusion is greatly suppressed, due to transient formation of a signaling molecular complex (oligomerization-induced trapping). Formation of such a signaling complex is likely to be facilitated by Ras-activation-dependent scaffolding proteins. The complex may then interact with effector molecules, such as Raf-1 and the Ras-deactivating protein RasGAP

a large signaling complex on the plasma membrane would induce its trapping in and/or binding to the actin-based membrane skeleton mesh, described in previous sections as "oligomerization-induced trapping" (Kusumi and Sako 1996; Iino et al. 2001; Fujiwara et al. 2002). The diffusion of the activated-Ras signaling complex is greatly reduced because the complex formation would enhance the binding to the membrane skeleton due to the avidity effect of the molecular complex, and enhance its corralling by the membrane skeleton "fences" and the anchored-protein "pickets", due to the increased size of the diffusant upon molecular complex formation.

5.11 Recruitment of GAP334, the Ras-Binding, Catalytic Domain of p120RasGAP, on the Cell Membrane

To test the model for the formation of the transient signaling complex of activated Ras and the subsequent confinement within and/or binding to the membrane skeleton mesh, the recruitment and movement of GFP-tagged GAP334 (the catalytic domain of p120RasGAP that binds to activated Ras) on the plasma membrane was examined. GAP334-GFP, which is normally in the cytoplasm, binds to activated Ras on the plasma membrane, and greatly accelerates the hydrolysis of the GTP molecule bound to Ras for its deactivation. As expected, individual GAP334-GFP molecules suddenly appear on the membrane from the cytoplasm after EGF stimulation (Fig. 5.11). Representative trajectories of GAP334-GFP on the membrane (Fig. 5.9C) indicated that the majority of the GAP334-GFP recruited to the plasma membrane is rather stationary, consistent with the activated Ras being corralled or bound by the membrane skeleton mesh. Latrunculin

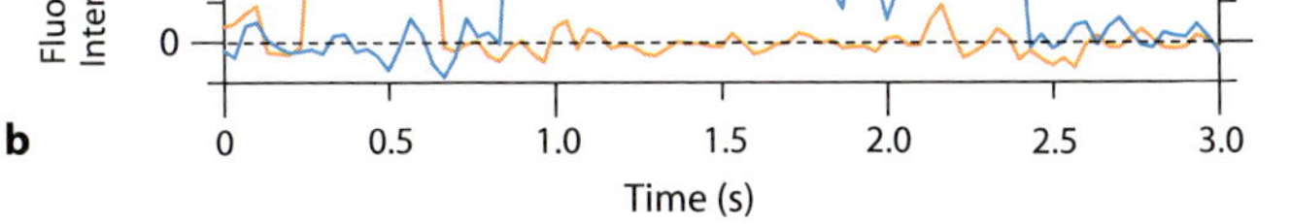

Fig. 5.11. GAP334-GFP molecules recruited to the membrane after EGF stimulation are stationary. Images of single molecules of GAP334-GFP (**a**) and their fluorescence intensities (**b**). GAP334-GFP molecules are recruited to the plasma membrane at ~ 0.25 s (*green line*) and 0.9 s (*blue line*) after EGF stimulation, and most of them are immobile (*arrowheads*). At ~0.6 s (*green line*) and 2.4 s (*blue line*), GAP334-GFP molecules were either photobleached or dissociated from the membrane. *Scale bar* 1 µm

treatment mobilized GAP334-GFP without affecting the recruitment of GAP334-GFP to the cell membrane. These results are consistent with a model in which the activated-Ras-induced signaling complex is confined in and/or bound to the membrane skeleton mesh.

5.12 Conclusions Regarding the FRET Method for Monitoring the Activation of Single Ras Molecules

The single molecule FRET method to detect the binding of BodipyTR-GTP to a G protein fused with YFP in real time is a new, useful technique to study the activation of G proteins and the dynamics of activated G proteins at the level of single molecules, and can be applied to many other G proteins. Taken together, we envisage that simultaneous observations of the movement, recruitment, and activation of signaling molecules at the level of single molecules will become a particularly useful method for investigation of the signaling mechanisms in live cells.

References

Bussell SJ, Hammer DA, Koch DL (1994) The effect of hydrodynamic interactions on the tracer and gradient diffusion of integral membrane-proteins in lipid bilayers. J Fluid Mech 258:167–190

Bussell SJ, Koch DL, Hammer DA (1995) Effect of hydrodynamic interactions on the diffusion of integral membrane proteins: diffusion in plasma membranes. Biophys J 68:1836–1849

De Brabander M, Geuens G, Nuydens R, Moeremans M, De Mey J (1985) Probing microtubule-dependent intracellular motility with nanometre particle video ultramicroscopy (nanovid ultramicroscopy). Cytobios 43:273–283

De Brabander M, Nuydens R, Geerts H, Hopkins CR (1988) Dynamic behavior of the transferrin receptor followed in living epidermoid carcinoma (A431) cells with nanovid microscopy. Cell Motil Cytoskeleton 9:30–47

De Brabander M, Nuydens R, Ishihara A, Holifield B, Jacobson K, Geerts H (1991) Lateral diffusion and retrograde movements of individual cell surface components on single motile cells observed with nanovid microscopy. J Cell Biol 112:111–124

Draganescu A, Hodawadekar SC, Gee KR, Brenner C (2000) Fhit-nucleotide specificity probed with novel fluorescent and fluorogenic substrates. J Biol Chem 275:4555–4560

Feder TJ, Brust-Mascher I, Slattery JP, Baird B, Webb WW (1996) Constrained diffusion or immobile fraction on cell surfaces: a new interpretation. Biophys J 70:2767–2773

Fujiwara T, Ritchie K, Murakoshi H, Jacobson K, Kusumi A (2002) Phospholipids undergo hop diffusion in compartmentalized cell membrane. J Cell Biol 157:1071–1081

Gamberucci A, Innocenti B, Fulceri R, Banhegyi G, Giunti R, Pozzan T, Benedetti A (1994) Modulation of Ca^{2+} influx dependent on store depletion by intracellular adenine-guanine nucleotide levels. J Biol Chem 269:23597–23602

Gelles J, Schnapp BJ, Sheetz MP (1988) Tracking kinesin-driven movements with nanometre-scale precision. Nature 331:450–453

Ghosh RN, Webb WW (1994) Automated detection and tracking of individual and clustered cell surface low density lipoprotein receptor molecules. Biophys J 66:1301–1318

Harms GS, Cognet L, Lommerse PH, Blab GA, Kahr H, Gamsjager R, Spaink HP, Soldatov NM, Romanin C, Schmidt T (2001) Single-molecule imaging of l-type $Ca^{(2+)}$ channels in live cells. Biophys J 81:2639–2646

Hegener O, Prenner L, Runkel F, Baader SL, Kappler J, Haberlein H (2004) Dynamics of beta2-adrenergic receptor-ligand complexes on living cells. Biochemistry 43:6190–6199

Hibino K, Watanabe TM, Kozuka J, Iwane AH, Okada T, Kataoka T, Yanagida T, Sako Y (2003) Single- and multiple-molecule dynamics of the signaling from H-Ras to cRaf-1 visualized on the plasma membrane of living cells. Chem Phys Chem 4:748–753

Iino R, Kusumi A (2001) Single-fluorophore dynamic imaging in living cells. J Fluorescence 11:187–195

Iino R, Koyama I, Kusumi A (2001) Single molecule imaging of green fluorescent proteins in living cells: E-cadherin forms oligomers on the free cell surface. Biophys J 80: 2667–2677

Kobayashi T, Storrie B, Simons K, Dotti CG (1992) A functional barrier to movement of lipids in polarized neurons. Nature 359:647–650

Krengel U, Schlichting L, Scherer A, Schumann R, Frech M, John J, Kabsch W, Pai EF, Wittinghofer A (1990) Three-dimensional structures of H-ras p21 mutants: molecular basis for their inability to function as signal switch molecules. Cell 62:539–548

Kucik DF, Elson EL, Sheetz MP (1989) Forward transport of glycoproteins on leading lamellipodia in locomoting cells. Nature 340:315–317

Kusumi A, Sako Y (1996) Cell surface organization by the membrane skeleton. Curr Opin Cell Biol 8:566–574

Kusumi A, Sako Y, Yamamoto M (1993) Confined lateral diffusion of membrane receptors as studied by single particle tracking (nanovid microscopy). Effects of calcium-induced differentiation in cultured epithelial cells. Biophys J 65:2021–2040

Li W, Han M, Guan KL (2000) The leucine-rich repeat protein SUR-8 enhances MAP kinase activation and forms a complex with Ras and Raf. Genes Dev 14:895–900

Lommerse PH, Blab GA, Cognet L, Harms GS, Snaar-Jagalska BE, Spaink HP, Schmidt T (2004) Single-molecule imaging of the H-ras membrane-anchor reveals domains in the cytoplasmic leaflet of the cell membrane. Biophys J 86:609–616

Mashanov GI, Tacon D, Peckham M, Molloy JE (2004) The spatial and temporal dynamics of pleckstrin homology domain binding at the plasma membrane measured by imaging single molecules in live mouse myoblasts. J Biol Chem 279:15274–15280

McEwen D, Gee K, Kang H, Neubig R (2001) Fluorescent BODIPY-GTP analogs: real-time measurement of nucleotide binding to G proteins. Anal Biochem 291:109–117

Mineo C, James GL, Smart EJ, Anderson RG (1996) Localization of epidermal growth factor-stimulated Ras/Raf-1 interaction to caveolae membrane. J Biol Chem 271:11930–11935

Murakoshi H, Iino R, Kobayashi T, Fujiwara T, Ohshima C, Yoshimura A, Kusumi A (2004) Single-molecule imaging analysis of Ras activation in living cells. Proc Natl Acad Sci USA 101: 7317–7322

Murase K, Fujiwara T, Umemura Y, Suzuki K, Iino R, Yamashita H, Saito M, Murakoshi H, Ritchie K, Kusumi A (2004) Ultrafine membrane compartments for molecular diffusion as revealed by single molecule techniques. Biophys J 86:4075–4093

Nagle JF (1992) Long tail kinetics in biophysics? Biophys J 63:366–370

Nakada C, Ritchie K, Oba Y, Nakamura M, Hotta Y, Iino R, Kasai RS, Yamaguchi K, Fujiwara T, Kusumi A (2003) Accumulation of anchored proteins forms membrane diffusion barriers during neuronal polarization. Nat Cell Biol 5:626–632

Nelson S, Horvat RD, Malvey J, Roess DA, Barisas BG, Clay CM (1999) Characterization of an intrinsically fluorescent gonadotropin-releasing hormone receptor and effects of ligand binding on receptor lateral diffusion. Endocrinology 140:950–957

Niv H, Gutman O, Henis YI, Kloog Y (1999) Membrane interactions of a constitutively active GFP-Ki-Ras 4B and their role in signaling. Evidence from lateral mobility studies. J Biol Chem 274:1606–1613

Niv H, Gutman O, Kloog Y, Henis YI (2002) Activated K-Ras and H-Ras display different interactions with saturable nonraft sites at the surface of live cells. J Cell Biol 157:865–872

Ormo M, Cubitt AB, Kallio K, Gross LA, Tsien RY, Remington SJ (1996) Crystal structure of the *Aequorea victoria* green fluorescent protein. Science 273:1392–1395

Paller MS (1994) Lateral mobility of Na,K-ATPase and membrane lipids in renal cells. Importance of cytoskeletal integrity. J Membr Biol 142:127–135

Parton RG, Hancock JF (2004) Lipid rafts and plasma membrane microorganization: insights from Ras. Trends Cell Biol 14:141–147

Paz A, Haklai R, Elad-Sfadia G, Ballan E, Kloog Y (2001) Galectin-1 binds oncogenic H-Ras to mediate Ras membrane anchorage and cell transformation. Oncogene 20:7486–7493

Peters R, Cherry RJ (1982) Lateral and rotational diffusion of bacteriorhodopsin in lipid bilayers: experimental test of the Saffman–Delbrück equations. Proc Natl Acad Sci USA 79:4317–4321

Prior IA, Harding A, Yan J, Sluimer J, Parton RG, Hancock JF (2001) GTP-dependent segregation of H-ras from lipid rafts is required for biological activity. Nat Cell Biol 3:368–375

Prior IA, Muncke C, Parton RG, Hancock JF (2003) Direct visualization of Ras proteins in spatially distinct cell surface microdomains. J Cell Biol 160:165–170

Rebollo A, Martinez AC (1999) Ras proteins: recent advances and new functions. Blood 94:2971–2980

Rotblat B, Prior IA, Muncke C, Parton RG, Kloog Y, Henis YI, Hancock JF (2004) Three separable domains regulate GTP-dependent association of H-ras with the plasma membrane. Mol Cell Biol 24:6799–6810

Rothberg KG, Heuser JE, Donzell WC, Ying YS, Glenney JR, Anderson RG (1992) Caveolin, a protein component of caveolae membrane coats. Cell 68:673–682

Roy S, Luetterforst R, Harding A, Apolloni A, Etheridge M, Stang E, Rolls B, Hancock JF, Parton RG (1999) Dominant-negative caveolin inhibits H-Ras function by disrupting cholesterol-rich plasma membrane domains. Nat Cell Biol 1:98–105

Saffman PG, Delbrück M (1975) Brownian motion in biological membranes. Proc Natl Acad Sci USA 72:3111–3113

Sako Y, Kusumi A (1994) Compartmentalized structure of the plasma membrane for receptor movements as revealed by a nanometer-level motion analysis. J Cell Biol 125: 1251–1264

Sako Y, Kusumi A (1995) Barriers for lateral diffusion of transferrin receptor in the plasma membrane as characterized by receptor dragging by laser tweezers: fence versus tether. J Cell Biol 129:1559–1574

Sako Y, Nagafuchi A, Tsukita S, Takeichi M, Kusumi A (1998) Cytoplasmic regulation of the movement of E-cadherin on the free cell surface as studied by optical tweezers and single particle tracking: corralling and tethering by the membrane skeleton. J Cell Biol 140:1227–1240

Sako Y, Sato SB, Ohnishi S (1990) Subpopulations of endosomes generated at sequential stages in the endocytic pathway of asialoganglioside-containing ferrite ligands in rat liver. J Biochem (Tokyo) 107:846–853
Sako Y, Minoghchi S, Yanagida T (2000) Single-molecule imaging of EGFR signalling on the surface of living cells. Nat Cell Biol 2:168–172
Satoh T, Nakafuku M, Kaziro Y (1992) Function of Ras as a molecular switch in signal transduction. J Biol Chem 267:24149–24152
Saxton MJ (1989) The spectrin network as a barrier to lateral diffusion in erythrocytes. A percolation analysis. Biophys J 55:21–28
Saxton MJ (1990) The membrane skeleton of erythrocytes. A percolation model. Biophys J 57: 1167–1177
Saxton MJ (1994) Single-particle tracking: models of directed transport. Biophys J 67: 2110–2119
Saxton MJ (1996) Anomalous diffusion due to binding: a Monte Carlo study. Biophys J 70: 1250–1262
Saxton MJ, Jacobson K (1997) Single-particle tracking: applications to membrane dynamics. Annu Rev Biophys Biomol Struct 26:373–399
Schmidt K, Nichols BJ (2004) A barrier to lateral diffusion in the cleavage furrow of dividing mammalian cells. Curr Biol 14:1002–1006
Schmidt T, Schütz GJ, Baumgartner W, Grüber HJ, Schindler H (1995) Characterization of photophysics and mobility of single molecules in a fluid lipid-membrane. J Phys Chem 99: 17662–17668
Schnapp BJ, Gelles J, Sheetz MP (1988) Nanometer-scale measurements using video light microscopy. Cell Motil Cytoskeleton 10:47–53
Schütz GJ, Kada G, Pastushenko VP, Schindler H (2000a) Properties of lipid microdomains in a muscle cell membrane visualized by single molecule microscopy. EMBO J 19:892–901
Schütz GJ, Sonnleitner M, Hinterdorfer P, Schindler H (2000b) Single molecule microscopy of biomembranes (review). Mol Membr Biol 17:17–29
Sheetz MP (1983) Membrane skeletal dynamics – role in modulation of red-cell deformability, mobility of transmembrane proteins, and shape. Semin Hematol 20:175–188
Sheetz MP, Schindler M, Koppel DE (1980) Lateral mobility of integral membrane proteins is increased in spherocytic erythrocytes. Nature 285:510–511
Sheetz MP, Turney S, Qian H, Elson EL (1989) Nanometre-level analysis demonstrates that lipid flow does not drive membrane glycoprotein movements. Nature 340:284–288
Sieburth DS, Sun Q, Han M (1998) SUR-8, a conserved Ras-binding protein with leucine-rich repeats, positively regulates Ras-mediated signaling in *C. elegans*. Cell 94:119–130
Singer SJ, Nicolson GL (1972) The fluid mosaic model of the structure of cell membranes. Science 175:720–731
Takeuchi M, Miyamoto H, Sako Y, Komizu H, Kusumi A (1998) Structure of the erythrocyte membrane skeleton as observed by atomic force microscopy. Biophys J 74:2171–2183
Tank DW, Wu ES, Webb WW (1982) Enhanced molecular diffusibility in muscle membrane blebs: release of lateral constraints. J Cell Biol 92:207–212
Tomishige M, Sako Y, Kusumi A (1998) Regulation mechanism of the lateral diffusion of band 3 in erythrocyte membranes by the membrane skeleton. J Cell Biol 142:989–1000
Tsuji A, Ohnishi S (1986) Restriction of the lateral motion of band 3 in the erythrocyte membrane by the cytoskeletal network: dependence on spectrin association state. Biochemistry 25:6133–6139

Tsuji A, Kawasaki K, Ohnishi S, Merkle H, Kusumi A (1988) Regulation of band 3 mobilities in erythrocyte ghost membranes by protein association and cytoskeletal meshwork. Biochemistry 27:7447–7452

Vrljic M, Nishimura SY, Brasselet S, Moerner WE, McConnell HM (2002) Translational diffusion of individual class II MHC membrane proteins in cells. Biophys J 83:2681–2692

Wakatsuki T, Schwab B, Thompson NC, Elson EL (2001) Effects of cytochalasin D and latrunculin B on mechanical properties of cells. J Cell Sci 114:1025–1036

Wakioka T, Sasaki A, Kato R, Shouda T, Matsumoto A, Miyoshi K, Tsuneoka M, Komiya S, Baron R, Yoshimura A (2001) Spred is a Sprouty-related suppressor of Ras signalling. Nature 412:647–651

Winckler B, Forscher P, Mellman I (1999) A diffusion barrier maintains distribution of membrane proteins in polarized neurons. Nature 397:698–701

Wu C, Butz S, Ying Y, Anderson RG (1997) Tyrosine kinase receptors concentrated in caveolae-like domains from neuronal plasma membrane. J Biol Chem 272:3554–3559

Wu ES, Tank DW, Webb WW (1982) Unconstrained lateral diffusion of concanavalin A receptors on bulbous lymphocytes. Proc Natl Acad Sci USA 79:4962–4966

Yang F, Moss LG, Phillips GN Jr (1996) The molecular structure of green fluorescent protein. Nat Biotechnol 14:1246–1251

Chemokine Signaling: The Functional Importance of Stabilizing Receptor Conformations

José Miguel Rodríguez-Frade, Antonio Serrano, Carlos Martínez-A, Mario Mellado

6.1 Chemokines and Their Receptors

The family of low molecular weight pro-inflammatory cytokines termed chemokines has been the focus of exceptional interest over the last 25 years (Mackay 2001). Originally described as specific mediators of leukocyte directional movement, current perspectives implicate the chemokines in a wide variety of cell types and functions. The chemokines have thus been linked to lymphocyte trafficking (Baggiolini 1998), regulation of T cell differentiation (Sallusto et al. 1998), HIV-1 infection (Berger et al. 1999), angiogenesis (Belpario et al. 2000), development (Zou et al. 1998; Raz 2003), and tumor metastasis (Müller et al. 2001).

Nearly 50 different chemokines have been described, which have remarkably comparable three-dimensional structures despite their limited amino acid sequence similarity (Rollins 1997). The original classification of chemokines based on structural criteria (C, CC, CXC, and CX3C chemokines) is being abandoned, and replaced by a functional classification by which chemokines are grouped into two main categories, constitutive and inducible (Baggiolini 1998). In general, constitutive chemokines are usually regulated during development, whereas inducible chemokine expression is regulated mainly during inflammatory processes. In addition, several viruses encode highly selective chemokine receptor ligands that serve as agonists or antagonists, and may thus have a role in viral dissemination or evasion of host immune response (Alcami 2003).

Based on their broad range of functions, it is easy to deduce that chemokines must be central players in a variety of diseases characterized by inflammation and cell infiltration. They have become a major focus of interest as therapeutic targets, as there is a clear correlation between the expression of specific chemokines and the orchestrated recruitment of cell populations during the course of certain disease processes (Proudfoot 2002).

The chemokines act by binding to class A, rhodopsin-like, seven-transmembrane, G protein-coupled receptors (GPCR) (Horuk 2001). Interaction with the receptor is thought to involve the receptor N-terminal domain and third extracellular loop, while its intracellular regions bear motifs that allow participation in different signaling cascades (Rodriguez-Frade et al. 1997; Rossi and Zlotnik 2000). The 18 receptors characterized to date are classified as CCR, CXCR, CX3CR, and XCR, based on the ligand they bind (Rossi and Zlotnik 2000). Most chemokine receptors are able to interact with more than one chemokine (shared receptors), although there are examples of specific chemokine-receptor pairs (specific recep-

Springer Series in Biophysics
S. Damjanovich (Ed.)
Biophysical Aspects of Transmembrane Signaling

tors) (Horuk 2001). Expression of these receptors is finely regulated by several factors that include cytokines, growth factors, and cell cycle status (Loetscher et al. 1996; Parks et al. 1998; Papadopoulus et al. 1999). It is therefore not surprising that cells respond differently to a chemokine, depending on the microenvironment. In addition, some chemokine receptors are used by the HIV-1 virus as co-receptors for cell entry, a fact that has greatly motivated the search for chemokine receptor-based AIDS pharmaceuticals (Berger et al. 1999).

The contribution of chemokines to the control of cell movement assigns these molecules a role in physiological or pathological processes in which cells are recruited to specific sites. Most diseases involve the coordinated recruitment of cell populations, which correlates with the expression of specific chemokines, as shown in the murine model of asthma (Gerard and Rollins 2001; Lukacs et al. 2003). In another case, leukocyte migration to synovial tissue in rheumatoid arthritis (Yang et al. 2002; Bruhl et al. 2004), CCR1, CCR2, CCR3, CCR4, and CCR5 participate in monocyte recruitment from the circulation, CXCR4 contributes to T cell accumulation in the synovium, and CCR3 and CCR5 are implicated in leukocyte retention in the joint. Other well-established examples are experimental autoimmune encephalitis (EAE) (Rajan et al. 2000; Godessart and Kunkle 2001), nephritis (Zernecke et al. 2001), inflammatory bowel disease (Ulbrich et al. 2003), multiple sclerosis (Sorensen et al. 1999), diabetes (Kim et al. 2002; Carvalho-Pinto et al. 2004), and HIV-1 infection (Berger et al. 1999). Of particular interest is the role of some chemokine receptors in tumor progression and metastasis; here, chemokines can act as growth factors, attracting dendritic cells to the tumor, or to induce angiogenesis, migration, and invasion by increasing integrin expression or by triggering expression of mediators such as TNF, uPAR, PAI-1, and MMP1 (Müller et al. 2001; Homey et al. 2002).

Animal models are important tools for correlating a given chemokine with a specific disease, although conclusions from animals cannot always be extrapolated to man. Mice have been developed that lack (knockout, KO) or overexpress (transgenic) nearly every chemokine and chemokine receptor described to date. Although this strategy has not clarified the role of chemokines in all cases, some KO mice have indicated additional chemokine functions. This was the case of the CXCR4/CXCL12, and CXCR5/CXCL13 KO mice, whose study led to recognition of the role of chemokines in development (Förster et al. 1996; Zou et al. 1998).

6.2 Signaling Through Chemokine Receptors

6.2.1 Background

Chemokine receptor function was originally assumed to be mediated entirely through G_i-mediated processes (Bokoch 1995), in analogy with classic studies of other G protein-coupled receptor families. More recent studies of GPCR signaling broadened this view, as the GPCR were found to activate G proteins other than their originally assigned "partners" (Soede et al. 2001). In addition, GPCR can induce G protein-independent signaling, as occurs in some tyrosine kinase-related pathways (Ali et al. 1997; Park et al. 2000; Showkat et al. 2000).

The chemokine receptors, which should be considered as a discrete GPCR family, integrate numerous signaling pathways (Mellado et al. 2001a; Soriano et al. 2003). These receptors can form homo- or heterodimers, thus enhancing their signaling possibilities (Rodriguez-Frade et al. 2001). The conformational changes induced by ligands on these receptor complexes allow coupling of JAK (Janus kinases) that, by phosphorylating the receptors, activate STAT (signal transducers and transactivators of transcription) and G proteins. Thus, chemokine functions are blocked in cells lacking JAK or in cells treated with JAK inhibitors (Soriano et al. 2003; Stein et al. 2003). These two major signaling cascades are responsible for most known chemokine functions, including gene transcription and cytoskeletal reorganization, as well as receptor internalization and recycling.

6.2.2 The Use of Biophysical Techniques to Measure Chemokine Receptor Oligomerization

The importance of GPCR dimerization in signaling has only recently attracted the attention of the scientific community, although it was first proposed 25 years ago and corroborative evidence has accumulated since then. Chemokine receptors were initially thought to function as monomers, but in analogy to other GPCR, chemokine receptor dimers and/or oligomers have been described, as has their importance in signaling (Lee et al. 2003; Milligan et al. 2003). Until recently, most assays to demonstrate receptor oligomerization have been based on biochemical approaches such as immunoprecipitation or cross-linking. Alternatively, dominant negative receptor mutants abrogated wild-type receptor functions by forming non-functional complexes (Rodriguez-Frade et al. 1999).

Although chemokine receptor homo- and heterodimer formation can be observed, these approaches only allow imprecise definition of the role of ligands in complex formation. Chemokines could promote receptor oligomerization or stabilize pre-existing dimers, and the results from biochemical and genetic studies showed that the dimerization process is chemokine-triggered. Nonetheless, recent data obtained using new technologies have modified this outlook. These new approaches, based on energy transfer between fluorochromes, followed by confocal microscopy, take advantage of important technological advances such as laser light sources, fluorescent probes, and advances in computer science that allow digital imaging and image analysis.

Modern optical microscopy allows us not only to visualize organelles and molecules, but also to study their function. In living cells, we can analyze how a molecule moves, changes location, or associates with other molecules. Such phenomena were originally evaluated using colocalization assays, which detect light from two different fluorophores and evaluate a digital image for the presence of the same pixel in two distinct channels. Signal colocalization indicates the adjacency of fluorophores, and thus of the molecules they label. A high-numerical aperture microscope lens permits resolution near 300 nm, sufficient to locate molecules in different cellular compartments, but not to demonstrate molecular association.

Fluorescence resonance energy transfer (FRET) is another new approach. This quantum mechanical process is based on energy transmission from a donor fluorophore to a nearby acceptor without photon emission. The use of a donor emission spectrum that overlaps acceptor absorption, and a distance between donor and acceptor in the 2- to 10-nm range, make FRET a highly sensitive technique for the study of protein–protein interactions. Energy transfer requires very short distances between donor and acceptor, as efficiency varies inversely with the distance between fluorophores (Sekar and Periasamy 2003).

Several methods determine and quantify FRET; the most commonly used are (1) sensitized acceptor fluorescence, (2) increase in donor fluorescence after acceptor photobleaching, and (3) decrease in donor fluorophore lifetime. In the case of sensitized acceptor fluorescence, the donor fluorescent dye is excited and the acceptor signal is quantitated (Fig. 6.1). An appropriate filter set is used to excite donor and detect acceptor emission, and a factor is included to correct for autofluorescence, photobleaching, and crossover between donor and acceptor emission (Sorkin et al. 2000).

The acceptor photobleaching method is based on quenching donor fluorescence. Some donor photons are used to excite the acceptor, decreasing the emission energy detected. Photobleaching of the acceptor abolishes FRET, increasing donor light emission (Kenworthy 2001; Fig. 6.2).

Fluorescence lifetime imaging microscopy (FLIM) measures a chromophore's fluorescence lifetime, allowing spatial resolution of biochemical processes. The fluorescence lifetime of a donor dye decreases under FRET conditions, independently of fluorophore concentration or of excitation intensity. Fluorescence lifetime imaging on a pseudocolor scale allows analysis of molecule localization and the intermolecular associations triggered (Elangovan et al. 2002; Fig. 6.3).

For all these approaches, donor/acceptor choice is critical in determining FRET. Ideally, donor emission spectra should have maximum overlap with acceptor absorption spectra, although acceptor and donor emissions should be clearly separable in order to minimize background interference. Fluorochrome incorporation in the protein to be analyzed is another feature to be considered. To study fixed cells, specific antibodies can also be used, which are conjugated to appropriately selected fluorescent dyes. For example, although Cy3 dye as donor and Cy5 as acceptor form a suitable fluorochrome pair, the Cy2 donor/Cy3 acceptor pair is often more convenient, as it permits use of the widely available 488-nm argon laser line. Use of secondary antibodies is occasionally necessary; although the increased distance between fluorophores complicates FRET detection; this can be resolved using dye-labeled F(ab′) fragments.

To evaluate FRET in living cells, the protein of interest is modified to incorporate a fluorochrome. Chimeric constructs are generated in which each of the proteins of interest is fused to a different, modified form of the green fluorescent protein (GFP); the most commonly used variants are cyan (CFP) as donor and yellow (YFP) as acceptor (Pollok and Heim 1999).

Bioluminescence resonance energy transfer (BRET) is another method that makes use of non-radiative energy transfer between a light donor and a fluores-

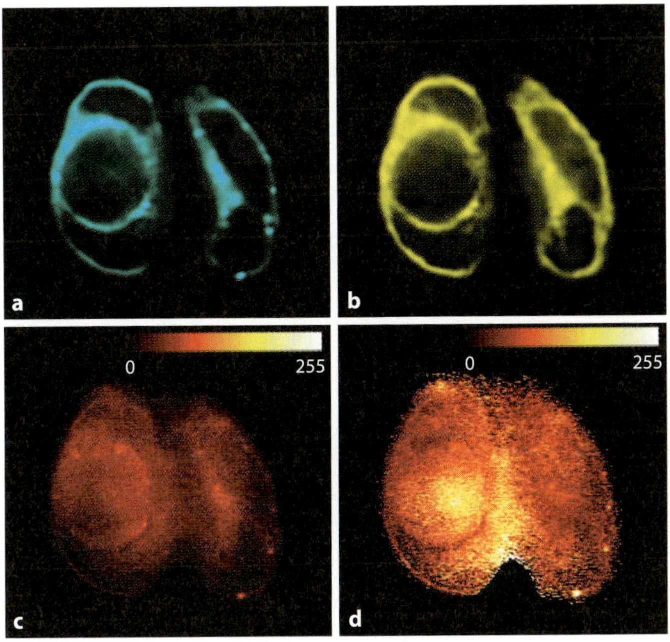

Fig. 6.1. FRET evaluation of chemokine receptor dimerization using a sensitized acceptor fluorescence method. Unstimulated CCR5-Cyan/CCR5-Yellow HEK 293 cells were evaluated by fluorescent microscopy. **a** CFP was detected using a specific filter set (435/10-nm excitation filter, 455-nm dichroic beam splitter, and 480/20-nm emission filter). **b** YFP was viewed under a filter set with a 500/20-nm excitation filter, 515-nm dichroic beam splitter, and 535/30-nm emission filter. **c** FRET was detected with a 435/10-nm excitation filter, 455-nm dichroic beam splitter, and 535/30-nm emission filter. **c** shows the corrected FRET (FRETC) image obtained by subtracting autofluorescence and bleed-through from original FRET images, displayed in quantitative pseudocolor (arbitrary linear units of fluorescence intensity). **d** Normalized FRET (FRETN), calculated on the CCR5-CFP and CCR5-YFP colocalization areas using pixel-by-pixel analysis of images **a** and **c** and displayed in quantitative pseudocolor

cent acceptor. In BRET, the bioluminescent energy resulting from the catalytic degradation of coelenterazine by luciferase is transferred to an acceptor fluorophore, which in turn emits a fluorescent signal (McVey et al. 2001).

6.2.3 Characterization of Chemokine Receptor Homo- and Heterodimerization

Chemokine receptors are found on the cell membrane in various conformational states that include monomers, homodimers, heterodimers, and higher molecular weight oligomers. Ligands stabilize the conformation that initiates sub-

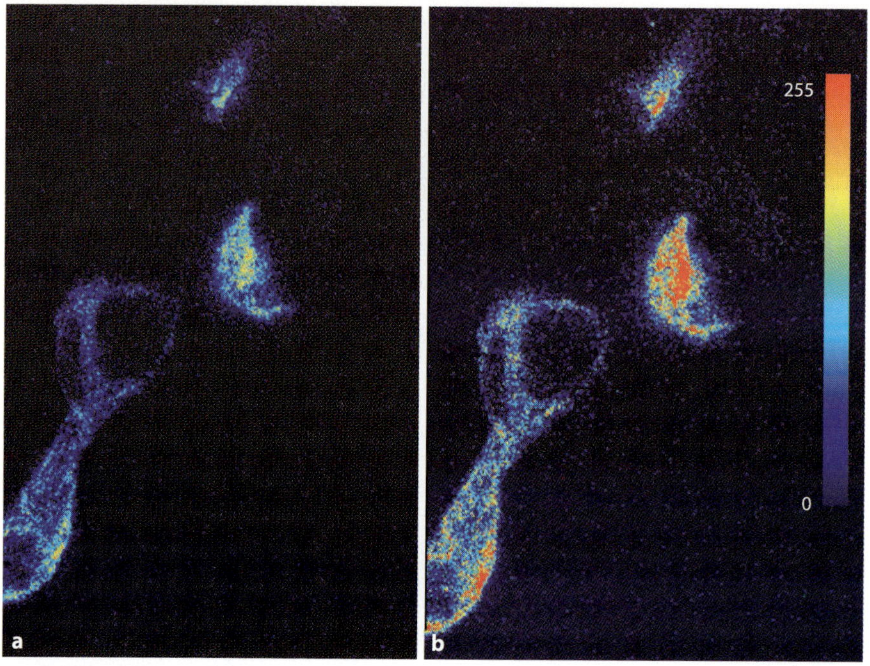

Fig. 6.2. FRET evaluation of chemokine receptor heterodimerization using an acceptor photobleaching method. Unstimulated CCR2/CCR5 HEK 293 cells were fixed and stained with anti-CCR2 (IgG) and anti-CCR5 (IgM) mAb, followed by Cy2-mIgGFab or Cy3-mIgMFab (excitation at 488 and 543 nm, respectively). **a** Cy2 fluorescence determined before Cy3 photobleaching. **b** Cy2 fluorescence determined after Cy3 photobleaching. In both cases, images are displayed in quantitative pseudocolor

sequent signaling cascades. Ligand-induced homodimerization was shown for CCR2, CCR5, and CXCR4, and its functional relevance has been described (Rodriguez-Frade et al. 1999; Vila-Coro et al. 1999; Mellado et al. 2001b). The role of these receptor complexes was described for AIDS, as HIV-1 entry into cells through CCR5 can be inhibited by inducing receptor dimerization (Vila-Coro et al. 2000; Rodriguez-Frade et al. 2004).

Bioinformatic analysis of the chemokine receptor family predicts TM1, TM2, and TM4 as the principal transmembrane (TM) domains implicated in the interaction between receptor monomers; mutation of these regions gives rise to nonfunctional receptors unable to form dimers (Hernanz-Falcon et al. 2004). Energy transfer techniques (FRET and FLIM) were used to show that a CCR5 with TM1 and TM4 point mutations did not dimerize even after ligand stimulation, although membrane expression and ligand binding capacity were indistinguishable from those of wild-type CCR5. In other FRET experiments, ligand binding induced only a modest alteration of FRET values, suggesting that the ligand does not drive dimerization, but rather stabilizes the dimeric conformation.

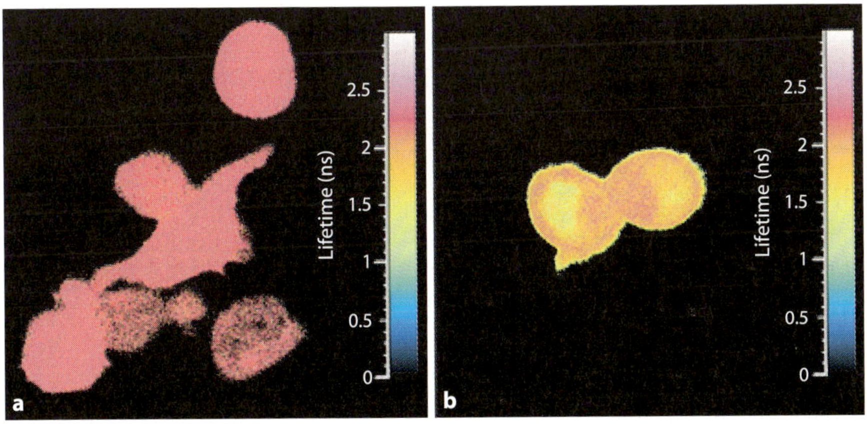

Fig. 6.3. FRET evaluation of chemokine receptor dimerization using the fluorescence lifetime imaging method (FLIM). CFP fluorescence lifetime images calculated from the phase shift of HEH 293 cells expressing CCR5-CFP (**a**) or CCR5-CFP/CCR5-YFP (**b**). The pseudocolor scale ranges from 0 ns (*black*) to 3.0 ns (*white*). *Arrowheads* indicate Golgi localization, showing that FLIM detection is not dependent on the level of receptor expressed

Chemokine receptor oligomers are not detected in unstimulated cells using Western blot assays, however, indicating that many factors affect dimer stability; these include the detergents used to lyse cells for immunoprecipitation, which could explain the current controversy regarding the effect of the ligand on GPCR dimerization. These data support a model of GPCR structural flexibility for activation-associated conformational changes, with the monomeric conformation as an inactive state, unable to trigger function or to activate the steps that initiate chemokine signaling.

In most physiological and pathological situations, several chemokines and receptors coexist in the same environment (Gerard and Rollins 2001). Under these conditions, specific chemokine receptor heterodimers can be formed, which mediate functions that cannot be implemented by the homodimer alone. This is the case with CCR2 and CCR5, whose heterodimerization results in coupling of pertussis toxin (Ptx)-insensitive G proteins, in a reduced response threshold to the ligand, and in distinct phosphoinositide-3 kinase (PI3 K) activation kinetics compared to homodimers (Mellado et al. 2001b). Chemokine receptor heterodimerization also affects HIV-1 infection. A CCR2-specific monoclonal antibody (mAb) that induces CCR2/CCR5 and CCR2/CXCR4 heterodimer formation blocks HIV-1 entry (Rodriguez-Frade et al. 2004).

Chemokines are produced within specific tissues in response to a variety of mediators; once they enter the circulation, they are immobilized by low-affinity binding to heparin-bearing proteoglycans on the vascular endothelial barrier (Culley et al. 2003). Chemokine availability dictates the formation of signaling domain complexes, regulating receptor sensitivity (Fig. 6.4); this affects the ability of a ligand to stabilize homo- or heterodimer conformations. Receptor homo-

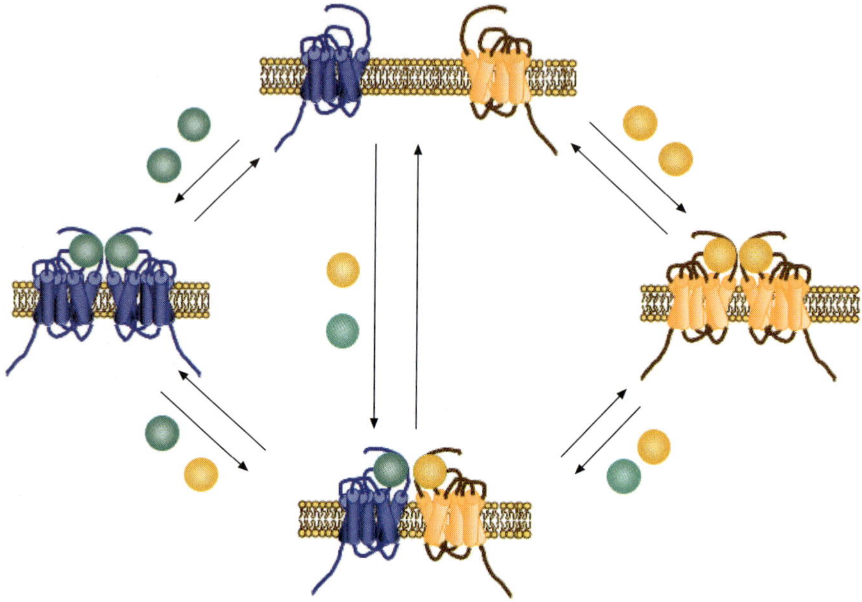

Fig. 6.4. Chemokine availability stabilizes different receptor conformations. Schematic representation of the equilibrium between chemokine receptors and showing that availability of specific chemokines stabilizes receptor homo- or heterodimers

or heterodimerization increases immune system plasticity and specificity. The range of chemokine responses is thus modulated by the spread of activity through a receptor array, explaining the abundance and variety of chemokine receptors and receptor conformations observed (Rossi and Zlotnik 2000).

Further complexity is added by chemokine receptor interaction with other GPCR such as the opioid receptors (Lee et al. 2003). Opium-induced inhibition of chemotaxis was first reported over a century ago. More recent indications of this cross-talk include increased opioid receptor ligand-mediated susceptibility to HIV-1 infection, as well as altered humoral or cell-mediated immune responses when both receptor types are coexpressed (Suzuki et al. 2002; Szabo et al. 2002). This cross-talk was previously explained as heterologous desensitization, whose signaling pathways remain poorly defined. Nonetheless, the mechanism underlying this GPCR cross-talk might also be the formation of heterologous heterodimers that trigger specific responses; in fact, coimmunoprecipitation assays showed ligand-independent complexes formed by δ, μ or κ opioid receptors and CCR5 (Suzuki et al. 2002). Once again, energy transfer technologies may help define the precise role of chemokines and opioids in stabilizing these complexes.

Several functional implications have been proposed for chemokine receptor dimerization. Oligomeric assembly is an important endoplasmic reticulum (ER) control mechanism for the GPCR, as it modulates specific retention signals and hydrophobic patches to retain or release proteins from the ER (Reddy and Corley 1998). An example for the chemokine receptors is CCR5, whose oligomerization with the CCR5Δ32 variant impedes its cell-surface expression (Benkirane et al. 1997).

The importance of GPCR dimerization for the generation of pharmacological diversity was first indicated by Jordan and Devi in studies of opioid receptors (Jordan and Devi 1999). Dimerization could affect ligand binding affinity (Jordan and Devi 1999) or modulate signal transduction (Terrillon and Bouvier 2004). We observed that specific chemokine receptor heterodimers trigger specific signaling events which in turn promote specific cell functions, for example, cell adhesion.

The role of the ligand in promoting or inhibiting receptor dimerization continues to be debated. Whereas several studies suggest that ligand binding regulates dimer formation (Rodriguez-Frade et al. 1999), others conclude that dimerization is a constitutive process that is not modulated by ligand binding (Issafras et al. 2002; Trettel et al. 2003). This discrepancy may result from differences in the analytic techniques employed, although it can also mirror the diversity of individual receptor behaviors, such as preferential localization in specific cell membrane domains.

6.2.4 Chemokines Trigger JAK/STAT Pathway Activation

JAK/STAT signaling was traditionally associated with single-transmembrane domain cytokine family receptors (Liu et al. 1998). GPCR have been only recently shown to activate the JAK/STAT pathway. In *Dictyostelium*, G protein-independent GPCR activation of STAT has been reported. Other examples of GPCR-mediated activation of this pathway include the receptors for angiotensin II, serotonin, α-MSH, and chemokines such as CCL5, CCL2, and CXCL12 (Araki et al. 1998; Buggy 1998; Park et al. 2000; Showkat 2000; Mellado et al. 2001a; Pelletier et al. 2003). In contrast to cytokine receptors, GPCR ligands and thus chemokines mediate JAK association with the receptor in a transient process (Mellado et al. 1998). Experiments using mutated chemokine receptor forms, JAK inhibitors and JAK-deficient cells showed that JAK activation is earlier than that of the G protein (Soriano et al. 2003). Lack of JAK activation abrogates G protein coupling to the chemokine receptor and receptor functions. These data indicate clearly that chemokines activate at least two distinct pathways: one involving G proteins, and the other, JAK family members.

There are several steps in JAK/STAT activation through chemokine receptors. First, ligand binding promotes or stabilizes receptor dimers, giving rise to a new receptor conformation that permits JAK interaction. This interaction may be direct or involve an adaptor protein, but it requires the conserved DRYLAIV motif in the second intracellular loop of the receptor. Once associated, JAK is activated by transphosphorylation, which in turn phosphorylates the receptor. This phosphorylated receptor/JAK complex can then bind G protein, which triggers functional responses (Mellado et al. 2001a).

6.2.5
Chemokine Signaling Involves G Protein Activation

In addition to the role of JAKs in chemokine signaling, G protein activation is critical in most chemokine-driven events (Kuang et al. 1996). Signaling studies showed potent, chemokine-dependent inhibition of adenylyl cyclase and mobilization of intracellular calcium, consistent with receptor coupling to $G\alpha_i$ (Locati et al. 1994). $G\alpha_i$ also associates with several chemokine receptors in response to ligand binding (Mellado et al. 1998; Rodriguez-Frade et al. 1999; Vila-Coro et al. 1999); consistent with G_i association, the majority of these responses are inhibited by Ptx treatment. Under certain conditions, however, chemokine receptors also associate with other, Ptx-insensitive G proteins (Kuang et al. 1996; Wright et al. 2003). One of these cases is the activation of chemokine receptor heterodimerization, which associates with $G_{q/11}$, indicating that chemokine signaling is much more versatile than previously thought. Although by stabilizing a specific receptor conformation, the ligand may determine the G protein that associates with the receptor, we cannot discard an effect of cell type on G protein availability (Mellado et al. 2001b).

G proteins mediate activation of several signaling molecules, among them PI3 K, phospholipase Cβ (PLCβ), and small GTPases, which cooperate to mediate chemokine-associated functions, including actin cytoskeleton reorganization, polarization, cell adhesion, and detachment (Fukata et al. 2003). Although the PI3 Kγ isoform is directly associated with GPCR, the role of PI3 K in cell migration is controversial. Whereas PI3 $K\gamma^{-/-}$ mice show impaired neutrophil migration, PI3 K blockade affected T cell migration only slightly (Hirsch et al. 2000; Ward 2004). The influence of other proteins on cell migration may differ depending on cell type. Through PIP3 generation, PI3 K activates Rho small GTPases including Rac1, RhoA, and Cdc42, which connect receptors with the cytoskeleton, modulating the changes needed for acquisition of a migratory phenotype (Hall 1998; Ridley 2001).

Thus for chemokines as well as for other GPCR, ligand availability conditions cell responses. Depending on the receptor conformation stabilized, chemokines activate distinct signaling pathways that trigger specific cell functions.

6.3
General Considerations on the Biophysical Aspects of Signaling

Transmembrane signaling is affected by biophysical considerations that are not often taken into account, and which vary depending on the cell system analyzed. In the case of chemokine signaling, several variables should be kept in mind, including receptor expression, recycling, and cell polarization.

6.3.1 Receptor Expression

Chemokines and their receptors are classified as constitutive or inducible according to functional criteria. The former group is comprised of cell-surface receptors with little fluctuation in expression levels, whereas expression of inducible receptors is altered by factors such as stimulation or cell cycle status (Baggiolini 1998; Papadopoulus et al. 1999).

The chemokine receptor C-terminal region is of particular importance in receptor expression. It has several serine and threonine residues, which are targets for the molecules that trigger receptor desensitization and internalization, such as G protein-coupled kinases (GRK) and arrestins (Franci et al. 1996; Aragay et al. 1998). Chemokine receptor internalization requires receptor phosphorylation, β-arrestin association, and clathrin vesicle formation (Yang et al. 1999). Certain signaling events, such as initiation of the MAPK cascade, are activated by receptor internalization (Pouysségur 2000).

When receptor expression is debated, the intracellular pool of chemokine receptors should not be overlooked, even though their function remains to be elucidated. The sum of these observations demonstrates that chemokine receptor expression is a finely regulated phenomenon. This must be recalled when both planning and evaluating biophysical methods that require receptor overexpression or the use of fluorescence-labeled receptor constructs, as these manipulations may interfere with correct receptor localization or function.

6.3.2 Receptor Localization

Cell migration is an essential part of physiological and pathological processes such as inflammatory responses, lymphocyte homing, development, and tumor metastasis (Baggiolini 2000; Homey et al. 2002; Ara et al. 2003). Forward cell movement requires coordination between substrate adhesion at the cell front and detachment at the rear, that is, cell polarity (Sanchez-Madrid and del Pozo 1999). Many molecules, including chemokines and cytokines, induce lymphocyte polarization; during this Rho GTPase-associated process, a cell leading edge and a uropod are developed, with redistribution of signaling molecules and receptors (del Pozo et al. 1999). For example, adhesion molecules and ezrin-radixin-moesin (ERM) complex proteins accumulate in the uropod of chemoattractant-stimulated neutrophils and T lymphocytes (Serrador et al. 1997). In migrating T lymphocytes, β-actin and chemokine receptors concentrate at the leading edge, while myosin II, the Golgi apparatus, and a microtubule meshwork are packed into the uropod (Sanchez-Madrid and del Pozo 1999). This asymmetric organization serves the specialized functions of each compartment, as the leading edge directs cell migration and the uropod sustains cell–cell interactions and lymphocyte recruitment.

Chemokine receptors and signaling molecules associate with rafts, which are lipid-based plasma membrane domains that appear to have a role in receptor redistribution (Manes et al. 1999). Their condensed cholesterol- and sphingolipid-rich structure dictates specific physical properties, which condition protein inclusion or exclusion. Glycosylphosphatidylinositol (GPI)-anchored proteins (Hooper 1999), double-acylated proteins such as the G protein α subunit (Resh 1999), and palmitoylated proteins (Brown and London 1998) are examples of molecules with raft affinity. It is not yet clear why some transmembrane proteins are included in rafts, but mutational analysis has shown that amino acids in the transmembrane domains near the exoplasmic leaflet are critical (Scheiffele et al. 1997). In any case, raft domains may serve as platforms for interaction between receptors and signal transduction molecules, thus increasing, restricting, or modulating signaling efficiency. In some models, rafts define the signaling molecules able to bind a receptor, but also facilitate interaction between different receptors at the cell surface. This has been observed for certain HIV-1 chemokine coreceptors and CD4, which localize to the same raft domains and whose physical association is necessary for HIV-1 infection; manipulation of these platforms may thus be of interest for blocking HIV-1 entry (Manes et al. 2000). Furthermore, rafts might influence chemokine receptor homo- and heterodimer stabilization.

Receptor localization also affects formation of the immunological synapse. This structure is a contact face formed during T cell interaction with antigen-presenting cells (APC). Polarity is inherent to crawling lymphocytes prior to APC contact; APC recognition and binding take place at the T cell leading edge (Kupfer and Singer 1989). Ongoing research is attempting to determine how these supramolecular assemblies are formed, and progress has been made in defining the order in which several molecules are recruited to signaling centers (Jacobelli et al. 2004). Although the role of the chemokine receptors has yet to be clarified, active cytoskeleton-driven mechanisms are thought to be responsible for synapse assembly (Wulfing and Davis 1998).

6.4 Concluding Remarks

Chemokines are the principal chemotactic factors implicated in the regulation of leukocyte traffic as well as in establishing lymphoid organ architecture. They regulate lymphocyte precursor entry into primary lymphoid organs as well as mature lymphocyte migration to secondary lymphoid organs, where, following activation, they are responsible for triggering functional immune responses. The chemokines implicated in these activities constitute the homeostatic chemokines. Another large subset of chemokines appears to provoke inflammatory cell migration into tissues: the so-called inflammatory chemokines. Both chemokine subsets mediate their function by interacting with specific receptors belonging to the family of seven-transmembrane, G protein-coupled receptors, expressed on the leukocyte surface.

Much information is available on the biochemical pathways activated by this large receptor family. In addition, recent studies have shown that chemokines and other ligands activate a tyrosine kinase pathway that shares many components with the biochemical cascade activated by cytokine receptors. As in cytokine responses, this activation depends on ligand-mediated receptor homo- and heterodimerization. These findings indicate a new way to intervene in chemokine responses, especially those involving tissue infiltration, and link chemokine and cytokine function.

In the near future, new experiments need to be designed to elucidate the precise role of the ligand in promoting or stabilizing receptor dimerization. The use of classical biochemical techniques, such as cross-linking, immunoprecipitation, or Western blot, must be implemented with new technologies such as those based on energy transfer between fluorochromes and microscopy analysis. For these techniques, methodological questions will need to be clarified, including use of different cell types, overexpression of proteins, or use of chemical inhibitors that can alter in vitro distribution, availability, and/or function of chemokine receptors compared to their in vivo behavior. Correctly used, biophysical analysis techniques will clearly be of value in disentangling the complex web of ligands, receptors, and their signals.

Acknowledgements. We thank the members of the DIO chemokine group, who contributed to some of the work described in this chapter. We also thank C. Bastos and C. Mark for secretarial support and helpful editorial assistance, respectively. This work was partially funded by grants from the Spanish MCyT. The Department of Immunology and Oncology was founded and is supported by the Spanish Council for Scientific Research (CSIC) and by Pfizer.

References

Alcami A (2003) Viral mimicry of cytokines, chemokines and their receptors. Nat Rev Immunol 3:36–50

Ali M, Sayeski PP, Dirksen L, Hayzer D, Marrero M, Bernstein K (1997) Dependence on the motif YIPP for the physical association of JAK2 kinase with the intracellular carboxyl tail of the angiotensin II AT1 receptor. J Biol Chem 272:23382–23388

Ara T, Itoi M, Kawabata K, Egawa T, Tokoyoda K, Sugiyama T, Fukii N, Amagau T, Nagasawa T (2003) A role of CXC chemokine ligand 12/stromal cell-derived dactor-1/pre-B cell growth stimulating factor and its receptor CXCR4 in fetal and adult T cell development in vivo. J Immunol 170:4649–4655

Aragay A, Mellado M, Rodriguez-Frade JM, Martin AM, Jimenez-Sainz MC, Martínez-A C, Mayor F Jr (1998) Monocyte chemoattractant protein-1-induced CCR2B receptor desensitization mediated by the G protein-coupled receptor kinase 2. Proc Natl Acad Sci USA 95: 2985–2990

Araki T, Gamper M, Early A, Fukuzawa M, Abe T, Kawata T, Kim E, Firtel RA, Williams JG (1998) Developmentally and spatially regulated activation of a *Dictyostelium* STAT protein by a serpentine receptor. EMBO J 17:4018–4028

Baggiolini M (1998) Chemokines and leukocyte traffic. Nature 392:565–568

Baggiolini M, Loetscher P (2000) Chemokines in inflammation and immunity. Immunol Today 21:418–420

Belpario J, Keane M, Arenberg D, Addison C, Ehlert J, Burdick M, Strieter R (2000) CXC chemokines in angiogenesis. J Leuk Biol 68:1–8

Benkirane M, Jin D-Y, Chun RF, Koup RA, Jeang KT (1997) Mechanism of transdominant inhibition of CCR5-mediated HIV-1 infection by ccr5D32. J Biol Chem 272:30603–30606

Berger EA, Murphy PM, Farber JM (1999) Chemokines as HIV-1 coreceptors: roles in viral entry, tropism and disease. Annu Rev Immunol 17:657–700

Bokoch GM (1995) Chemoattractant signaling and leukocyte activation. Blood 86:1649–1660

Brown DA, London E (1998) Functions of lipid rafts in biological membranes. Annu Rev Cell Dev Biol 14:111–136

Bruhl H, Cihak J, Schneider MA, Plachy J, Rupp T, Wenzel I, Shakarami M, Milz S, Ellwart JW, Stangassinger M, Schlondorff D, Mack M (2004) Dual role of CCR2 during initiation and progression of collagen-induced arthritis: evidence for regulatory activity of CCR2+ T cells. J Immunol 172:890–898

Buggy JJ (1998) Binding of α-melanocyte-stimulating hormone to its G-protein-coupled receptor on B-lymphocytes activates the Jak/STAT pathway. Biochem J 331:211–216

Carvalho-Pinto C, Garcia MI, Gomez L, Ballesteros A, Zaballos A, Flores JM, Mellado M, Rodriguez-Frade JM, Balomenos D, Martinez-A C (2004) Leukocyte attraction through the CCR5 receptor controls progress from insulitis to diabetes in non-obese diabetic mice. Eur J Immunol 34:548–557

Culley F, Fadlon EJ, Kirchem A, Williams TJ, Jose PJ, Pease JE (2003) Proteoglycans are potent modulators of the biological responses of eosinophils to chemokines. Eur J Immunol 33:1302–1310

del Pozo MA, Vicente-Manzanares M, Tejedor R, Serrador JM, Sanchez-Madrid F (1999) Rho GTPases control migration and polarization of adhesion molecules and cytoskeletal ERM components in T lymphocytes. Eur J Immunol 29:3609–3620

Elangovan M, Day RN, Periasamy A (2002) Nanosecond fluorescence resonance energy transfer – fluorescence lifetime imaging microscopy to localize the protein interactions in a single living cell. J Microsc 205:3–14

Förster R, Mattis AE, Kremmer E, Wolf E, Brem G, Lipp M (1996) A putative chemokine receptor, BRL1, directs B cell migration to defined lymphoid organs and specific anatomic compartments of the spleen. Cell 87:1037–1047

Franci C, Gosling J, Tsou CL, Coughlin SR, Charo IF (1996) Phosphorylation by a G-protein-coupled kinase inhibits signalling and promotes internalization of the monocyte chemoattractant protein-1 receptor. Critical role of carboxyl-tail serines/threonines in receptor function. J Immunol 157:5606–5612

Fukata M, Nakagawa M, Kaibuchi K (2003) Roles of Rho-family GTPases in cell polarisation and directional migration. Curr Opin Cell Biol 15:590–597

Gerard C, Rollins BJ (2001) Chemokines and disease. Nat Immunol 2:108–115

Godessart N, Kunkle SL (2001) Chemokines in autoimmune diseases. Curr Opin Immunol 13:670–675

Hall A (1998) Rho GTPases and the actin cytoskeleton. Science 279:509–514

Hernanz-Falcon P, Rodriguez-Frade JM, Serrano A, Juan D, del Sol A, Soriano SF, Roncal F, Gomez L, Valencia A, Martinez-A C, Mellado M (2004) Identification of amino acid residues crucial for chemokine receptor dimerization. Nat Immunol 5:216–223

Hirsch E, Katanaev VL, Garlanda C, Azzolino O, Pirola L, Silengo L, Sozzani S, Mantovani A, Altruda F, Wymann MP (2000) Central role for G protein-coupled phosphoinositide 3-kinase gamma in inflammation. Science 287:1049–1053

Homey B, Müller A, Zlotnik A (2002) Chemokines: agents for the immunotherapy of cancer? Nat Rev Immunol 2:175–184

Hooper NM (1999) Detergent-insoluble glycosphingolipid/cholesterol-rich membrane domains, lipid rafts and caveolae. Mol Membr Biol 16:145–156

Horuk R (2001) Chemokine receptors. Cytokine Growth Factor Rev 12:313–335

Issafras H, Angers S, Bulenger S, Blanpain C, Parmentier M, Labbé-Jullié C, Bouvier M, Marullo S (2002) Constitutive agonist-independent CCR5 oligomerization and antibody-mediated clustering occurring at physiological levels of receptors. J Biol Chem 277:34666–34673

Jacobelli J, Andres PG, Boisvert J, Krummel MF (2004) New views of the immunological synapse: variations in assembly and function. Curr Opin Immunol 16:345–352

Jordan BA, Devi LA (1999) G-protein-coupled receptor heterodimerization modulates receptor function. Nature 399:697–700

Kenworthy AK (2001) Imaging protein–protein interactions using fluorescence resonance energy transfer microscopy. Methods 24:289–296

Kim SK, Cleary MM, Fox HS, Chantry D, Sarvetnick N (2002) CCR4-bearing T cells participate in autoimmune diabetes. J Clin Invest 110:1675–1686

Kuang Y, Wu Y, Jiang H, Wu D (1996) Selective G protein coupling by C-C chemokine receptors. J Biol Chem 271:3975–3978

Kupfer A, Singer SJ (1989) Cell biology of cytotoxic and helper T cell functions: immunofluorescence microscopic studies of single cells and cell couples. Annu Rev Immunol 7:309–337

Lee SP, O'Dowd BF, George SR (2003) Homo- and hetero-oligomerization of G protein-coupled receptors. Life Sci 74:173–180

Liu KD, Gaffen SL, Goldsmith MA (1998) JAK/STAT signalling by cytokine receptors. Curr Opin Immunol 10:271–278

Locati M, Zhou D, Luini W, Evangelista V, Mantovani A, Sozzani S (1994) Rapid induction of arachidonic acid release by monocyte chemotactic protein-1 and related chemokines. Role of calcium influx, synergism with platelet-activating factor and significance for chemotaxis. J Biol Chem 269:4746–4753

Loetscher P, Seitz M, Baggiolini M, Moser B (1996) Interleukin-2 regulates CC chemokine receptor expression and chemotactic responsiveness in T lymphocytes. J Exp Med 184:569–577

Lukacs NW, Miller AL, Hogaboam CM (2003) Chemokine receptors in asthma: searching for the correct immune targets. J Immunol 171:11–15

Mackay CR (2001) Chemokines: immunology's high impact factors. Nat Immunol 2:95–101

Manes S, Mira E, Gomez-Mouton C, Lacalle RA, Keller P, Labrador JP, Martinez-A C (1999) Membrane raft microdomains mediate front–rear polarity in migrating cells. EMBO J 18:6211–6220

Manes S, del Real G, Lacalle RA, Lucas P, Gomez-Mouton C, Sanchez-Palomino S, Delgado R, Alcami J, Mira E, Martinez AC (2000) Membrane raft microdomains mediate lateral assemblies required for HIV-1 infection. EMBO Rep 1:190–196

McVey M, Ramsay D, Kellet E, Rees S, Wilson S, Pope AJ, Milligan G (2001) Monitoring receptor oligomerization using time-resolved fluorescence resonance energy transfer and bioluminescence resonance energy transfer. J Biol Chem 276:14092–14099

Mellado M, Rodríguez-Frade JM, Aragay A, del Real G, Martín AM, Vila-Coro AJ, Serrano A, Mayor F Jr, Martínez-A C (1998) The chemokine monocyte chemotactic protein 1 triggers Janus kinase 2 activation and tyrosine phosphorylation of the CCR2B receptor. J Immunol 161:805–813

Mellado M, Rodríguez-Frade JM, Mañes S, Martínez-A C (2001a) Chemokine signaling and functional responses: the role of receptor dimerization and TK pathway activation. Annu Rev Immunol 19:397–421

Mellado M, Rodriguez-Frade JM, Vila-Coro AJ, Fernandez S, Martin de Ana A, Jones DR, Toran JL, Martinez AC (2001b) Chemokine receptor homo- or heterodimerization activates distinct signaling pathways. EMBO J 20:2497–2507

Milligan G, Ramsay D, Pascal G, Carrillo JJ (2003) GPCR dimerization. Life Sci 74:181–188

Müller A, Homey B, Soto H, Ge N, Catron D, Buchanan M, McClanahan T, Murphy E, Yuan W, Wagner SN, Barrera JL, Mohar A, Verástegui E, Zlotnik A (2001) Involvement of chemokine receptors in breast cancer metastasis. Nature 410:50–56

Papadopoulus EJ, Sassetti C, Saeki H, Yamada N, Kawamura T, Fitzhugh DJ, Saraf MA, Schall TJ, Blauvelt A, Rosen SD, Hwang ST (1999) Fractalkine, a CXCR3 chemokine is expressed by dendritic cells and is up-regulated upon dendritic cell maturation. Eur J Immunol 29:2551–2559

Park ES, Kim H, Suh JM, Park SJ, You SH, Chung HK, Lee KW, Kwon OY, Cho BY, Kim YK, Ro HK, Chung J, Shong M (2000) Involvement of JAK/STAT (Janus kinase/signal transducer and activator of transcription) in the thyrotropin signaling pathway. Mol Endocrinol 14:662–670

Parks E, Strieter RM, Lukacs NW, Gauldie J, Hitt M, Graham FL, Kunkel SL (1998) Transient gene transfer of IL-2 regulates chemokine expression and disease severity in experimental arthritis. J Immunol 160:4615–4619

Pelletier S, Duhamel F, Coulombe P, Popoff MR, Meloche S (2003) Rho family GTPases are required for activation of Jak/STAT signaling by G protein-coupled receptors. Mol Cell Biol 23:1316–1333

Pollok BA, Heim R (1999) Using GFP in FRET-based applications. Trends Cell Biol 9:57–60

Pouysségur J (2000) An arresting start for MAPK. Science 290:1515–1518

Proudfoot AEI (2002) Chemokine receptors: multifaceted therapeutic targets. Nat Rev Immunol 2:106–115

Rajan A, Asensio VC, Campbell IL, Brossnan CF (2000) Experimental autoimmune encephalomyelitis on the SJL mouse: effect of $\delta\gamma$ T cell depletion on chemokine and chemokine receptor expression in the central nervous system. J Immunol 164:2120–2130

Raz E (2003) Primordial germ-cell development: the zebrafish perspective. Nat Rev Genet 4:690–700

Reddy PS, Corley RB (1998) Assembly, sorting, and exit of oligomeric proteins from the endoplasmic reticulum. Bioessays 20:546–554

Resh MD (1999) Fatty acylation of proteins: new insights into membrane targeting of myristoylated and palmitoylated proteins. Biochem Biophys Acta 1451:1–16

Ridley A (2001) Rho family proteins: coordinating cell responses. Trends Cell Biol 11:471–477

Rodriguez-Frade JM, Llorente M, Mellado M, Alcamí J, Gutierrez-Ramos JC, Zaballos A, del Real G, Martínez-A C (1997) The amino-terminal domain of the CCR2 chemokine receptor acts as coreceptor for HIV-1 infection. J Clin Invest 100:1–6

Rodriguez-Frade JM, Vila-Coro AJ, de Ana AM, Albar JP, Martinez-A C, Mellado M (1999) The chemokine monocyte chemoattractant protein-1 induces functional responses through dimerization of its receptor CCR2. Proc Natl Acad Sci USA 96:3628–3633

Rodriguez-Frade JM, Mellado M, Martinez-A C (2001) Chemokine receptor dimerization: two are better than one. Trends Immunol 22:612–617

Rodriguez-Frade JM, Del Real G, Serrano A, Hernanz-Falcon P, Soriano SF, Vila-Coro AJ, De Ana AM, Lucas P, Prieto I, Martinez-A C, Mellado M (2004) Blocking HIV-1 infection via CCR5 and CXCR4 receptors by acting in *trans* on the CCR2 chemokine receptor. EMBO J 23:66–76

Rollins B (1997) Chemokines. Blood 90:909–928

Rossi D, Zlotnik A (2000) The biology of chemokines and their receptors. Annu Rev Immunol 18:217–242

Sallusto F, Lanzavecchia A, Mackay C (1998) Chemokines and chemokine receptors in T-cell priming and Th1/Th2-mediated responses. Immunol Today 19:568–574

Sanchez-Madrid F, del Pozo MA (1999) Leukocyte polarization in cell migration and immune interactions. EMBO J 18:501–511

Scheiffele P, Roth MG, Simons K (1997) Interaction of influenza virus haemagglutinin with sphingolipid-cholesterol membrane domains via its transmembrane domain. EMBO J 16: 5501–5508

Sekar RB, Periasamy A (2003) Fluorescence resonance energy transfer (FRET) microscopy imaging of live cell protein localizations. J Cell Biol 160:629–633

Serrador JM, Alonso-Lebrero JL, del Pozo MA, Furthmayr H, Schwartz-Albiez R, Calvo J, Lozano F, Sanchez-Madrid F (1997) Moesin interacts with the cytoplasmic region of intercellular adhesion molecule-3 and is redistributed to the uropod of T lymphocytes during cell polarization. J Cell Biol 138:1409–1423

Showkat AM, Sayeski PP, Bernstein KE (2000) Jak2 acts both as STAT1 kinase and as a molecular bridge linking STAT1 to the angiotensin II AT1 receptor. J Biol Chem 275:15586–15593

Soede RDM, Zeelenberg IS, Wijnands YM, Kamp M, Roos E (2001) Stromal cell-derived factor-1-induced LFA-1 activation during in vivo migration of T cell hybridoma cells requires Gq/11, RhoA, and myosin, as well as Gi and Cdc42. J Immunol 166:4293–4301

Sorensen TL, Tani M, Jensen J, Pierce V, Lucchinetti C, Folcik VA, Qin S, Rottman J, Sellebjerg F, Strieter RM, Frederiksen JL, Ransohoff RM (1999) Expression of specific chemokines and chemokine receptors in the central nervous system of multiple sclerosis patients. J Clin Invest 103:807–815

Soriano SF, Serrano A, Hernanz-Falcon P, Martin de Ana A, Monterrubio M, Martinez-A C, Rodriguez-Frade JM, Mellado M (2003) Chemokines integrate JAK/STAT and G protein pathways during chemotaxis and calcium flux responses. Eur J Immunol 33:1328–1333

Sorkin A, McClure M, Huang F, Carter R, (2000) Interaction of EGF receptor and grb2 in living cells visualized by fluorescence resonance energy transfer (FRET) microscopy. Curr Biol 10:1395–1398

Stein JV, Soriano SF, M'rini C, Nombela-Arrieta C, de Buitrago GG, Rodriguez-Frade JM, Mellado M, Girard JP, Martinez-A C (2003) CCR7-mediated physiological lymphocyte homing involves activation of a tyrosine kinase pathway. Blood 101:38–44

Suzuki S, Chuang LF, Yau P, Doi RH, Chuang RY (2002) Interactions of opioid and chemokine receptors: oligomerization of mu, kappa, and delta with CCR5 on immune cells. Exp Cell Res 280:192–200

Szabo I, Chen XH, Xin L, Adler MW, Howard OM, Oppenheim JJ, Rogers TJ (2002) Heterologous desensitization of opioid receptors by chemokines inhibits chemotaxis and enhances the perception of pain. Proc Natl Acad Sci USA 99:10276–10281

Terrillon S, Bouvier M (2004) Roles of G-protein-coupled receptor dimerization. EMBO Rep 5:30–34

Trettel F, Di Bartolomeo S, Lauro C, Catalano M, Ciotti MT, Limatola C (2003) Ligand-independent CXCR2 dimerization. J Biol Chem 278:40980–40988

Ulbrich H, Eriksson EE, Lindbom L (2003) Leukocyte and endothelial cell adhesion molecules as targets for therapeutic interventions in inflammatory disease, Trends Pharmacol Sci 24:640–647

Vila-Coro AJ, Rodriguez-Frade JM, Martin de Ana A, Moreno-Ortiz MC, Martinez-A C, Mellado M (1999) The chemokine SDF-1a triggers CXCR4 receptor dimerization and activates the JAK/STAT pathway. FASEB J 13:1699–1710

Vila-Coro AJ, Mellado M, Martin de Ana A, Lucas P, del Real G, Martinez-A C, Rodriguez-Frade JM (2000) HIV-1 infection through the CCR5 receptor is blocked by receptor dimerization. Proc Natl Acad Sci USA 97:3388–3393

Ward SG (2004) Do phosphoinositide 3-kinases direct lymphocyte navigation? Trends Immunol 25:67–74

Wright N, de Lera TL, Garcia-Moruja C, Lillo R, Garcia-Sanchez F, Caruz A, Teixido J (2003) Transforming growth factor-beta1 down-regulates expression of chemokine stromal cell-derived factor-1: functional consequences in cell migration and adhesion. Blood 102:1978–1984

Wulfing C, Davis MM (1998) A receptor/cytoskeletal movement triggered by costimulation during T cell activation. Science 282:2266–2269

Yang W, Wang D, Richmond A (1999) Role of clathrin-mediated endocytosis in CXCR2 sequestration, resensitization, and signal transduction. J Biol Chem 274:11328–11333

Yang YF, Mukai T, Gao P, Yamaguchi N, Ono S, Iwaki H, Obika S, Imanishi T, Tsujimura T, Hamaoka T, Fujiwara H (2002) A non-peptide CCR5 antagonist inhibits collagen-induced arthritis by modulating T cell migration without affecting anti-collagen T cell responses. Eur J Immunol 32:2124–2132

Zernecke A, Weber KSC, Erwig LP, Kluth DC, Schröppel B, Rees AJ, Weber C (2001) Combinatorial model of chemokine involvement in glomerular monocyte recruitment: role of CXC chemokine receptor 2 in infiltration during nephrotoxic nephritis. J Immunol 166:5755–5762

Zou Y, Kottmann A, Kuroda M, Taniuchi I, Littman D (1998) Function of the chemokine receptor CXCR4 in haematopoiesis and in cerebellar development. Nature 393:595–599

Signaling and Reverse Signaling in the Tumor Necrosis Factor/TNF Receptor System

Ernő Duda, Tünde Nagy, Daniela N. Männel, Günther Eissner

7.1 The TNF/TNFR System

An anti-tumor activity elicited by severe bacterial infections was first noted more than 100 years ago. When modern scientists revisited the observation, they discovered two closely related cytokines, tumor necrosis factor (TNF) and lymphotoxin (LT), that caused hemorrhagic necrosis of tumor tissues and lysis of certain types of cells, including certain tumor cells (Granger et al. 1969; Carswell et al. 1975).

These factors, produced by macrophages and lymphocytes, turned out to be multifunctional proteins that coordinate innate and adaptive immune responses, regulate differentiation and activation of different types of immune cells and control developmental processes, such as the organogenesis of peripheral lymphoid tissues, morphogenesis of exocrine glands, hair and teeth, and bone metabolism. Prolonged or inappropriate production of these factors or abnormal activation of their signaling pathways has been implicated in the pathogenesis of a wide spectrum of human diseases, including allograft rejection, autoimmune diseases (such as multiple sclerosis, rheumatoid arthritis and inflammatory bowel diseases), septic shock, cachexia and wasting, bone resorption and osteoporosis, atherosclerosis, cerebral malaria, diabetes and cancer metastasis (for reviews see Locksley et al. 2001; Khosla 2003; Pfeffer 2003; Tumanov et al. 2003).

Intensive research on the regulation of immune functions and recent genome projects have identified close to two dozen TNF homologues, establishing the TNF superfamily of proteins (Fig. 7.1). Corresponding receptors constitute the TNFR superfamily with even larger numbers of proteins (Locksley et al. 2001; Pfeffer 2003). The receptors of the TNF family have unique structural attributes that couple them directly to sophisticated signaling pathways for apoptosis, survival, proliferation, differentiation and activation (Fig. 7.2). TNFR superfamily members, just like TNF receptors themselves, can be divided into three classes: death domain (DD)-containing receptors, signaling receptors without death domain and decoy receptors. Decoys are soluble or membrane-anchored molecules, which bind their ligands with high specificity and affinity but elicit no intracellular signals (Locksley et al. 2001; Pfeffer 2003).

There are eight death domain receptors: the smaller (p55 or p60) receptor of TNF (TNFR1), Fas, death receptor 3 (DR3), DR4/TRAILR1, DR5/TRAILR2, DR6, nerve growth factor receptor (NGFR) and XEDAR. Better characterized members of the expanding family of TNF receptors *without death domain* include: 4-1BB,

Springer Series in Biophysics
S. Damjanovich (Ed.)
Biophysical Aspects of Transmembrane Signaling

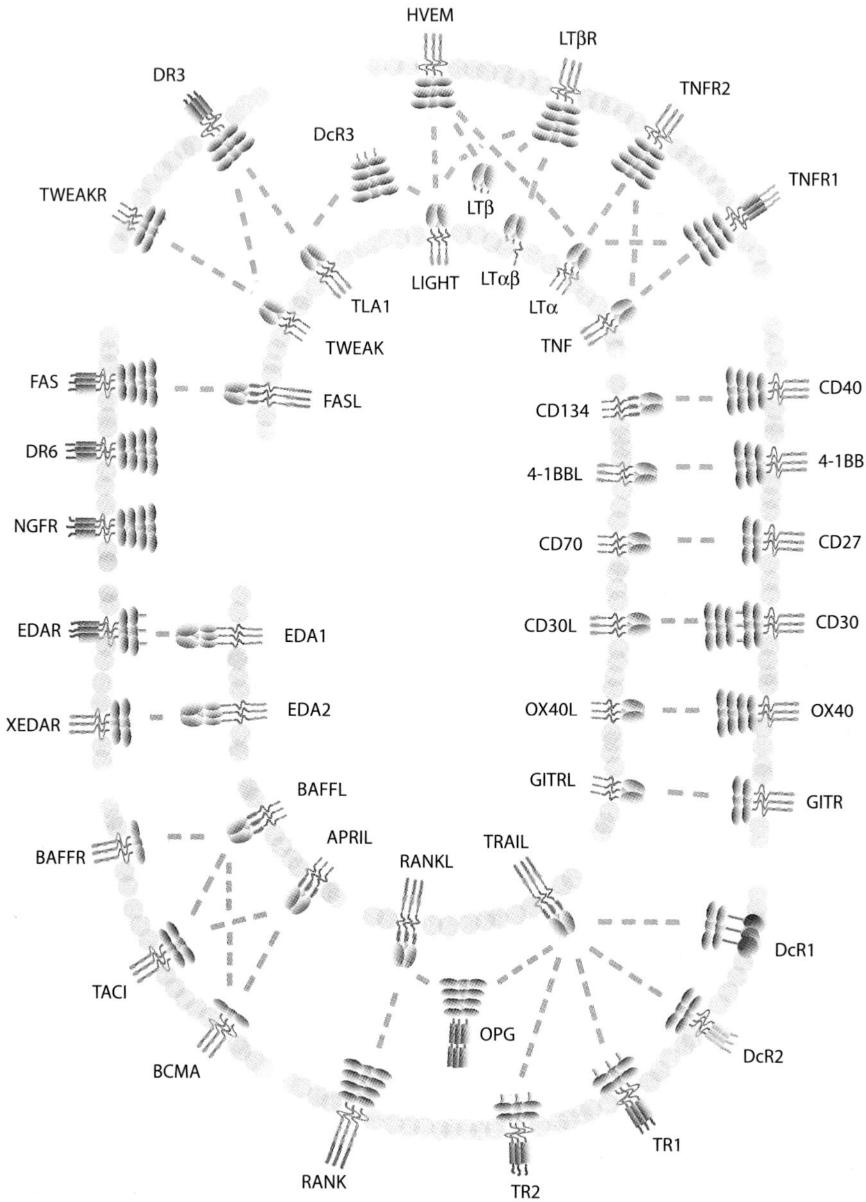

Fig. 7.1. The TNF/TNFR superfamilies. Members of the TNF superfamily of proteins are pictured in the *inner circle*, while receptor molecules are located in the *external circle*. Soluble decoy receptors take place in between. *Dotted lines* represent interactions between the molecules. The numerous soluble forms of transmembrane receptors and ligands are not shown

AITR/GITR, BAFFR, BCMA, CD27, CD30, CD40, EDAR, HVEM, LTβR/TNFR3, OX40, RANK, TACI, TNFR2, TWEAKR and XEDAR. The decoy receptors are DcR1 and DcR2, which bind only TRAIL, DcR3, interacting with LIGHT and TLA1, the soluble form of CD40, osteoprotegerin (OPG), which interferes with the activity of RANK and TRAIL receptors, and finally the soluble forms of the two TNF receptors. Both DcR1 and -2 are membrane-associated proteins, and DcR2 is the only decoy receptor that possesses a large, inactive intracellular domain.

Detailed information on TNF/TNFR superfamily members can be found online at http://www. gene.ucl.ac.uk/users/hester/tnfinfo.htm and www.copewithcytokines.de. The TNF Pathway Connections Map at Science's Signal Transduction Knowledge Environment presents current knowledge of the pathway's components and connections between them (http://stke.sciencemag.org/cgi/cm/cmp_7107).

7.1.1 Evolution of the TNF/TNFR System

The first invertebrate tumor necrosis factor (TNF) superfamily ligand, Eiger, that can induce cell death was described in *Drosophila* (Igaki et al. 2002). Eiger is a type II transmembrane protein with a C-terminal TNF homology domain. It is predominantly expressed in the nervous system. Overexpression of Eiger is sufficient to induce apoptosis. Eiger acts on its receptor Wengen (Kanda et al. 2002); however, it induces cell death indirectly by triggering JNK signaling, which, in turn, causes activation of the apoptosome. A direct mode of action via the apical FADD (*Fas-associated death domain*)/caspase-8 pathway may have been coopted by some TNF signaling systems only at subsequent stages of evolution. (Moreno et al. 2002).

Elements of the vertebrate signaling pathways used by members of the TNFR family predate the emergence of TNF/TNFR systems. In insects, not only Toll receptor signals the presence of infectious non-self. The intracellular components of the *Drosophila* and mosquito Imd pathways are fairly similar to the molecules utilized by the death receptors of the mammalian TNF family (Hoffmann and Reichhart 2002). The receptor is still not known, but Imd is a death domain protein with significant sequence homology with the TNFR *receptor interacting protein* (RIP). In *Drosophila* downstream of Imd we find dFADD, a homologue of FADD, and dTAK, a protein kinase, which is the most likely activator of the DmIKKβ–DmIKKγ complex, an equivalent of the mammalian signalosome. The *Drosophila* caspase DREDD together with dFADD forms a complex that seems to process Relish, liberating the DNA-binding N-terminal part clipping off the inhibitory domain (Hu and Yang 2000; Khush et al. 2001; Leulier et al. 2002). The pathway is almost identical in mosquito (Christophides et al. 2002).

The death domain found in many of the above proteins is related to the ankyrin domain, an oligomerization domain found – among others places – in many signaling proteins. The death domain is always encoded by a single exon, a property that has been conserved from *Drosophila* to mammals.

Most proteins of the TNF and TNFR superfamilies regulate the development, proliferation, activation or death of B and T lymphocytes. As the adaptive immune system became more and more sophisticated during evolution, more and more regulatory mechanisms were needed. The number of genes of these superfamilies multiplied parallel with the development of the adaptive immune systems (Locksley et al. 2001). Gene duplication and multiplication of whole gene clusters participated in this process. Many of the TNF and TNFR genes are linked in discrete loci, reflecting their evolutionary derivation. The TNF homologue found in fish seems to be a common ancestor of TNF and LT (Hirono et al. 2000; Bobe and Goetz 2001; Laing et al. 2001). In humans and other mammals, TNF and the two LT genes (LTα and LTβ) are neighbors. The conserved chromosomal location, transcriptional orientation and high amino acid homology prove their common origin, just as the shared receptor usage. Duplication of this chromosome segment might have given rise to another set of three ligand genes (4–1BB, CD27L and LIGHT). All these genes are still located on the same chromosome in mice (Ch 17). Translocations to other chromosomes multiplied the number of ligand genes. In human, the four clusters are located on separate chromosomes – OX-40L, Fas and GITP on Ch 1, LTα, LTβ and TNF on Ch 6, 4–1BB, CD27L and LIGHT on Ch 19, and finally CD40L, EDA1 and 2 on Ch X – while the remaining genes are scattered over the genome as singles or pairs (Pfeffer 2003).

A similar evolution of the receptor genes must have taken place. TNF receptors found on human chromosome 12 (TNFR1, LTR and CD27) might have duplicated and translocated to Ch 1 (TNFR2, HVEM, OX40, CD30, AITR, 4–1BB and DR3). Another cluster of five receptor genes (receptors and decoy receptors of TRAIL) are located on Ch 8. Interestingly, homologues of these five genes are scattered over three different chromosomes in mice (Pfeffer 2003).

The odd member of the TNFR superfamily is the NGFR, which does not interact with TNF-like ligands. It binds neurotropin ligands and interacts with the Trk family of receptor tyrosine kinases. It is the only death domain receptor that forms dimers, not trimers, and even the exon–intron structure of its gene is unique, suggesting that its evolution followed a different path from very early times onwards (Rabizadeh and Bredesen 2003)

One of the unusual features of the TNF superfamily is that most ligands are transmembrane proteins. They exert juxtacrine effects, typically observed in development and morphogenesis. Frequently, transmembrane and soluble forms have distinct biological roles. A number of receptors have soluble forms to modulate the effect of their ligand(s). An increasing number of publications suggest that the interaction with certain members of the "receptor family" induces intra-

Fig. 7.2. Signaling pathways of TNF receptor prototypes. Fas and TRAIL receptors (*left*) usually elicit death signals, activation of the caspase cascade and release of cytochrome *c* from mitochondria. Death-inducing signaling complexes (DISCs) are formed around the tails of these receptors and around the cytochrome *c* molecule. TNFR1- and DR3-type receptors (*middle*) are able to trigger both death signals and survival signals. TNFR2- and CD40-type receptors (*right*) generally induce proliferation and differentiation through activation of MAPK and NF-κB signaling pathways

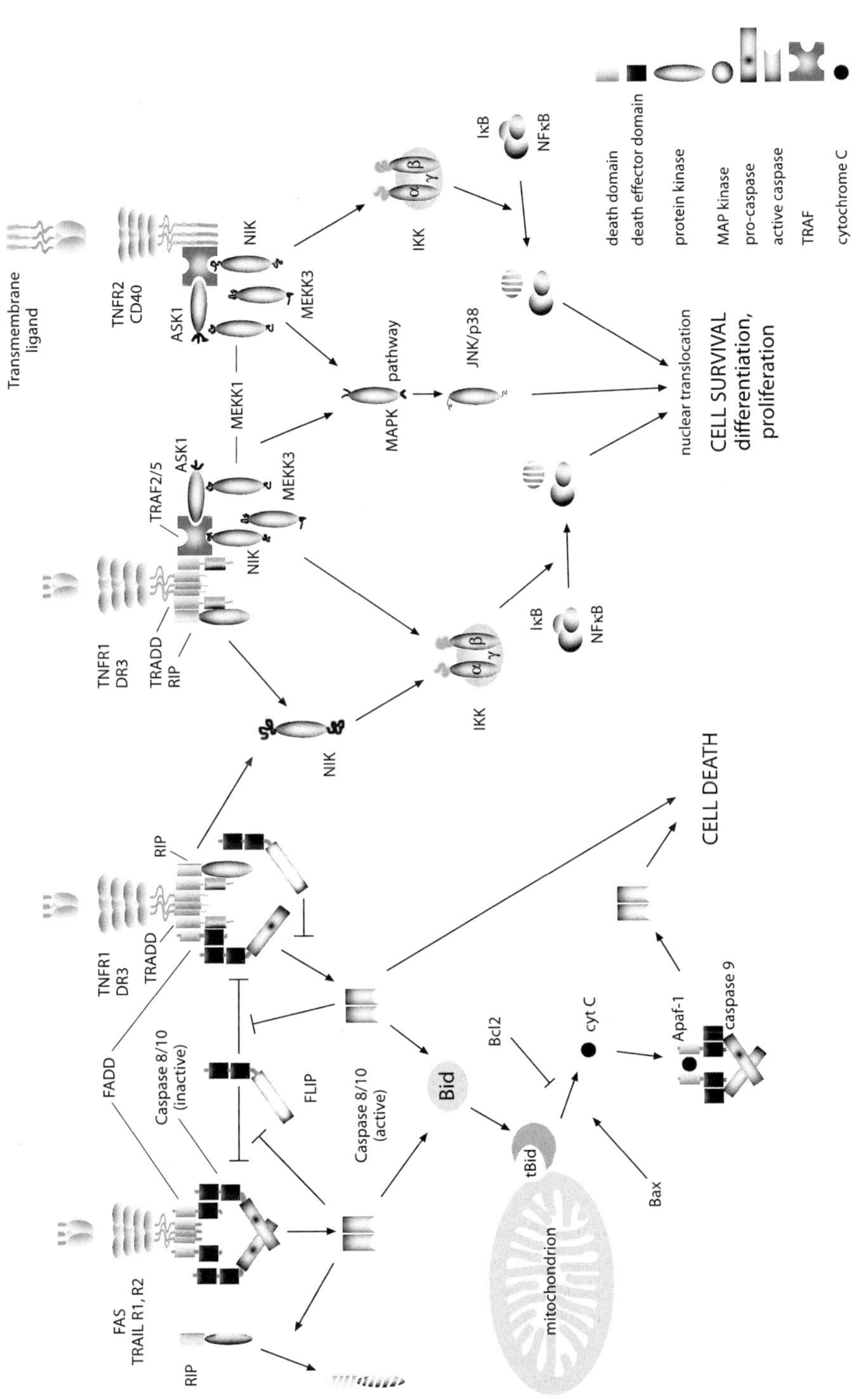

Transmembrane ligand
TNFR2
CD40
ASK1
NIK
MEKK3
MEKK1
IKK
IκB
NFκB
death domain
death effector domain
protein kinase
MAP kinase
pro-caspase
active caspase
TRAF
cytochrome C
TNFR1
DR3
TRADD
RIP
TRAF2/5
MAPK
pathway
JNK/p38
nuclear translocation
CELL SURVIVAL
differentiation,
proliferation
FAS
TRAIL R1, R2
FADD
Caspase 8/10
(inactive)
FLIP
Caspase 8/10
(active)
Bid
tBid
mitochondrion
Bcl2
cyt C
Bax
Apaf-1
caspase 9
CELL DEATH

cellular signaling in cells exhibiting the corresponding ligand. The phenomenon, called "reverse signaling", might be a further tool to modulate functions of the immune system (Ferran et al. 1994).

Considering the enormous importance of TNF/TNFR proteins in the coordination of the immune response and in the proliferation and death of immune cells, it is not surprising that pathogenic organisms – from viruses to multicellular parasites – carry genes that correspond to (parts of) cellular genes of TNFR family members. These sequences were obviously captured from the host genome and code for decoy receptors, constitutive intracellular signaling domains or molecules participating in the signaling pathway (Locksley et al. 2001; Pfeffer 2003). The function of these truncated genes is either attenuation of the inflammatory immune response (poxvirus, *Schistosoma*), prevention of apoptosis (vIAP, baculoviruses) or immortalization of the infected cells (LMP-1 of EBV herpesvirus).

7.1.2 General Structure

7.1.2.1 Receptors

TNFR family proteins are trimers of type I transmembrane subunits that form elongated structures stabilized by three intrachain disulfide bridges between six highly conserved cysteines. The disulfide bonds form "cysteine-rich domains" (CRDs), characteristic 40 amino acid pseudorepeats with a core motif of CXXCXXC (Harless Smith and Cancro 2003). Although four CRDs can be regarded as typical among family members, significant variations occur from BAFFR with only a partial CDR to six in CD30 (Hsu and Chao 1993; Gordon et al. 2003). Although the low affinity nerve growth factor receptor (NGFR) has the classic CRD structure, it binds proteins structurally and functionally unrelated to TNF family members (Gruss and Dower 1995).

The elongated receptor chains fit in the "grooves" formed between the subunits of the trimeric ligands. The crystal structure of LT-alpha in complex with the TNFR1 extracellular domain reveals no direct contact between the individual chains of the receptor (Banner et al. 1993). Comparing the structure of the LT/TNFR1 complex with that of DR5 receptor engaged with its ligand, TRAIL (Hymowitz et al. 1999), a striking conservation of structural details can be seen, though the protein sequences exhibit a rather low degree of sequence homology.

Homotrimeric structures are not frequent among proteins and unusual among receptors. However, trimers require more contacts than dimers and may cause an exponential increase in avidity and specificity. In the case of TNF receptors, both ligand binding and signal complex formation involve stoichiometrically defined protein complexes with three-fold symmetry. TNFR family proteins bind their ligands with high affinity (K_d 5×10^{-9}–10^{-10} M) (Vandenabeele et al. 1995).

The characteristic structure in the intracellular tail of one type of the receptors is the death domain (Tartaglia et al. 1993). The death domain is a globular bundle of six α-helices (60–65 amino acids). The death domains found in the receptors and adaptor molecules share little sequence homology (only 10–20%), but the structure is highly conserved. Very similar domains, called *death effector domains* (DEDs) and *caspase recruitment domains* (CARDs), share the same overall six α-helical structure (Martin 2001). The other type of receptors contain a *TRAF (TNFR-associated factor) interacting motif* (TIM). This structural element is short, with less than a dozen contact residues in the receptor tail. During signaling it is clutched by a pocket in the globular head group of the TRAF molecules through charged residues (Fesik 2000).

Ligand-induced clustering of receptor subunits is not typical for TNF receptors; at least several receptors in the TNFR family self-assemble in the absence of ligand, and signaling involves rearrangement or conformational changes of the preassembled chains (Chan et al. 2000; Siegel et al. 2000).

7.1.2.2 Ligands

The characteristic structural element of the TNF-related ligands is the extracellular TNF homology domain (THD) and the trimeric configuration (Gruss and Dower 1995). The THD folds into an anti-parallel "jellyroll" beta-sandwich. The individual subunits interact at hydrophobic interfaces and self-assemble into inverted bell-shaped, non-covalent trimers (Fesik 2000).

Sequence homology of different ligands is largely confined to internal aromatic residues responsible for oligomerization. The external surfaces of the molecules exhibit little sequence conservation, which accounts for receptor selectivity (Gruss and Dower 1995).

Not members of the TNF superfamily, but structurally related proteins with TNF homology domains and trimeric configuration are complement protein C1q (its globular domain) and a number of C1q-like proteins (Kishore and Reid 2000). TNF and other family members tend to form multimeric structures. This and the fact that a large group of structurally related proteins that form highly organized oligomers are viral capsid proteins indicate that ancestors of TNF receptors might have served as virus-detecting pattern recognition molecules (Locksley et al. 2001).

7.2 TNFR1 and the Apoptotic Signal

A prototype of the TNFR superfamily of proteins is the smaller (55–60 kDa) TNFR1. Although it contains an intracellular death domain, it can also signal survival and proliferation, depending on the balance of several signaling pathways activated by its interaction with its ligands (TNF and LT) (Bhardwaj and Aggarwal 2003). Experimental evidence shows that – unlike subunits of growth factor receptors – subunits of the trimeric TNFR are clustered in the absence of ligands. The *pre-ligand assembly domain* (PLAD) structure is responsible for the aggregation and seems to be essential for ligand binding (Chan et al. 2000). However, this clustering of the subunits does not lead to activation; actually this structure of the trimeric receptor seems to prevent accidental oligomerization of the intracellular death domains. Clustering of death domains – which can be facilitated by over-production of the receptor subunits or expressing fusion proteins of death domain and proteins of trimeric structure – activates the caspase cascade (Vandevoorde et al. 1997). Deletion of the death domain results in a non-functional receptor, indicating that the death domain is required for all the different biological functions of the receptor (Devergne et al. 1996).

Since simple overexpression of death receptors leads to apoptosis, there must clearly be further mechanisms to prevent ligand-independent activation of death receptors. By the yeast two hybrid method, a protein was found that constitutively associates with TNFR1 (and DR3). This protein is a *silencer of death domain* (SODD), which inhibits receptor association of adaptor molecules of the apoptotic pathway [FADD and TRADD (*TNFR-associated death domain*)], competing for the binding site on the death domain of TNFR1 (and DR3) (Jiang et al. 1999).

In nature, ligand binding-dependent conformational changes are responsible for activation. Interaction with specific ligands can lead to the recruitment of intracellular death domain-containing adaptors such as FADD and TRADD. TNFR1 (and DR3) can recruit both FADD and TRADD, while death domains of Fas, TRAIL-R1 and TRAIL-R2 are restricted to interacting with only FADD (Chinnaiyan et al. 1996; Kischkel et al. 2000).

FADD is a 26-kDa protein with a C-terminal death domain and a homologous death effector domain (DED). Within minutes of ligand binding, the conformational change of the receptor triggers the assembly of a death-inducing signaling complex (DISC), containing FADD and caspase-8 (Kischkel et al. 1995; Boldin et al. 1996). Caspase-8 has two DEDs that interact with the homologous domain of FADD. Caspase-8 is a proenzyme, a zymogen, which can be activated by crowding in the death-inducing complex. A two-step activation removes the prodomain and the smaller, 10-kDa subunit, liberating the active protease of 20 kDa (Medema et al. 1997). The activated caspase-8 can process its precursor and downstream procaspases. Caspase activation is an amplification process, which proceeds in a hierarchical manner, resulting in the serial proteolytic activation of effector caspase-3, caspase-6 and caspase-7. Activation of executor caspases leads to apoptosis. Experimentally induced aggregation of caspase-8 elicits proteolytic activation of the molecule and results in cell death (the induced proximity model of caspase activation) (Yang et al. 1998). In type I apoptosis, robust DISC association and processing of caspase-8 activates the effector caspases directly, committing the cells to death with characteristic fragmentation of DNA.

Activated caspase-8 cleaves not only downstream caspases, but also Bid, releasing a potent pro-apoptotic fragment, tBid. tBid is targeted to the outer mitochondrial membrane (Li et al. 1998), inducing type II apoptosis. Type II apoptosis relies on organelle dysfunction, on the failure of anti-apoptotic activity of Bcl-2 family proteins. In living cells, anti-apoptotic family members Bcl-2 or Bcl-XL, containing a *Bcl-2 homology domain-4* (BH-4), balance the pro-apoptotic members of the same family, such as Bax or Bak [which contain *Bcl-2 homology domains-1 to -3* (BH-1 to -3) but not BH-4] (Hsu et al. 1999).

Caspase-generated tBid induces the oligomerization of Bak, which forms a pore complex in the membrane, leading to dissipation of the mitochondrial transmembrane potential and allowing the release of cytochrome *c* (Wei et al. 2000). In the cytoplasm, cytochrome *c* drives oligomerization of the *apoptotic protease activating factor-1* (Apaf-1) (Zou et al. 1997), which contains a *caspase activation and recruitment domain* (CARD). In a reaction resembling the formation of DISC, cytochrome *c* and Apaf-1 interact with the CARD containing caspase-9, forming a functional apoptosome (Shiozaki et al. 2002). The assembly requires the presence of ATP. The apoptosome can cleave and activate procaspase-3, thereby completing the same cycle of activation as the initiator caspase-8. The balance between caspase-8-mediated cleavage of Bid and inhibition of tBid by the presence of Bcl-2 and other anti-apoptotic members of the Bcl-2 family regulates cytochrome *c* release and cell fate in type II apoptosis.

There is competition between anti-apoptotic and pro-apoptotic pathways and several mechanisms have evolved to inhibit different steps leading to apoptosis. TNFR1-induced anti-apoptotic signaling through NF-kappaB (NF-κB) activation upregulates the synthesis of proteins (Bcl-x and Bcl-2) that can sequester Bak and prevent pore formation of mitochondrial membranes (Gedrich et al. 1996; Galibert et al. 1998).

A gene duplication of the initiator caspase-8 (previously also called FLICE) gene could have generated the gene of caspase-8 inhibitor, *FLICE-inhibitory protein* (FLIP). Like caspase-8, FLIP has two DED and a caspase-like domain, which cannot be converted into an active protease. FLIP forms complexes with FADD and caspase-8, inhibiting auto-activation of caspase-8. FLIP inhibits the apoptotic signaling pathways of all known human death receptors (Irmler et al. 1997). As expected, cells from FLIP knockout (KO) animals exhibit very high sensitivity to stimulation of death receptors.

Vertebrate cells are programmed to perform apoptosis in case of viral infection. Obviously, viruses use different strategies to block this process. *Inhibitor of apoptosis protein* (IAP) was described as a viral protein, which specifically inactivated the effector caspases, but soon after, a family of related cellular proteins – c-IAP-1, c-IAP-2 and X-linked IAP (XIAP) – was discovered (Uren et al. 1996). These proteins contain three *baculovirus IAP repeats* (BIR) and an N-terminal ring finger motif. XIAP, the best known member of the group, inhibits three different caspases, i. e. caspase3 and caspase 7 which act in concert as well as caspase 9, through binding to their intermediate and fully cleaved forms. Its ring finger domain exhibits a ubiquitin–protein ligase activity, which facilitates the ubiquitination and proteosomal degradation of caspase-3, providing additional protection against apoptosis (Suzuki et al. 2001). However, if the pro-apoptotic signal is strong, Smac/DIABLO is released from the mitochondria of apoptotic cells and accelerates cell death activation by displacing XIAP from the caspases (Takeuchi et al. 1996; Wang et al. 2001).

In *Drosophila*, caspases are produced in the active form and need no proteolytic processing. They are counterinhibited by IAPs, and relief of IAP-mediated caspase inhibition triggers cell death.

7.3 TNF Receptors and the Survival Signal

Activation of Fas almost always leads to cell death. On the other hand, TNFR1 only signals for cell death in certain types of cells or in special circumstances (inhibition of protein synthesis) and TNFR2 does not activate apoptotic pathways. TNFR1 ligation induces several pathways that can synergize or antagonize each other. In most cases TNFR1 signaling provides a mechanism to suppress the apoptotic stimulus and induce the synthesis of pro-inflammatory cytokines. Fas directly recruits FADD, but TNFR1 frequently associates with TRADD. TRADD can in turn associate with the death domain of FADD, thereby initiating the caspase-8 activation pathway.

Alternatively, TRADD can recruit TRAF2, TRAF1 and *receptor-interacting protein* (RIP) to activate the NF-κB and JNK pathways, protecting cells from apoptosis and initiating inflammatory responses (Stanger et al. 1995; Hsu et al. 1996). RIP is the primary mediator of NF-κB activation: animals with a targeted deletion for RIP are unable to activate NF-κB in response to TNFR1 stimulation, inevitably leading to TNF-induced apoptosis (Kelliher et al. 1998). In contrast, TRAF2 deficiency prevents activation of the MAP kinase JNK in response to TNF stimulation (Yeh et al. 1997), also resulting in an apoptotic response to TNF stimulation.

Ligand-induced activation of TNFR1 leads to the recruitment of a signaling complex that can activate both the apoptotic and the anti-apoptotic pathways. The balance is regulated at several levels, including strength of signal, expression of signaling molecules and protective enzymes and cross talk with other signals. Experiments on TNFR-mediated signaling suggest that strength of signal and complex stability are controlled by receptor–ligand interactions. Soluble ligands can transiently activate a few receptors. Interaction between receptor-producing cells and cells with thousands of surface-bound ligand molecules obviously produces longer lasting and stronger signals (Krippner-Heidenreich et al. 2002).

The balance between death and survival pathways is influenced by other factors. RIP protein inhibits the interaction between TRADD and FADD. However, caspase-8-dependent cleavage of RIP prevents the RIP-mediated activation of NF-κB (Lin et al. 1999). The death domain-containing cleavage product RIPc serves as a dominant negative inhibitor of NF-κB activation. In addition, binding of RIPc to TRADD enhances the interaction between TRADD and FADD. RIP cleavage is also seen in TRAIL- and Fas-stimulated cells, implying that RIP is a general molecular switch between apoptosis and activation in death receptor signaling.

In a similar way, caspase-3 can specifically cleave the *I-κB-specific kinase* (IKK) during TNF-induced apoptosis. This proteolytic step prevents degradation of the inhibitor, tilting the balance towards caspase-mediated apoptosis (Tang et al. 2001).

7.3.1 TNF Receptor-Associated Factors (TRAF1–7)

Interacting partners of TRADD include survival factors RIP and TRAFs (Stanger et al. 1995; Hsu et al. 1996). Those members of the TNF receptor superfamily that do not contain intracellular death domains possess a short TIM sequence within their cytoplasmic tail that serves as a direct recognition site for TRAF proteins (Rothe et al. 1994; Chen et al. 1995; Pullen et al. 1998).

As adapter proteins, TRAFs elaborate receptor signal transduction by serving as both a convergent and a divergent platform. Different TRAFs evolved with individual biological roles. Without any detectable enzyme activity these proteins are able to mediate the activation of protein kinases, leading to signaling through several pathways. Their characteristic feature, the large TRAF domain (200 amino acids), is highly conserved; proteins with similar domains were found in *Dictyostelium*, plants, worms and insects (*Drosophila* has three TRAF homologues, dTRAF1, 2 and 3). TRAF proteins are also known to be associated with IL-1 and Toll-like receptors. The TRAF domain, which is located at the C terminus of mammalian TRAF proteins, is responsible for receptor binding. Its distal TRAF-C part is a coiled-coil domain, able to interact with downstream signaling molecules, such as NIK, while the proximal TRAF-N domain associates with anti-apoptotic molecules (cIAP1 and 2).

TRAFs are homo- and heterotrimers, different combinations serving individual receptors. TRAF1 – unlike other TRAFs – lacks a ring finger domain, which seems to be essential for NF-κB activation, but TRAF1 can form activating trimers with other TRAFs. The intact TRAF domain is required for trimerization, and TRAF domains spontaneously form trimers in solution.

TRAFs also have a number (5–7) of zinc finger domains, which are essential for downstream signaling through JNK and NF-κB. The significance of the isoleucine zipper domain found in some TRAFs is not known.

7.3.2 Signaling Pathways Activated by TRAFs

TRAF2 is the prototypical member of the TRAF family. A close relative of TRAF3 and 5, TRAF2 plays a cytoprotective role, through activation of the JNK/SAPK pathway. Deficiency of the factor caused severe runting and premature death of the KO mice. Lack of TRAF2 or the expression of its dominant-negative form resulted in a severe reduction in JNK/SAPK activation (Lee et al. 1997; Yeh et al. 1997).

TRAF2 can associate with several upstream MAP kinases to induce NF-κB and AP-1 activation. These include NIK (Malinin et al. 1997; Song et al. 1997), MEKK1 and MEKK3 (Baud et al. 1999) for IKK activation, and ASK1, MEKK1 and MEKK3 for initiating MAP kinase pathways and AP-1 activation (Nishitoh et al. 1998, Baud et al. 1999, Hoeflich et al. 1999, Shi et al. 1999).

TRAF1 is a negative regulator of signaling by certain receptors of the TNF family; it has the most restricted expression among TRAFs, and is found almost exclusively in activated lymphocytes and dendritic cells. TRAF1 protects cells from apoptosis and cooperates with TRAF2 and cIAPs to fully suppress TNF-induced apoptosis. TRAF1 is induced by NF-κB and plays a role in the feedback regulation of receptor signaling (Speiser et al. 1997; Wang et al. 1998; Carpentier and Beyaert 1999; Schwenzer et al. 1999; Nolan et al. 2000).

Interestingly, clustering of the receptor subunits – at least in the case of CD40 – does not lead to direct interaction of their intracellular domains. Crystallographic analyses suggest that individual tails of the receptor subunits may contact TRAF trimers, each multiplying the numbers of molecules participating in the process and amplifying the signal (McWhirter et al. 1999; Park et al. 1999).

TRAF4 might interact with some of the TNFR family members (Krajewska et al. 1998; Ye et al. 1999); its biological importance stems from its morphogenetic role: TRAF4-deficient mice (Shiels et al. 2000) exhibited severe tracheal malformation, suggesting a parallel function with the *Drosophila* Toll pathway.

TRAF5 is considered to be a close functional and structural homologue of TRAF2, and overexpression of TRAF5 can also activate NF-κB and AP-1 transcription factors (Ishida et al. 1996; Nakano et al. 1996). However, deletion of TRAF5 did not cause perinatal lethality, perhaps owing to the more restricted expression pattern of TRAF5 compared with TRAF2 (Ishida et al. 1996; Nakano et al. 1996).

TRAF6 has a unique position among TRAFs, as the signaling mediator for both the TNF receptor superfamily and the IL-1R/TLR superfamily, a central player in innate immunity against pathogens. TRAF6 also has a crucial role in TRANCE-R-mediated osteoclast activation and participates in the adaptive immune response as the signaling partner of CD40 in B-cells (Lomaga et al. 1999; Naito et al. 1999; Wong et al. 1999). TRAF7 is the most recently identified member of the family, found to interact with MEKK3, a MAP3 K needed for TNF-induced activation of NF-κB. TRAF7 activated JNK and p38 but had little effect on IKK activation (Bouwmeester et al. 2004).

Since TRAFs can form homo- and heterodimers of different compositions, it is conceivable that different forms regulate the specificity of signaling. The expression of TRAF genes depends on the cell's lineage and degree of differentiation, and consequently the activation of the same receptor could lead to diverse results in different cells.

In certain cases, TNFR family members seem to compete for TRAFs: CD30 activation induces the formation of receptor–TRAF complexes in the cells, decreasing the level of available TRAF2 molecules. As a consequence, these cells exhibit an increased sensitivity to TNF, since TRAF2 depletion results in the association of less anti-apoptotic cIAP molecules to TNFR trimers, shifting the balance toward cell death. On receptor activation, TRAFs tend to redistribute into membrane microdomains (see later) which leads to a more sustained signaling of the activated receptor. The simultaneous depletion of TRAFs in the cytoplasm leads to the downregulation of subsequent TRAF-dependent signals elicited by other receptors (Arch et al. 2000).

Interestingly, several TRAFs have been shown to interact with proteins of the cytoskeleton, such as the p62 nucleoporin, a component of the nuclear pore central plug (Gamper et al. 2000), the *microtubule-binding protein* MIP-T3 (Ling and

Goeddel 2000) and filamin (Leonardi et al. 2000). Activation of TNFR2 triggers the binding of TRAF2 to caveolin-1, a protein characteristic of caveoli and associated with receptor-mediated endocytosis. The receptor-associated complex is needed for the activation of NF-κB and the anti-apoptotic effect of TNFR2 activation (Feng et al. 2001).

7.3.3 The NF-kappaB Pathway

The mammalian Rel family is composed of five structurally related members, RelA/p65, RelB, c-Rel, p52 and p50, which form various homo- and heterodimeric combinations. The dimers bind to κB transcriptional elements and are called *nuclear factor kappaB* (NF-κB). κB sites are present in a large number of genes involved in cell survival, stress response, inflammation and other immune responses; therefore NF-κB activates the transcription of an impressive number of genes.

All five members share a large (300 amino acids) *Rel homology domain* (RHD), which mediates dimerization, nuclear transport and DNA recognition (unexpectedly, RelB homodimers do not bind to κB sites). RelA, RelB and c-Rel also contain a C-terminal domain responsible for transcriptional activation. In most cells, p50 and p52, lacking this domain, cannot transactivate [the thymus seems to be an exception (Ishikawa 1998)], but repress the expression of target genes (Li and Verma 2002). They are produced as precursors of 105 and 100 kDa, respectively.

The function of NF-κB dimers is inhibited by a third partner, one of the *inhibitor of kappaB* (IκB) isoforms. IκBα inhibits the interaction of NF-κB with DNA transcription elements, and with the help of its two nuclear export sequences promotes the export of NF-κB from the nucleus. It covers only one of the two nuclear localization sequences (NLS) of NF-κB dimers, allowing the NF-κB/IκB complex to shuttle between the nucleus and the cytoplasm.

IκBβ has a higher affinity to NF-κB, and by masking both NLSs of NF-κB, it restricts its presence to the cytoplasm. Degradation of IκBβ by the proteasome is needed to allow nuclear translocation of NF-κB. TNFs mediate a rapid activation of *I-kappaB kinases* (IKK) which phosphorylate IκB (leading to ubiquitinylation and degradation) and the p65 subunit of the liberated NF-κB (activating it in several steps). IKK is composed of a heterodimer of the catalytic subunits, IKKα and β and the regulatory γ subunit (also called NEMO). NEMO is catalytically inactive; still, it is essential for activation of the complex. The two other subunits – a catalytic and a leucine-zipper domain – share approx. 50% sequence homology.

Different subunits of NF-κB are activated by two distinct pathways. RelA, RelB and c-Rel are activated by the above, canonical (type 1) mechanism, leading to the degradation of IκB. The β subunit plays a dominant role in the TNFR-induced "canonical" activation of NF-κB (Li Q et al. 1999b; Li ZW et al. 1999; Tanaka et al. 1999). The alternative, type 2 pathway controls the processing of the p100 precursor of NF-κB subunit p52 and is governed by the α subunit.

There were two models of TNFR-induced activation of the IKK complex. One suggested that autophosphorylation was the result of the recruitment of IKK to the signaling complex formed at the cytoplasmic tail of the receptor after TNF-mediated receptor activation. According to the other model, members of the *mitogen-activated protein kinase kinase kinase* (MAP3 K) family were responsible for the activation of IKK. *NF-κB-inducing kinase* (NIK), *NF-κB-activating kinase* (NAK), *receptor-proximal kinase* (RIPK1), *TGF-β-activating kinase* (TAK1) and *MAP kinase/ERK kinase kinases* (MEKK1 and 3) were found to be involved in phosphorylation of IKK.

Recent data indicate that RIP and TRAF2 have roles in recruiting IKK to the receptor complex. RIP directly interacts with the regulatory subunit (NEMO) of IKK, although its enzyme activity is not required for IKK activation. (Devin et al. 2001). The ubiquitin ligase activity of TRAF6 and 7 is important in the activation of TAK1 and MEKK3, respectively. These MAP3 kinases then stimulate IKK activity (Wang et al. 2001). TAK1 and MEKK3 seem to be the key players in the activation of NF-κB – and also in the activation of AP1 (Bouwmeester et al. 2004). Although genetic experiments have argued against the essential role of the other above-mentioned enzymes in the canonical NF-κB pathway, the negative results cannot rule out their participation in an activation process, with redundant kinases compensating for the loss of individual members.

The non-canonical pathway leading to NF-κB activation regulates the processing of p100, modulating the levels (and activity) of the RelB/p52 heterodimer. p100 contains ankyrin repeats, homologous to those present in IκB. Therefore the RelB/p100 complex binds Iκ with low affinity, but remains in the cytoplasm due to the shielding effect of p100 sequences on the NLS of RelB. p100 is processed by a signal-dependent post-translational mechanism, which is dependent on the activity of the NF-κB-inducing kinase, NIK and IKKα. NIK starts the processing of p100 by inducing IKKα to phosphorylate the C-terminal part of p100 (Senftleben et al. 2001). Recombinant IKKα can phosphorylate p100, which induces ubiquitination (Amir et al. 2004). Production of p52 leads to the accumulation of RelB/p52 dimers in the nuclear compartment of the cells.

As targeted deletion of the gene encoding the IKKα subunit proved, IKKα plays a unique role in epidermal differentiation, mammary gland development and lactation (Hu et al. 1999, 2001; Li Q et al. 1999a; Takeda et al. 1999; Cao et al. 2001; Sil et al. 2004). This phenomenon is related to RANK signaling (Cao et al. 2001; Hu et al. 2001). BAFFR, CD40 and LTβR-elicited signals also involve this subunit, and malfunctions of the pathway result in defects in secondary lymphoid organogenesis and germinal center formation (Kaisho et al. 2001; Matsushima et al. 2001).

A further role of the IKKα subunit has been revealed recently: IKKα proved to be a histone H3 kinase, which activated TNF-induced genes binding to the promoter regions of target genes. IKKα interacted with the transactivation domain of *CREB-binding protein* (CBP), and by phosphorylating H3 it enhanced CBP-dependent histone acetylation and remodeling of the chromatin structure (Yamamoto et al. 2003). Nuclear IKKα helped to make promoter and transcriptional factor binding sequences available for DNA binding proteins and synergized with the transcription-activating effect of NF-κB.

7.3.4 The Role of Membrane Microdomains (Rafts) in the Formation of Signaling Complexes

Before we discuss the activation of the MAP kinase pathway, we have to look at the role of membranes in signaling. The chemical composition of the membrane lipids does not allow homogenous distribution of the different molecular species. Spatial separation of lipid mixtures leads to the dynamic formation of membrane (micro)domains. Since the physical state and thickness of these membrane domains are different they can accommodate diverse sets of membrane proteins with different transmembrane domain structures.

Sphingolipid and cholesterol-rich microdomains, also called “rafts”, constitute only a small fraction of the plasma membrane; therefore they can serve as platforms to facilitate the interaction of raft-associated molecules. In receptor-mediated signaling events, relatively large numbers of different proteins should interact with each other, and thus rafts seem to be the ideal location for signaling molecules. The importance of these microdomains in the initiation of signaling pathways has only been recognized during the last decade.

Rafts have a lower density than other cellular membranes and at low temperatures they are resistant to non-ionic detergents. These facts allow relatively easy isolation of the rafts and microdomain-associated proteins. Recruitment of TRAF2 and 3 to detergent-insoluble membrane fractions was first shown to occur during the interaction of CD40 with its ligand in intact B cells (Hostager et al. 2000). Simultaneously with TRAF recruitment, JNK and Lyn (a kinase of the Src family) were also translocated into rafts and became phosphorylated, activating new signaling pathways. The ring finger and zinc finger domains of TRAFs were shown to be required for communication between CD40 and microdomain-associated signaling molecules.

TNFR signaling leads to activation of sphingomyelinases (SM) liberating ceramide. Rafts are modified by ceramide to form larger domains, which serve to cluster receptor molecules. The generation of a high receptor density might be required for the amplification of receptor-specific signaling and explain the function of the SM and ceramide in multiple signaling pathways.

More recently, rafts have been implicated in the signal transduction by several other members of the TNF receptor family. Raft microdomains proved to have a crucial role in RANK signaling (Ha et al. 2003a). The majority of the ectopically expressed RANK and a substantial portion of endogenous TRAF2 and 6 were detected in the low-density membrane fractions. RANKL stimulation increased the association of TRAF6 with rafts. Disruption of rafts blocked the RANK ligand-induced translocation of TRAF6 and impeded the interaction between RANK and TRAF6. These observations demonstrated that proper RANK signaling requires the function of raft membrane microdomains.

Further studies (Ha et al. 2003a) proved that the raft microdomain also played an essential role in osteoclast function and differentiation. Expression of raft component flotillin greatly increased during osteoclast differentiation, whereas engagement of RANK induced the translocation of TRAF6 to rafts where Src was constitutively resident. Disruption of rafts blocked TRAF6 translocation and Akt activation by RANK ligand in osteoclasts and further reduced the survival of osteoclasts. Actin ring formation and bone resorption by osteoclasts were also found to require the integrity of rafts (Ha et al. 2003b).

A number of TNF/TNFR superfamily members are expressed both as transmembrane and as soluble proteins. Proteolytic processing (shedding) of the precursor forms by metalloproteinases is controlled since the soluble forms frequently have regulatory functions (e.g., soluble CD30 is able to suppress the Th1-type immune response). Microdomains were found to influence this process. Cholesterol depletion obviously leads to the disruption of rafts. Treatment of human lymphoma-derived cell lines with cholesterol-binding methyl-beta-cyclodextrin or drugs interfering with cholesterol metabolism, such as lovastatin, cholesterol oxidase and filipin, led to the increased release of sCD30. Since TNF-alpha-converting enzyme (TACE), the enzyme responsible for CD30 shedding, is not raft-associated, while CD30 is partially present in lipid rafts, disintegration of lipid rafts converts raft-associated CD30 into a better substrate, resulting in enhanced shedding of CD30 (von Tresckow et al. 2004).

No Fas-associated proteins can be detected prior to Fas-L binding. However, stimulation with the ligand induces clustering and the rapid recruitment of FADD and caspase-8 to the Fas DISC (Medema et al. 1997; Scaffidi et al. 1999). Interestingly, the DISC has been shown to assemble in the glycosphingolipid-rich microdomains. This localization is dependent on an intact death domain. Indeed, raft localization can be transferred from TNF-R1 to TNF-R2 by recombinantly grafting the TNF-R1 death domain (Cottin et al. 2002).

7.3.5 Activation of JNK

The N-terminal zinc fingers of TRAFs seem to be essential for the activation of the MAP kinase pathway, which eventually leads to activation of the AP1 transcription factor (a dimer of c-Fos and c-Jun), through the activation of *c-Jun N-terminal kinase* (JNK). The MAP kinase pathway consists of four levels of Ser/Thr kinases, which are sequentially activated, proceeding downstream from the fourth level.

A significant proportion of death domain receptors and TRAF molecules are associated with detergent-resistant membrane domains. Overexpression of most TRAFs (2, 4, 5 or 6) activated the MAPK pathway. TRAF3 was the only exception. TRAF3 was also the only one that proved to be detergent soluble. Myristoylation of the molecule changed both subcellular localization and the ability to activate MAP kinases (Dadgostar and Cheng 2000).

Apoptosis signal-regulating kinase 1 (ASK1) seems to be the key upstream activator of JNK and p38 MAPK signaling cascades, and is activated after TNFR engagement or in case of TRAF overexpression. Its deletion or negative dominant mutations lead to loss of JNK activation and inhibition of apoptosis (Ichijo et al. 1997; Nishitoh et al. 1998). Evidence shows that the *ASK1-interacting protein*, AIP1, plays an important role in TNF-induced ASK1 activation by facilitating dissociation of ASK1 from its inhibitor.

Several other MAP4Ks (germinal center kinases) were also shown to play a role in TNFR-mediated activation of JNK, but their importance is limited to embryonic stem cells (Yujiri et al. 2000).

7.4 Signaling of Other Family Members

7.4.1 4-1BB

4-1BB (CD137) is one of the costimulatory members of the TNFR family. Activation of these receptors generates costimulatory signals, which have been defined as signals that fortify of modify signals elicited by the interaction of TCR and MHC-peptide complexes on antigen-presenting cells. In most cases, T cells are ineffective, become unresponsive or undergo apoptosis in the absence of costimulatory signals. 4-1BB is expressed on activated CD4(+) and CD8(+) T cells, on a series of malignant cell lines as well as on dendritic cells and monocytes. 4-1BB-mediated T cell costimulation involves recruitment of TNFR-associated factor 2 (TRAF2) and activation of the stress-activated protein kinase cascade. In normal T cells, p38 MAPK is activated synergistically by immobilized anti-CD3 plus immobilized 4-1BB ligand. T cells from TRAF2 dominant-negative mice are impaired in 4-1BB-mediated p38 MAPK activation. A link between TRAF2 and the p38 cascade is provided by ASK-1. A T cell hybrid expressing a kinase-dead ASK-1 fails to activate p38 MAPK in response to 4-1BB signaling.

In yeast two-hybrid screening, a novel leucine-rich repeat (LRR)-containing protein, LRR-1, was found to specifically interact with the cytoplasmic domain of 4-1BB. Overexpression of LRR-1 suppressed the activation of NF-κB induced by 4-1BB or TRAF2. In addition, LRR-1 downregulated JNK1 activity. These results indicate that LRR-1 negatively regulates the 4-1BB-mediated signaling cascades and the activation of NF-κB and JNK1.

4-1BB is induced on CD4+ and CD8+ T cells upon engagement of the TCR/CD3 complex with the antigen bound to MHC. The transient nature of its expression on activated CD4+ T cells is the result of proteolytic shedding catalyzed by metalloproteinases.

4-1BB plays an important role during T cell activation, enhances clonal expansion and rescues T cells from apoptosis. The survival of the activated CD8+ T cells following expansion is the result of the upregulation of anti-apoptotic genes bcl-x(L) and bfl-1 via 4-1BB-mediated NF-κB activation, but no increase is detected in Bcl-2 expression. 4-1BB enhances cell cycle progression and proliferation of CD8(+) T cells in both an IL-2-dependent and -independent manner. 4-1BB upregulates the expression of cyclins D2, D3 and E, and the translation of cyclin D2 via the *phosphatidylinositol 3-kinase* (PI3K)-Akt pathway. The enhanced cyclin D2 and D3 expression inhibits the expression of the *cyclin-dependent kinase inhibitor* p27(Kip1).

4-1BB is the regulator of regulators: signaling through the 4-1BB receptor inhibits the suppressive function of CD4+CD25+ regulatory T cells (Treg), especially when these cells are preactivated. However, *proliferation* of Treg cells is only marginally influenced by 4-1BB-elicited signaling.

7.4.2 BAFFR

BAFFR (Bcmd/BR3) is one of the receptors of the *B-cell activation factor* (BAFF), also known as BlyS. BAFFR signaling plays a critical role in the maturation, selection and survival of B cells. BAFFR activation induces B cell survival, at least in part, through the regulation of Bcl-2 family members via NF-κB-mediated mechanisms. Experimental evidence suggests that BAFFR expression is controlled by B cell antigen receptor signaling, linking specificity-based selection and BLyS-mediated survival. BAFFR and BCMA have the shortest ligand-binding modules within the family. The extracellular domain of BAFFR comprises a minimal TNF receptor-like module that encodes a highly focused ligand-binding site. This is analogous to the first half of a canonical TNFR CRD but is stabilized by an additional non-canonical disulfide bond. Several of the key BAFF-binding residues form a beta-turn that was shown to be sufficient for ligand binding when transferred to a structured beta-hairpin scaffold (Kayagaki et al. 2002).

7.4.3 BCMA

The ligand of *B-cell maturation antigen* (BCMA) is a rather new member of the TNF family: *A proliferation-inducing ligand* (APRIL, also called TRDL-1 or TALL-2) is thus named for its capacity to stimulate the proliferation of tumor cells in vitro. Its expression is low in normal tissues but elevated in several types of tumors and transformed cell lines. In immunodeficient mice, APRIL increases tumorigenicity of tumor grafts. Obviously, this is not its physiological role.

APRIL and BAFF [also termed as *B-lymphocyte stimulator* (BLyS), TALL-1, THANK and zTNF4) form a new subfamily of TNF-like ligands that are expressed in hematopoietic cells. Both ligands can bind their cognate receptors and the *transmembrane activator and calcium modulator cyclophilin ligand interactor* (TACI).

As a special twist in the story of transmembrane and soluble forms of ligands of the TNF superfamily of proteins, Pradet-Balade et al. (2002) described a chimaeric molecule, TWE-PRIL. Genes of TWEAK and APRIL are located very close on human Ch 17. TWEAK is a transmembrane protein expressed on the cell surface, whereas APRIL acts solely as a secreted factor. A hybrid transcript between TWEAK and APRIL was identified in human primary T cells and monocytes, and TWE-PRIL protein composed of cytoplasmic and transmembrane domains of TWEAK fused to the C-terminal domain of APRIL was detected in primary human T lymphocytes and monocytic cell lines. TWE-PRIL proved to be a biologically active ligand, as it stimulated cycling in T- and B-lymphoma cell lines.

7.4.4 CD27

CD27 is a lymphocyte-specific member of the TNF receptor (TNFR) family, another costimulatory molecule for peripheral T cells. In a yeast two-hybrid screen, CD27 directly associated with TRAF2 and 3 (Gravestein et al. 1998; Yamamoto et al. 1998). Transfection experiments using dominant-negative TRAF2 indicated that CD27 communicates with Jun N-terminal kinase (JNK) via TRAF2 (Gravestein et al. 1998). CD27 augments TCR-induced JNK activity in primary murine lymph node T cells. Since TRAF proteins initiate anti-apoptotic signaling, CD27 regulates not only proliferation but also survival of T lymphocytes. CD27 and its ligand, CD70, are crucial molecules to direct the differentiation of CD27+ memory B cells toward plasma cells in cooperation with IL-10, while preventing IL-10-mediated apoptosis (Gravestein et al. 1998).

CD27 is unique in this family for its disulfide-linked homodimerization of 55-kDa monomers. Overexpression of CD27 in 293 cells induces a low level of NF-κB activation, and the ligation of the receptor by its ligand dramatically augments this signal. The NF-κB activation signal is inhibited by dominant-negative TRAF2 or intact TRAF3, indicating that in this case TRAF2 and 3 have antagonistic effects (Yamamoto et al. 1998).

The induction of immunoglobulin isotype switching in B cells requires several signals given by cytokines and cell contact-delivered cosignals. Interaction of CD27 with its ligand increased IgE synthesis synergistically in the presence of IL-4 plus anti-CD40 mAb. CD27 negative B cells do not produce IgE (Nagumo et al. 1998). CD27/CD70 interaction upregulates the *positive regulatory domain I-binding factor-1*, resulting in enhanced B cell proliferation and promotion of differentiation into plasma cells in the presence of IL-4 or IL-4 plus anti-CD40 (Nagumo et al. 1998).

A death domain-containing pro-apoptotic protein, Siva-1, was identified as an intracellular partner of CD27 and GITR. Overexpression of Siva protein triggers a typical apoptotic process mediated by the activation of both initiator and effector caspases. This pathway involves a mitochondrial step, evidenced by activation of Bid and cytochrome *c* release (Spinicelli et al. 2002).

7.4.5 CD30

CD30 is a characteristic cell-surface receptor for activated lymphocytes and the malignant cells of Hodgkin's disease (HD), anaplastic large cell lymphoma (ALCL) and a few other non-Hodgkin's lymphomas. CD30 functions include participation in negative selection of thymocytes, costimulation of activated T cells, isotype switching of B cells and regulation of the effector activity of cytotoxic lymphocytes. CD30-mediated signal transduction is capable of promoting cell proliferation and cell survival as well as anti-proliferative effects and cell death, depending on cell type and costimulatory effects.

Some data indicate the opposite signaling of CD30 in HD or ALCL cells, while other information points to pleiotropic signaling pathways in both malignancies. Most Hodgkin and Reed-Sternberg (H-RS) cells are derived from germinal center B-cells. Ligand-independent signaling by overexpressed CD30 was shown to be a common mechanism that induced constitutive NF-κB activation in these cells (Horie et al. 2003).

7.4.6 CD40

The CD40 receptor is expressed constitutively on B cells and also found on macrophages and dendritic cells. In B lymphocytes, CD40 provides essential signals controlling clonal expansion, antibody production and isotype switching. In monocytes and CDs, its signals enhance antigen presentation (Grewal et al. 1995, 1996; van Essen et al. 1995; Soong et al. 1996; Stout et al. 1996; Maruo et al. 1997; Moodycliffe et al. 2000). CD40 was found to be expressed transiently on CD8+ T cells during an immune response, providing important signals for the generation of T cell memory (Bourgeois et al. 2002).

During complex interactions between B and T cells, CD40 and its ligand provide a number of signals that play important roles in regulating the adaptive immune response. Defective signaling due to a mutation of CD154 leads to the rare human disease *X-linked hyper-IgM syndrome* (HIGM), a block in isotype switching of the immunoglobulin synthesis (Callard et al. 1993; DiSanto et al. 1993; Fuleihan et al. 1993; Korthauer et al. 1993).

CD40 upregulates surface molecules involved in T cell–B cell interactions (costimulatory and adhesion molecules, cytokines and their receptors). Defects in or absence of CD40 signals lead to defects in antigen presentation by B cells, macrophages and dendritic cells, causing severe deficiencies in cell-mediated as well as humoral immunity (Campbell et al. 1996; Grewal et al. 1995, 1996; van Essen et al. 1995; Grewal and Flavell 1996; Hollander et al. 1996; Kamanaka et al. 1996). Studies with CD40- and CD154-deficient mice have revealed the importance of these molecules in T cell-dependent B cell activation.

As mentioned above, the CD40 signaling complex is assembled in membrane rafts. Before stimulation, CD40 is present in detergent-sensitive domains of the plasma membrane of B cells, exhibiting no interaction with TRAF molecules. Straight after engagement with ligands on the surface of T cells, CD40 aggregates in detergent-resistant rafts recruiting TRAF molecules. CD40 binds at least four distinct TRAFs, and three of these (TRAFs 1–3) have overlapping binding sites (Bishop et al. 2002). Additional microdomain-associated signaling molecules interact with the TRAFs, forming a signaling complex, or become activated by its assembly. CD40 engagement has been shown to activate the Src family kinase Lyn, and to induce the phosphorylation of phosphatidylinositol-3 kinase (PI3 K), the Ser/Thr kinase Pim-1 and phospholipase C2 in B cells (Ren et al. 1994; Zhu et al. 2002).

CD40 signaling induces rapid ubiquitination and subsequent degradation of TRAF molecules (both TRAF2 and 3, but not TRAF1 or 6) (Hostager et al. 2000;

Brown et al. 2001). Ubiquitination-dependent degradation of the TRAF molecules is a negative regulatory feedback ending the signaling process. Blocking CD40-mediated TRAF2 degradation leads to intensive and long-lasting CD40 signals in B cells (Brown et al. 2002). At the same time, ubiquitination of other signaling molecules seems to be required for the initiation of signaling.

CD40 signaling leads to the activation of JNK, stress-activated protein kinase, p38 and NF-κB. *Germinal center kinase* (GCK) and related enzymes (Yuasa et al. 1998; Shi et al. 1999) can interact with the TRAF domain of TRAF2 in B cells, and by interaction with MEKK1 may contribute to CD40-induced JNK activation (Sanchez et al. 1994; Yan et al. 1994; McAdam et al. 2001). ASK1 seems to be involved also with CD40 signal, participating in the activation of JNK and p38 (Ichijo et al. 1997).The NF-κB-inducing MAP3-kinase, NIK (Malinin et al. 1997; Song et al. 1997), turned out to be essential only for LTβR, but dispensable in the CD40-mediated activation of NF-κB (Yin et al. 2001).

B cell receptors and CD40 do not exhibit homologous structures; the signaling pathways induced by ligation of these two receptors are able to cooperate in a synergistic way (Wheller et al. 1993; Bishop et al. 1995). Activation of both types of receptors activates similar targets: *extracellular signal-regulated kinase* (ERK), JNK, p38 and similar transcription factors, such as NF-B, NF-AT and AP-1 (Francis et al. 1995; Li et al. 1996; Sutherland et al. 1996).

The relationship between CD40 and Fas is extremely interesting. During activation of mature B cells, CD40 signals upregulate Fas, which renders activated B cells more susceptible to apoptosis induction by the Fas ligand present on activated T cells. However, the Fas signaling complex is disabled by CD40 signals, upregulating the expression of the anti-apoptotic caspase homologue c-FLIP (Hennino et al. 2000; Wang et al. 2000). Defects resulting in an absence of Fas signals in vivo result in dramatic dysregulation of humoral immune responses, resulting in autoimmunity, splenomegaly and lymphadenopathy (Krammer 2000).

The human herpesvirus, Epstein-Barr virus (EBV), codes for a membrane protein, LMP1, which participates in the immortalization of B cells. (EBV is the etiological agent for mononucleosis and Burkitt lymphoma.) The intracellular domain of LMP1 assembles a signaling complex quite similar to that of CD40, but its activation is constitutive. The strong and sustained B cell signaling induced by this virus-encoded protein correlates with its ability to prevent ubiquitination and degradation of TRAF proteins (Brown et al. 2001).

7.4.7 DR3/TRAMP and DcR3

DR3 is a death domain-containing receptor that is upregulated during T cell activation. Its ligand, TL1A, is produced by endothelial cells. In T cells, TL1A acts as a costimulator of gamma IFN synthesis, and increases IL-2 responsiveness and secretion of proinflammatory cytokines both in vitro and in vivo. TL1A induces NF-κB activation and apoptosis in DR3-expressing cell lines; the same effects can be elicited by DR3 overexpression. TL1A also binds to decoy receptor DcR3 (which interacts with LIGHT, too). The interaction of TL1A with DR3 promotes T cell expansion during an immune response, whereas DcR3 has an opposing effect.

7.4.8 DR4 and DR5 (TRAILR1 and R2) and Decoy Receptors DcR1 and DcR2

It seems that nature judged the ligand of these receptors simply too dangerous to regulate at the transcriptional and translational levels. The ligand, *tumor necrosis factor-related apoptosis-inducing ligand* (TRAIL) or Apo2 ligand (Apo2L), interacts with a complex system of receptors: two pro-apoptotic death receptors and three anti-apoptotic decoys. TRAIL/Apo2L is able to trigger apoptosis through engagement of its death receptors, TRAILR1 and 2, and its activity is attenuated by the presence of two membrane-associated receptors and a soluble decoy, DcR1, DcR2 and osteoprotegerin (OPG), the last one carrying non-functional death domains.

TRAILR1 and 2 contain cytoplasmic death domains and signal apoptosis. Many cancer cell lines preferentially express TRAILR1 and TRAILR2, suggesting differential regulation of the death and decoy receptors. As TNF did two decades earlier, TRAIL raised high hopes among tumor immunologists and oncologists as a promising anti-tumor factor, since it can selectively induce apoptosis of a variety of transformed and malignant cells, without killing most types of normal diploid cells.

TRAIL is expressed as a transmembrane protein on the surface of different cells of the immune system and plays a role in both T-cell- and natural killer cell-mediated tumor surveillance and suppression of tumor metastasis. One of the key regulators of TRAIL expression in lymphocytes is the NF-κB transcription factor. Transcriptional activation of Apo2L/TRAIL by interferons and a number of other cytokines suggests a possible involvement of this protein in the activation of natural killer cells, cytotoxic T lymphocytes and dendritic cells.

TRAILR1 is broadly expressed in many tissues including the spleen and lung, and TRAILR2 level is high in the lung, heart, spleen and peripheral blood lymphocytes. At first sight it is difficult to understand the puzzling TRAIL resistance of these organs. The presence of decoy receptors that express TRAIL-binding ectodomains and either minimal cytoplasmic domains (DcR2) or GPI anchors (DcR1) provides one explanation of how apoptosis can be avoided (Pan et al. 1997).

7.4.9 EDAR

The pathway activated by the ligation of this receptor was discovered through the cloning of human genes that cause *hypohidrotic ectodermal dysplasia* (HED) syndrome. All components of this system, the ligand, ectodysplasin or EDA, the death domain-containing receptor EDAR and the death domain adapter molecule

EDARADD, were previously unknown proteins. Their exact roles were investigated in mice by the analysis of mutation in the corresponding genes (*crinkled*, *downless* and *Tabby*). These mutants showed defects in the development of certain exocrine glands and morphogenesis of hair and teeth.

Ectodysplasin-EDAR has unique, important functions in the regulation of embryonic development, but its effects are mediated by the activation of the well-known NF-κB (and probably other, unidentified pathways). EDAR mediates cell–cell interactions within the ectoderm and regulates the initiation and growth of hair and teeth. In evolutionary respect it is interesting to note that EDAR is also necessary for the development of *fish scales*, indicating that this pathway and its function have been conserved during the evolution of ectodermal organs.

The development of hair follicles and the sebaceous glands is the process whereby NGFR, the odd member of the TNFR family, works together with the EDA/EDAR, XEDAR, Troy system. Hair follicle formation in the mouse begins with intense NGFR expression. During hair follicle development, NGFR expression is attenuated and the receptors EDAR, XEDAR and Troy, as well as the ligand EDA, help form the hair follicle.

7.4.10 Fas Signaling

Fas (Apo-1, CD95) and its ligand (FasL, CD95L) are extremely important members of the TNFR and TNF families, respectively, with a crucial role in the regulation of activation-induced cell death, clonal selection of lymphocytes, T-cell-induced cytotoxicity, immune privilege, tumor surveillance and immune escape. Malfunction of the FasL/Fas system has been implicated in lymphoproliferative and autoimmune diseases, fulminant hepatitis and immune deficiency. Until recently, the FasL/Fas system was mainly appreciated with respect to its potential to kill, and the apoptotic features of the FasL/Fas system and the corresponding signaling pathways have been well investigated. There is increasing evidence that activation of Fas can lead to cell proliferation or NF-κB activation; however, the pathways that are utilized by Fas to transduce proliferative and activating signals are not very well understood.

7.4.11 GITR/AITR

Glucocorticoid-induced TNFR family-related gene product (GITR) is expressed in different cell types, including T lymphocytes, and its cytoplasmic region shows high homology with that of other costimulatory receptors of the TNFRSF. GITR stimulation potentiated the proliferative response of the anti-CD3 stimulus in CD8+ and CD4+ peripheral T cell subpopulations, and increased the production of IL-2 and IFN-gamma and expression of IL-2 receptor alpha. GITR is also involved in MAPK-pathway activation, since GITR-elicited signaling enhanced the anti-CD3-induced ERK phosphorylation.

GITR has a stimulatory effect on CD4+CD25+ regulatory T cell proliferation, simultaneously downregulating suppressor activity of these cells. GITR expression is probably dispensable for the development of Treg cells, since GITR–/– CD4+CD25+ cells are capable of exerting suppressor activity on CD4+CD25– responder cells.

7.4.12 HVEM

HVEM belongs to a rather complex system: it has three ligands, LIGHT, lymphotoxin (LT) alpha and LTαβ2, all of which interact with at least one more receptor. HVEM also serves as a receptor protein for human herpes simplex virus. The physiological role of HVEM is the regulation of T cell immune responses and it shares its ligand, LIGHT, with the lymphotoxin-beta receptor (LTβR) and one of the soluble decoy receptors, DcR3. LIGHT has emerged as a potent initiator of T cell costimulation signals, effecting CTL-mediated tumor rejection, allograft rejection and graft versus host disease. Constitutive activation of HVEM through overexpression of LIGHT leads to tissue destruction and autoimmune-like disease syndromes. In contrast to LTαβ2, LIGHT plays a minimal role in lymphoid tissue development, yet some evidence indicates a role in negative selection in the thymus.

7.4.13 Lymphotoxin-β Receptor

The lymphotoxin-β receptor (LTβR) is expressed on most cell types including cells of epithelial and myeloid lineages. This receptor is most closely related to TNFR2 and CD40, not only in sequence but also in its signaling capacity by belonging to the family of non-death-domain receptors (Hehlgans and Männel 2002). LTβR plays a critical role in inflammation and lymphoid organogenesis through activation of NF-κB (Futterer et al. 1998; Rennert et al. 1998).The first evidence in vivo has been observed in mice homozygous for the *aly* point mutation in the *NF-κB-inducing kinase* (NIK), resulting in alymphoplasia, a phenotype similar to that found in mice that have been genetically deleted for LTβR.

Although wild-type and *aly*-NIK do not differ in their in vitro binding to TRAF1, 2, 3 or 6 or in their in vivo association with TRAF2 and do not differ in their poor binding to IκB kinase beta (IKKβ), only wild-type NIK is able to bind to IKK. These results demonstrate that activation of NF-κB by LTβR is mediated by an interaction of NIK with IKK that is abrogated in the *aly* mutation (Dejardin et al. 2002). In addition to this classical NF-κB activation which requires NIK and IKK, an IKKα-dependent activation is required for the expression of proinflammatory molecules including VCAM-1, MIP-2 and MIP-1. In contrast, IKK controls the LTβR-dependent expression of chemokines and cytokines involved in lymphoid organogenesis, including SLC, BLC and ELC. Thus LTβR signaling induces different patterns of gene expression via distinct NF-κB pathways (Luftig et al. 2001).

7.4.14 RANK and Osteoprotegerin (OPG)

The RANK/RANKL system is linked to the DR4 and DR5/TRAIL system through a common decoy receptor, OPG, which has affinity to both RANKL and TRAIL, two transmembrane ligands. The discovery of OPG was a very significant step in understanding bone metabolism. The molecule interacting with OPG [initially called OPG-ligand/osteoclast differentiating factor (ODF), now RANKL or TRANCE], was identified as the key mediator of osteoclastogenesis. It was expressed as a transmembrane protein on the surface of various cell types and particularly on pre-osteoblastic/stromal cells and activated T cells. RANKL can be cleaved and the soluble form is biologically active. RANKL induces the differentiation of osteoclast precursor cells and stimulates the resorption function and survival of mature osteoclasts. The decoy receptor, OPG, strongly inhibits bone resorption by binding with high affinity to RANKL, thereby preventing RANKL from engaging its receptor RANK. This system is regulated by the calciotropic hormones.

Mutations of mice affecting expression of the genes of this system resulted in abnormal skeletal phenotypes (osteoporosis and osteopetrosis), proving the decisive role played by these factors in regulating bone metabolism. Manipulation of the RANK/RANKL system could provide therapy for very common, debilitating human diseases (e. g., postmenopausal osteoporosis, osteolytic metastatic cancer, arthritis and periodontitis).

The pleiotropic RANKL has other activities as well: it is a survival factor for activated dendritic cells, and RANK signaling seems to be important for the maintenance of immunological tolerance. Activated T cells express RANKL, and knockout mice for RANKL acquire severe immunological abnormalities (and osteopetrosis). Also in the developmental processes leading to both lymph node formation and the expansion and function of mammary glands during pregnancy and lactation, RANKL has been shown to play essential roles.

7.4.15 TACI

Transmembrane activator and calcium modulator and cyclophilin ligand interactor (TACI) shares one of its ligands, BAFF, with BAFFR, while the other, APRIL, also binds to BCMA. Both ligands are usually expressed by myeloid cells. They are crucial survival factors and powerful B cell-activating molecules for B cells, attenuating apoptosis, increasing NF-κB activation, upregulating Bcl-2 and Bcl-x(L) and downregulating Bax.

7.4.16 TNFR2

Signaling information concerning TNFR2 is rather limited compared to signaling via TNFR1. For activation of TNFR2, soluble TNF can serve as a ligand;

however, the membrane-bound 26-kDa form of TNF is superior to soluble TNF (Grell et al. 1995). Different splice variants of TNFR2 have been described. Besides the full-length membrane-located TNFR2 (Lewis et al. 1991) and an intracellularly expressed splice variant which is identical to the full-length TNFR2 (Seitz et al. 2001), a soluble TNFR2 lacking the transmembrane and intracellular domain (Lainez et al. 2004) has been found. A PLAD domain is present in the extracellular part of TNFR2. The intracellular part of the molecule does not contain a death domain but still recruits adapter proteins such as TRAF2 and can be a target for kinases (MacEwan 2002). Since TNFR2 is readily cleaved by metalloproteases into a soluble form, which is still capable of TNF binding (Higuchi and Aggarwal 1994; Crowe et al. 1995), one obvious function of this decoy TNFR2 is the modulation of TNF activity by neutralization.

7.4.17 TWEAKR/Fn14

TWEAKR/Fn14 is the smallest member of the TNF receptor superfamily, with one single cysteine-rich domain (CRD) in its extracellular ligand binding structure. It is a type I transmembrane protein alternatively referred to as either TWEAK receptor or *fibroblast growth factor-inducible* 14 (Fn14). It has multiple biological activities, including stimulation of cell growth and angiogenesis, induction of inflammatory cytokines and – under some experimental conditions – stimulation of apoptosis. TWEAKR appears to signal via recruitment of several different TRAFs.

Its ligand, *tumor necrosis factor-like weak inducer of apoptosis* (TWEAK), is a relatively new, not very well-characterized member of the TNF superfamily. TWEAK is a cell-surface-associated type II transmembrane protein and, like several other members of the family, it is processed into a smaller, biologically active soluble form. According to our present knowledge, there is one receptor that binds TWEAK with high affinity.

7.5 Reverse Signaling

Reverse signaling, i.e. the activation of cells expressing transmembrane ligands by their receptors, is characteristic of the TNF/TNFR system (Eissner et al. 2004). Stimulation of the membrane-associated form of TNF (mTNF) has been described for T cells (Ferran et al. 1994; Chen et al. 2001; Chou et al. 2001) and monocytes, where in the latter it serves as a silencing mechanism rather than a stimulatory signal. Interaction of mTNF with TNF *receptor* positive cells, soluble receptor molecules or anti-TNF antibodies confers resistance to bacterial endotoxin (LPS) in monocytes. As a result, such prestimulated monocytes no longer respond to LPS with the release of soluble (yet unidentified) pro-apoptotic factors and cytokines such as TNF, IL-1 and IL-10 (Eissner et al. 2000). Thus, reverse signaling by mTNF causes an anergic state of these cells. Under pathological conditions, mTNF might additionally transmit apoptotic signals, as it has been shown for monocytes and T cells from patients with Crohn's disease (Lugering et al. 2001; ten Hove et al. 2002).

mTNF has receptor-like properties and its cytoplasmic domain was detected in the nuclei after anti-TNF stimulation (Domonkos et al. 2001). Very recently, we have begun to unravel the intracellular cascades that mediate reverse signaling by mTNF. Activation of mTNF triggered at least two independent pathways that could be distinguished by *protein kinase C* (PKC) inhibitors. The suppression of the LPS-induced Death Factor X (Eissner et al. 2000) was dependent on PKC, whereas the suppression of LPS-mediated cytokine release was not. Both LPS and reverse signaling stimulated the (MAPK)/extracellular signal-regulated kinase (ERK) pathway (Kirchner et al. 2004). Interestingly, the activation of reverse signaling by mTNF rendered monocytes and macrophages refractory to a subsequent activation of the MAPK/ERK pathway by LPS (Kirchner et al. 2004). It remains to be elucidated where the LPS and the mTNF signals separate, but preliminary data suggest that a differential engagement of PKC isozymes by the antagonistic stimuli LPS and mTNF, respectively, might, at least in part, do that. LPS is known to stimulate classical PKC isozymes, whereas reverse signaling rather activates non-classical PKCs.

Reverse signaling is not limited to TNF alpha. CD30, a lymphoid activation marker, is shed into the cell environment after endoproteolytic cleavage of its ectodomain. Soluble (s)CD30 is able to suppress the Th1-type immune response (von Tresckow et al. 2004). Recombinant 4-1BB that was immobilized on culture plates strongly inhibited M-CSF/RANKL-induced in vitro osteoclast formation from bone marrow cells. Antibody against 4-1BB ligand was also active (to a lesser extent), indicating involvement of reverse signaling through 4-1BBL during inhibition of osteoclast formation. Reverse signaling abolished RANKL-mediated induction of *nuclear factor of activated T cells* (NFAT)-2, and experimental results suggest that it probably interfered with the PI3 K/Akt pathway, and *casein kinase I* (CKI) was also involved in the signaling process (Saito et al. 2004).

Soluble GITR induced cell cycle arrest in murine macrophages by decreasing the activities of cdk2 and 4 as a result of lowering the expression of cyclins A and D2 and cdk4. Inhibition of G1/S phase transition provoked by sGITR acting on its membrane-bound ligand lead to apoptosis of the cells (Shin et al. 2004). The GITR ligand is constitutively expressed on macrophages, though its level is low.

Multiple myeloma (MM)-induced bone destruction requires increased RANK-L expression and is facilitated by a concurrent reduction in OPG, a natural decoy receptor for RANK-L. Administration of the RANK-L antagonist RANK-Fc limited MM-induced osteoclastogenesis, development of bone disease and MM tumor progression. RANK-Fc is a recombinant RANK-L antagonist that is formed by fusing the extracellular domain of RANK to the Fc portion of human IgG1. In vitro, addition of RANK-Fc virtually eliminated the formation of osteoclasts in cocultures of MM with bone marrow and osteoblast/stromal cells.

Clearly more experimental evidence is required to conclusively answer important questions as to what extent reverse signaling of TNF ligand family members is important in vivo. The data obtained so far strongly suggest that this signaling system contributes to the fine tuning of the cellular and functional interactions in the immune system. The fact that the majority of reverse signals so far described in B and T cells are of a stimulatory nature (Suzuki and Fink 1998; Harashima et al. 2001), whereas monocytes are mainly inhibited in their effector functions by reverse signaling (Eissner et al. 2000; Lugering et al. 2001), argues for a differential role of this signal type in adaptive and innate immunity. The bi-directional signaling of mTNF in monocytes could be a mechanism to control inflammatory responses, thus limiting the destructive potential of monocytes against other target cells. Furthermore, preliminary evidence suggests that experimental induction of reverse signaling via mTNF by antibodies or soluble receptors leads not only to resistance, but also to a generalized anergy towards non-specific inflammatory stimuli, and even to cell death under specific conditions (Lugering et al. 2001; ten Hove et al. 2002).

If such regulatory circuits are acting in vivo, one has to critically reconsider therapeutic and prophylactic anti-TNF strategies, which are used in inflammatory bowel diseases (for review see van Deventer 2001), rheumatoid arthritis and allogeneic stem cell transplantation (Holler et al. 1995). In all these clinical settings there is still a need for fighting secondary infections due to immunosuppression or other therapeutic modalities, and a total monocytic loss of function rather than controlled immune reactivity is undesirable. Indeed, infectious complications have been reported in anti-TNF-treated patients (Nahar et al. 2003). Another important aspect to be taken into account is the nature of the therapeutic agents. Recent data indicate a differential clinical effect of the humanized anti-TNF antibody infliximab versus the humanized TNF receptor etanercept (Scallon et al. 2002), which might be explained by a different reverse signaling of mTNF.

7.6 Therapy/Intervention

The TNF superfamily encompasses a number of biologically powerful members and knowledge of their function is steadily increasing. So far, only one molecule, namely TNF, has been found to be able to serve as a useful target for immunotherapy. For treatment of patients with chronic inflammatory diseases like rheumatoid arthritis or Crohn's disease, either monoclonal antibodies or soluble TNFR constructs have been used as TNF antagonists with benefit for the patients (Hanauer et al. 2002; Shanahan and St Clair 2002; ten Hove et al. 2002). However, since TNF has a central role in the defense of microbial infections, therapeutic neutralization of TNF not only reduces the inflammatory reaction in patients with chronic inflammatory diseases but also enhances the risk of infections such as tuberculosis (Gardam et al. 2003).

References

Amir RE, Haecker H, Karin M, Ciechanover A (2004) Mechanism of processing of the NF-kappa B2 p100 precursor: identification of the specific polyubiquitin chain-anchoring lysine residue and analysis of the role of NEDD8-modification on the SCF(beta-TrCP) ubiquitin ligase. Oncogene 23(14):2540–2547

Arch RH, Gedrich RW, Thompson CB (2000) Translocation of TRAF proteins regulates apoptotic threshold of cells. Biochem Biophys Res Commun 272(3):936–945

Banner DW, D'Arcy A, Janes W, Gentz R, Schoenfeld HJ, Broger C, Loetscher H, Lesslauer W (1993) Crystal structure of the soluble human 55 kD TNF receptor-human TNF beta complex: implications for TNF receptor activation. Cell 73(3):431–445

Baud V, Liu ZG, Bennett B, Suzuki N, Xia Y, Karin M (1999) Signaling by proinflammatory cytokines: oligomerization of TRAF2 and TRAF6 is sufficient for JNK and IKK activation and target gene induction via an amino-terminal effector domain. Genes Dev 13(10):1297–1308

Bhardwaj A, Aggarwal BB (2003) Receptor-mediated choreography of life and death. J Clin Immunol 23(5):317–332

Bishop GA, Warren WD, Berton MT (1995) Signaling via MHC class II molecules and antigen receptors enhances the B cell response to gp39/CD40 ligand. Eur J Immunol 25:1230–1238

Bishop GA, Hostager BS, Brown KD (2002) Mechanisms of tumor necrosis factor receptor associated factor (TRAF) regulation in B lymphocytes. J Leuk Biol 72:19–23

Bobe J, Goetz FW (2001) Molecular cloning and expression of a TNF receptor and two TNF ligands in the fish ovary. Comp Biochem Physiol B Biochem Mol Biol 129(2–3):475–481

Boldin MP, Goncharov TM, Goltsev YV, Wallach D (1996) Involvement of MACH, a novel MORT1/FADD-interacting protease, in Fas/APO-1- and TNF receptor-induced cell death. Cell 85:803–815

Bourgeois C, Rocha B, Tanchot C (2002) A role for CD40 expression on CD8+ T cells in the generation of CD8+ T cell memory. Science 297:2060–2063

Bouwmeester T, Bauch A, Ruffner H, Angrand PO, Bergamini G, Croughton K, Cruciat C, Eberhard D, Gagneur J, Ghidelli S, Hopf C, Huhse B, Mangano R, Michon AM, Schirle M, Schlegl J, Schwab M, Stein MA, Bauer A, Casari G, Drewes G, Gavin AC, Jackson DB, Joberty G, Neubauer G, Rick J, Kuster B, Superti-Furga G (2004) A physical and functional map of the human TNF-alpha/NF-kappa B signal transduction pathway. Nat Cell Biol 6(2):97–105

Brown KD, Hostager BS, Bishop GA (2001) Differential signaling and TRAF degradation by CD40 and the EBV oncoprotein LMP1. J Exp Med 193:943–954

Brown KD, Hostager BS, Bishop GA (2002) Regulation of TRAF2 signaling by self-induced degradation. J Biol Chem 277:19433–19438

Callard RE, Armitage RJ, Fanslow WC, Spriggs MK (1993) CD40 ligand and its role in X-linked hyper-IgM syndrome. Immunol Today 14:559–564

Campbell KA, Ovendale PJ, Kennedy MK, Fanslow WC, Reed SG, Maliszewski CR (1996) CD40L is required for protective cell-mediated immunity to *Leishmania major*. Immunity 4:283–289

Cao Y, Bonizzi G, Seagroves TN, Greten FR, Johnson R, Schmidt EV, Karin M (2001) IKKa provides an essential link between RANK signaling and cyclin D1 expression during mammary gland development. Cell 107:763–775

Carpentier I, Beyaert R (1999) TRAF1 is a TNF inducible regulator of NF-kappaB activation. FEBS Lett 460(2):246–250

Carswell EA, Old LJ, Kassel RL, Green S, Fiore N, Williamson B (1975) An endotoxin-induced serum factor that causes necrosis of tumors. Proc Natl Acad Sci USA 72:3666–3670

Chan FK, Chun HJ, Zheng L, Siegel RM, Bui KL, Lenardo MJ (2000) A domain in TNF receptors that mediates ligand-independent receptor assembly and signaling. Science 288(5475): 2351–2354

Chen NJ, Huang MW, Hsieh SL (2001) Enhanced secretion of IFN-gamma by activated Th1 cells occurs via reverse signaling through TNF-related activation-induced cytokine. J Immunol 166(1):270–276

Chen PC, DuBois GC, Chen MJ (1995) Mapping the domain(s) critical for the binding of human tumor necrosis factor-alpha to its two receptors. J Biol Chem 270(6):2874–2878

Chinnaiyan AM, O'Rourke K, Yu GL, Lyons RH, Garg M, Duan DR, Xing L, Gentz R, Ni J, Dixit VM (1996) Signal transduction by DR3, a death domain-containing receptor related to TNF-R1 and CD95. Science 274:990–992

Chou AH, Tsai HF, Lin LL, Hsieh SL, Hsu PI, Hsu PN (2001) Enhanced proliferation and increased IFN-gamma production in T cells by signal transduced through TNF-related apoptosis-inducing ligand. J Immunol 167(3):1347–1352

Christophides GK, Zdobnov E, Barillas-Mury C, Birney E, Blandin S, Blass C, Brey PT, Collins FH, Danielli A, Dimopoulos G, Hetru C, Hoa NT, Hoffmann JA, Kanzok SM, Letunic I, Levashina EA, Loukeris TG, Lycett G, Meister S, Michel K, Moita LF, Muller HM, Osta MA, Paskewitz SM, Reichhart JM, Rzhetsky A, Troxler L, Vernick KD, Vlachou D, Volz J, von Mering C, Xu J, Zheng L, Bork P, Kafatos FC (2002) Immunity-related genes and gene families in *Anopheles gambiae*. Science 298(5591):159–165

Cottin V, Doan JE, Riches DW (2002) Restricted localization of the TNF receptor CD120a to lipid rafts: a novel role for the death domain. J Immunol 168:4095–4102

Crowe PD, Walter BN, Mohler KM, Otten-Evans C, Black RA, Ware CF (1995) A metalloprotease inhibitor blocks shedding of the 80-kD TNF receptor and TNF processing in T lymphocytes. J Exp Med 181(3):1205–1210

Dadgostar H, Cheng G (2000) Membrane localization of TRAF3 enables JNK activation. J Biol Chem 275:2539–2544

Dejardin E, Droin NM, Delhase M, Haas E, Cao Y, Makris C, Li ZW, Karin M, Ware CF, Green DR (2002) The lymphotoxin-beta receptor induces different patterns of gene expression via two NF-kappaB pathways. Immunity 17(4):525–535

Devergne O, Hatzivassiliou E, Izumi KM, Kaye KM, Kleijnen MF, Kieff E, Mosialos G (1996) Association of TRAF1, TRAF2, and TRAF3 with an Epstein–Barr virus LMP1 domain important for B-lymphocyte transformation: role in NF-kappaB activation. Mol Cell Biol 16: 7098–7108

Devin A, Lin Y, Yamaoka S, Li Z, Karin M, Liu Zg (2001) The alpha and beta subunits of IkappaB kinase (IKK) mediate TRAF2-dependent IKK recruitment to tumor necrosis factor (TNF) receptor 1 in response to TNF. Mol Cell Biol 21(12):3986–3994

DiSanto JP, Bonnefoy JY, Gauchat JF, Fischer A, de Saint Basile G (1993) CD40 ligand mutations in X-linked immunodeficiency with hyper-IgM. Nature 361:541–543

Domonkos A, Udvardy A, Laszlo L, Nagy T, Duda E (2001) Receptor-like properties of the 26 kDa transmembrane form of TNF. Eur Cytokine Netw 12(3):411–419

Eissner G, Kirchner S, Lindner H, Kolch W, Janosch P, Grell M, Scheurich P, Andreesen R, Holler E (2000) Reverse signaling through transmembrane TNF confers resistance to lipopolysaccharide in human monocytes and macrophages. J Immunol 164(12):6193–6198

Eissner G, Kolch W, Scheurich P (2004) Ligands working as receptors: reverse signaling by members of the TNF superfamily enhance the plasticity of the immune system. Cytokine Growth Factor Rev 15(5):353–366

Feng X, Gaeta ML, Madge LA, Yang JH, Bradly JR, Pober JS (2001) Caveolin-1 associates with TRAF2 to form a complex that is recruited to tumor necrosis factor receptors. J Biol Chem 276(11):8341–8349

Ferran C, Dautry F, Merite S, Sheehan K, Schreiber R, Grau G, Bach JF, Chatenoud L (1994) Anti-tumor necrosis factor modulates anti-CD3-triggered T cell cytokine gene expression in vivo. J Clin Invest 93(5):2189–2196

Fesik SW (2000) Insights into programmed cell death through structural biology. Cell 103(2):273–282

Francis DA, Karras JG, Ke X, Sen R, Rothstein TL (1995) Induction of the transcription factors NF-κB, AP-1 and NF-AT during B cell stimulation through the CD40 receptor. Int Immunol 7:151–161

Fuleihan R, Ramesh N, Loh R, Jabara H, Rosen FS, Chatila T, Fu SM, Stamenkovic I, Geha RS (1993) Defective expression of the CD40 ligand in X chromosome-linked Ig deficiency with normal or elevated IgM. Proc Natl Acad Sci USA 90:2170–2173

Futterer A, Mink K, Luz A, Kosco-Vilbois MH, Pfeffer K (1998) The lymphotoxin beta receptor controls organogenesis and affinity maturation in peripheral lymphoid tissues. Immunity 9(1):59–70

Galibert L, Tometsko ME, Anderson DM, Cosman D, Dougall WC (1998) The involvement of multiple tumor necrosis factor receptor (TNFR)-associated factors in the signaling mechanisms of receptor activator of NF-kappaB, a member of the TNFR superfamily. J Biol Chem 273:34120–34127

Gamper C, van Eyndhoven WG, Schweiger E, Mossbacher M, Koo B, Lederman S (2000) TRAF-3 interacts with p62 nucleoporin, a component of the nuclear pore central plug that binds classical NLS-containing import complexes. Mol Immunol 37(1–2):73–84

Gardam MA, Keystone EC, Menzies R, Manners S, Skamene E, Long R, Vinh DC (2003) Anti-tumour necrosis factor agents and tuberculosis risk: mechanisms of action and clinical management. Lancet Infect Dis 3(3):148–155

Gedrich RW, Gilfillan MC, Duckett CS, Van Dongen JL, Thompson CB (1996) CD30 contains two binding sites with different specificities for members of the tumor necrosis factor receptor-associated factor family of signal transducing proteins. J Biol Chem 271:12852–12858

Gordon NC, Pan B, Hymowitz SG, Yin J, Kelley RF, Cochran AG, Yan M, Dixit VM, Fairbrother WJ, Starovasnik MA (2003) BAFF/BLyS receptor 3 comprises a minimal TNF receptor-like module that encodes a highly focused ligand-binding site. Biochemistry 42(20):5977–5983

Granger GA, Shacks SJ, Williams TW, Kolb WP (1969) Lymphocyte in vitro cytotoxicity: specific release of lymphotoxin like materials from tuberculin-sensitive lymphoid cells. Nature 221:1155–1157

Gravestein LA, Amsen D, Boes M, Calvo CR, Kruisbeek AM, Borst J (1998) The TNF receptor family member CD27 signals to Jun N-terminal kinase via Traf-2. Eur J Immunol 28(7):2208–2216

Grell M, Douni E, Wajant H, Lohden M, Clauss M, Maxeiner B, Georgopoulos S, Lesslauer W, Kollias G, Pfizenmaier K, Scheurich P (1995) The transmembrane form of tumor necrosis factor is the prime activating ligand of the 80 kDa tumor necrosis factor receptor. Cell 83(5):793–802

Grewal IS, Flavell RA (1996) A central role of CD40L in the regulation of CD4+ T-cell responses. Immunol Today 17:410–414

Grewal IS, Xu J, Flavell RA (1995) Impairment of antigen-specific T-cell priming in mice lacking CD40 ligand. Nature 378:617–620

Grewal IS, Foellmer HG, Grewal KD, Xu J, Hardardottir F, Baron JL, Janeway CA Jr, Flavell RA (1996) Requirement for CD40L in costimulation induction, T cell activation, and EAE. Science 273:1864–1867

Gruss HJ, Dower SK (1995) The TNF ligand superfamily and its relevance for human diseases. Cytokines Mol Ther 1(2):75–105

Ha H, Kwak HB, Lee SK, Na DS, Rudd CE, Lee ZH, Kim HH (2003a) Membrane rafts play a crucial role in receptor activator of nuclear factor kappaB signaling and osteoclast function. J Biol Chem 278(20):18573–18580

Ha H, Kwak HB, Le SW, Kim HH, Lee ZH (2003b) Lipid rafts are important for the association of RANK and TRAF6. Exp Mol Med 35(4):279–284

Hanauer SB, Feagan BG, Lichtenstein GR, Mayer LF, Schreiber S, Colombel JF, Rachmilewitz D, Wolf DC, Olson A, Bao W, Rutgeerts P, ACCENT I Study Group (2002) Maintenance infliximab for Crohn's disease: the ACCENT I randomised trial. Lancet 359(9317):1541–1549

Harashima S, Horiuchi T, Hatta N, Morita C, Higuchi M, Sawabe T, Tsukamoto H, Tahira T, Hayashi K, Fujita S, Niho Y (2001) Outside-to-inside signal through the membrane TNF-alpha induces E-selectin (CD62E) expression on activated human CD4+ T cells. J Immunol 166(1):130–136

Harless Smith S, Cancro MP (2003) BLyS: the pivotal determinant of peripheral B cell selection and lifespan. Curr Pharm Des 9(23):1833–1847

Hehlgans T, Männel DN (2002) The TNF–TNF receptor system. Biol Chem 383(10):1581–1585

Hennino A, Berard M, Casamayor-Pallejà M, Krammer PH, Defrance T (2000) Regulation of the Fas death pathway by FLIP in primary human B cells. J Immunol 165:3023–3030

Higuchi M, Aggarwal BB (1994) TNF induces internalization of the p60 receptor and shedding of the p80 receptor. J Immunol 152(7):3550–3558

Hirono I, Nam BH, Kurobe T, Aoki T (2000) Molecular cloning, characterization, and expression of TNF cDNA and gene from Japanese flounder *Paralychthys olivaceus*. J Immunol 165(8): 4423–4427

Hoeflich KP, Yeh WC, Yao Z, Mak TW, Woodgett JR (1999) Mediation of TNF receptor-associated factor effector functions by apoptosis signal-regulating kinase-1 (ASK1). Oncogene 18(42): 5814–5820

Hoffmann JA, Reichhart JM (2002) *Drosophila* innate immunity: an evolutionary perspective. Nat Immunol 3(2):121–126

Hollander GA, Castigli E, Kulbacki R, Su M, Burakoff SJ, Gutierrez-Ramos JC, Geha RS (1996) Induction of alloantigen-specific tolerance by B cells from CD40-deficient mice. Proc Natl Acad Sci USA 93:4994–4998

Holler E, Kolb HJ, Mittermuller J, Kaul M, Ledderose G, Duell T, Seeber B, Schleuning M, Hintermeier-Knabe R, Ertl B, Kempeni J, Wilmanns W (1995) Modulation of acute graft-versus-host-disease after allogeneic bone marrow transplantation by tumor necrosis factor alpha (TNF alpha) release in the course of pretransplant conditioning: role of conditioning regimens and prophylactic application of a monoclonal antibody neutralizing human TNF alpha (MAK 195F). Blood 86:890–899

Horie R, Higashihara M, Watanabe T (2003) Hodgkin's lymphoma and CD30 signal transduction. Int J Hematol 77(1):37–47

Hostager BS, Catlett IM, Bishop GA (2000) Recruitment of CD40 and tumor necrosis factor receptor-associated factors 2 and 3 to membrane microdomains during CD40 signaling. J Biol Chem 275(20):15392–15398

Hsu H, Huang J, Shu HB, Baichwal V, Goeddel DV (1996) TNF-dependent recruitment of the protein kinase RIP to the TNF receptor-1 signaling complex. Immunity 4:387–396

Hsu H, Lacey DL, Dunstan CR, Solovyev I, Colombero A, Timms E, Tan HL, Elliott G, Kelley MJ, Sarosi I, Wang L, Xia XZ, Elliott R, Chiu L, Black T, Scully S, Capparelli C, Morony S, Shimamoto G, Bass MB, Boyle WJ (1999) Tumor necrosis factor receptor family member RANK mediates osteoclast differentiation and activation induced by osteoprotegerin ligand. Proc Natl Acad Sci USA 96:3540–3545

Hsu KC, Chao MV (1993) Differential expression and ligand binding properties of tumor necrosis factor receptor chimeric mutants. J Biol Chem 268(22):16430–16436

Hu S, Yang X (2000) dFADD, a novel death domain-containing adapter protein for the *Drosophila* caspase DREDD. J Biol Chem 275(40):30761–30764

Hu Y, Baud V, Delhase M, Zhang P, Deerinck T, Ellisman M, Johnson R, Karin M (1999) Abnormal morphogenesis but intact IKK activation in mice lacking the IKKa subunit of IkB kinase. Science 284:316–320

Hu Y, Baud V, Oga T, Kim KI, Yoshida K, Karin M (2001) IKKa controls formation of the epidermis independently of NF-κB. Nature 410:710–714

Hymowitz SG, Christinger HW, Fuh G, Ultsch M, O'Connell M, Kelley RF, Ashkenazi A, de Vos AM (1999) Triggering cell death: the crystal structure of Apo2L/TRAIL in a complex with death receptor 5. Mol Cell 4(4):563–571

Ichijo H, Nishida E, Irie K, ten Dijke P, Saitoh M, Moriguchi T, Takagi M, Matsumoto K, Miyazono K, Gotoh Y (1997) Induction of apoptosis by ASK1, a mammalian MAPKKK that activates SAPK/JNK and p38 signaling pathways. Science 275:90–94

Igaki T, Kanda H, Yamamoto-Goto Y, Kanuka H, Kuranaga E, Aigaki T, Miura M (2002) Eiger, a TNF superfamily ligand that triggers the *Drosophila* JNK pathway. EMBO J 21(12):3009–3018

Irmler M, Thome M, Hahne M, Schneider P, Hofmann K, Steiner V, Bodmer JL, Schroter M, Burns K, Mattmann C, Rimoldi D, French LE, Tschopp J (1997) Inhibition of death receptor signals by cellular FLIP. Nature 388:190–195

Ishida TK, Tojo T, Aoki T, Kobayashi N, Ohishi T, Watanabe T, Yamamoto T, Inoue J (1996) TRAF5, a novel tumor necrosis factor receptor-associated factor family protein, mediates CD40 signaling. Proc Natl Acad Sci USA 93(18):9437–9442

Ishikawa H, Claudio E, Dambach D, Raventos-Suarez C, Ryan C, Bravo R (1998) Chronic inflammation and susceptibility to bacterial infections in mice lacking the polypeptide (p)105 precursor (NF-kappaB1) but expressing p50. J Exp Med 187(7):985–996

Jiang Y, Woronicz JD, Liu W, Goeddel DV (1999) Prevention of constitutive TNF receptor 1 signaling by silencer of death domains. Science 283:543–546

Kaisho T, Takeda K, Tsujimura T, Kawai T, Nomura F, Terada N, Akira S (2001) IkB kinase α is essential for mature B-cell development and function. J Exp Med 193:417–426

Kamanaka M, Yu P, Yasui T, Yoshida K, Kawabe T, Horii T, Kishimoto T, Kikutani H (1996) Protective role of CD40 in *Leishmania major* infection at two distinct phases of cell-mediated immunity. Immunity 4:275–281

Kanda H, Igaki T, Kanuka H, Yagi T, Miura M (2002) Wengen, a member of the *Drosophila* tumor necrosis factor receptor superfamily, is required for Eiger signaling. J Biol Chem 277(32):28372–28375

Kayagaki N, Yan M, Seshasayee D, Wang H, Lee W, French DM, Grewal IS, Cochran AG, Gordon NC, Yin J, Starovasnik MA, Dixit VM (2002) BAFF/BLyS receptor 3 binds the B cell survival factor BAFF ligand through a discrete surface loop and promotes processing of NF-kappaB2. Immunity 10:515–524

Kelliher MA, Grimm S, Ishida Y, Kuo F, Stanger BZ, Leder P (1998) The death domain kinase RIP mediates the TNF-induced NF-kappaB signal. Immunity 8:297–303

Khosla S (2003) The OPG/RANKL/RANK system. Endocrinology 142(12):5050–5055

Khush RS, Leulier F, Lemaitre B (2001) *Drosophila* immunity: two paths to NF-kappaB. Trends Immunol 22(5):260–264

Kirchner S, Boldt S, Kolch W, Haffner S, Kazak S, Janosch P, Holler E, Andreesen R, Eissner G (2004) LPS resistance in monocytic cells caused by reverse signaling through transmembrane TNF (mTNF) is mediated by the MAPK/ERK pathway. J Leukoc Biol 75(2):324–331

Kischkel FC, Hellbardt S, Behrmann I, Germer M, Pawlita M, Krammer PH, Peter ME (1995) Cytotoxicity-dependent APO-1 (Fas/CD95)-associated proteins form a death-inducing signaling complex (DISC) with the receptor. EMBO J 14:5579–5588

Kischkel FC, Lawrence DA, Chuntharapai A, Schow P, Kim KJ, Ashkenazi A (2000) Apo2L/TRAIL-dependent recruitment of endogenous FADD and caspase-8 to death receptors 4 and 5. Immunity 12:611–620

Kishore U, Reid KB (2000) C1q: structure, function, and receptors. Immunopharmacology 49(1–2):159–170

Korthauer U, Graf D, Mages HW, Briere F, Padayachee M, Malcolm S, Ugazio AG, Notarangelo LD, Levinsky RJ, Kroczek RA (1993) Defective expression of T-cell CD40 ligand causes X-linked immunodeficiency with hyper-IgM. Nature 361:539–541

Krajewska M, Krajewski S, Zapata JM, Van Arsdale T, Gascoyne RD, Berem K, McFadden D, Shabaik A, Hugh J, Reynolds A, Clevenger CV, Reed JC (1998) TRAF-4 expression in epithelial progenitor cells. Analysis in normal adult, fetal, and tumor tissues. Am J Pathol 152(6): 1549–1561

Krammer PH (2000) CD95's deadly mission in the immune system. Nature 407:789–795

Krippner-Heidenreich A, Tubing F, Bryde S, Willi S, Zimmermann G, Scheurich P (2002) Control of receptor-induced signaling complex (RISC) formation by the kinetics of ligand/receptor interaction. J Biol Chem 277(46):44155–44163

Lainez B, Fernandez-Real JM, Romero X, Esplugues E, Canete JD, Ricart W, Engel P (2004) Identification and characterization of a novel spliced variant that encodes human soluble tumor necrosis factor receptor 2. Int Immunol 16(1):169–177

Laing KJ, Wang T, Zou J, Holland J, Hong S, Bols N, Hirono I, Aoki T, Secombes CJ (2001) Cloning and expression analysis of rainbow trout *Oncorhynchus mykiss* tumour necrosis factor-α. Eur J Biochem 268:1315–1322

Lee SY, Reichlin A, Santana A, Sokol KA, Nussenzweig MC, Choi Y (1997) TRAF2 is essential for JNK but not NF-kappaB activation and regulates lymphocyte proliferation and survival. Immunity 7(5):703–713

Leonardi A, Ellinger-Ziegelbauer H, Franzoso G, Brown K, Siebenlist U (2000) Physical and functional interaction of filamin (actin-binding protein-280) and tumor necrosis factor receptor-associated factor 2. J Biol Chem 275(1):271–278

Leulier F, Vidal S, Saigo K, Ueda R, Lemaitre B (2002) Inducible expression of double-stranded RNA reveals a role for dFADD in the regulation of the antibacterial response in *Drosophila* adults. Curr Biol 12(12):996–1000

Lewis M, Tartaglia LA, Lee A, Bennett GL, Rice GC, Wong GH, Chen EY, Goeddel DV (1991) Cloning and expression of cDNAs for two distinct murine tumor necrosis factor receptors demonstrate one receptor is species specific. Proc Natl Acad Sci USA 88(7):2830–2834

Li H, Zhu H, Xu CJ, Yuan J (1998) Cleavage of BID by caspase-8 mediates the mitochondrial damage in the Fas pathway of apoptosis. Cell 94:491–501

Li Q, Verma IM (2002) NF-kappaB regulation in the immune system. Nat Rev Immunol 2(10):725–734

Li Q, Lu Q, Hwang JY, Buscher D, Lee KF, Izpusua-Belmonte JC, Verma IM (1999a) IKK1-deficient mice exhibit abnormal development of skin and skeleton. Genes Dev 13:1322–1328

Li Q, Van Antwerp D, Mercurio F, Lee KF, Verma IM (1999b) Severe liver degeneration in mice lacking the IkB kinase 2 gene. Science 284:321–325

Li YY, Baccam M, Waters SB, Pessin JE, Bishop GA, Koretzky GA (1996) CD40 ligation results in PKC-independent activation of ERK and JNK in resting murine splenic B cells. J Immunol 157:1440–1447

Li ZW, Chu W, Hu Y, Delhase M, Deerinck T, Ellisman M, Johnson R, Karin M (1999) The IKKb subunit of IkB kinase (IKK) is essential for nuclear factor kB activation and prevention of apoptosis. J Exp Med 189:1839–1845

Lin Y, Devin A, Rodriguez Y, Liu ZG (1999) Cleavage of the death domain kinase RIP by caspase-8 prompts TNF-induced apoptosis. Genes Dev 13:2514–2526

Ling L, Goeddel DV (2000) MIP-T3, a novel protein linking tumor necrosis factor receptor-associated factor 3 to the microtubule network. J Biol Chem 275(31):23852–23860

Locksley RM, Killeen N, Lenardo MJ (2001) The TNF and TNF receptor superfamilies: integrating mammalian biology. Cell 104:487–501

Lomaga MA, Yeh WC, Sarosi I, Duncan GS, Furlonger C, Ho A, Morony S, Capparelli C, Van G, Kaufman S, van der Heiden A, Itie A, Wakeham A, Khoo W, Sasaki T, Cao Z, Penninger JM, Paige CJ, Lacey DL, Dunstan CR, Boyle WJ, Goeddel DV, Mak TW (1999) TRAF6 deficiency results in osteo-petrosis and defective interleukin-1, CD40, and LPS signaling. Genes Dev 13(8):1015–1024

Luftig MA, Cahir-McFarland E, Mosialos G, Kieff E (2001) Effects of the NIK aly mutation on NF-kappaB activation by the Epstein-Barr virus latent infection membrane protein, lymphotoxin beta receptor, and CD40. J Biol Chem 276(18):14602–14606

Lugering A, Schmidt M, Lugering N, Pauels HG, Domschke W, Kucharzik T (2001) Infliximab induces apoptosis in monocytes from patients with chronic active Crohn's disease by using a caspase-dependent pathway. Gastroenterology 121(5):1145–1157

MacEwan DJ (2002) TNF receptor subtype signalling: differences and cellular consequences. Cell Signal 14(6):477–492

Malinin NL, Boldin MP, Kovalenko AV, Wallach D (1997) MAP3K-related kinase involved in NF-kappaB induction by TNF, CD95 and IL-1. Nature 385(6616):540–544

Martin SJ (2001) Dealing the CARDs between life and death. Trends Cell Biol 11(5):188–189

Maruo S, Oh-hora M, Ahn HJ, Ono S, Wysocka M, Kaneko Y, Yagita H, Okumura K, Kikutani H, Kishimoto T, Kobayashi M, Hamaoka T, Trinchieri G, Fujiwara H (1997) B cells regulate CD40L-induced IL-12 production in APC during T cell/APC interactions. J Immunol 158:120–126

Matsushima A, Kais ho T, Rennert PD, Nakano H, Kurosawa K, Uchida D, Takeda K, Akira S, Matsumoto M (2001) Essential role of nuclear factor (NF)-kB-inducing kinase and inhibitor of kB (IkB) kinase α in NF-κB activation through lymphotoxin B receptor, but not through tumor necrosis factor receptor I. J Exp Med 193:631–636

McAdam AJ, Greenwald RJ, Levin MA, Chernova T, Malenkovich N, Ling V, Freeman GJ, Sharpe AH (2001) ICOS is critical for CD40-mediated antibody class switching. Nature 409:102–105

McWhirter SM, Pullen SS, Holton JM, Crute JJ, Kehry MR, Alber T (1999) Crystallographic analysis of CD40 recognition and signaling by human TRAF2. Proc Natl Acad Sci USA 96(15): 8408–8413

Medema JP, Scaffidi C, Kischkel FC, Shevchenko A, Mann M, Krammer PH, Peter ME (1997) FLICE is activated by association with the CD95 death-inducing signaling complex (DISC). EMBO J 16:2794–2804

Moodycliffe AM, Shreedhar V, Ullrich SE, Walterscheid J, Bucana C, Kripke ML, Flores-Romo L (2000) CD40–CD40L interactions in vivo regulate migration of antigen-bearing dendritic cells from the skin to draining lymph nodes. J Exp Med 191:2011–2020

Moreno E, Yan M, Basler K (2002) Evolution of TNF signaling mechanisms: JNK-dependent apoptosis triggered by Eiger, the *Drosophila* homolog of the TNF superfamily. Curr Biol 12(14):1263–1268

Nagumo H, Agematsu K, Shinozaki K, Hokibara S, Ito S, Takamoto M, Nikaido T, Yasui K, Uehara Y, Yachie A, Komiyama A (1998) CD27/CD70 interaction augments IgE secretion by promoting the differentiation of memory B cells into plasma cells. J Immunol 161(12):6496–6502

Nahar IK, Shojania K, Marra CA, Alamgir AH, Anis AH (2003) Infliximab treatment of rheumatoid arthritis and Crohn's disease. Ann Pharmacother 37(9):1256–1265

Naito A, Azuma S, Tanaka S, Miyazaki T, Takasi S, Takatsu K, Nakao K, Nakamura K, Katsuki M, Yamamoto T, Inoue J (1999) Severe osteopetrosis, defective interleukin-1 signalling and lymph node organogenesis in TRAF6-deficient mice. Genes Cells 4(6):353–362

Nakano H, Oshima H, Chung W, Williams-Abbott L, Ware CF, Yagita H, Okumura K (1996) TRAF5, an activator of NF-kappaB and putative signal transducer for the lymphotoxin-beta receptor. J Biol Chem 271(25):14661–14664

Nishitoh H, Saitoh M, Mochida Y, Takeda K, Nakano H, Rothe M, Miyazono K, Ichijo H (1998) ASK1 is essential for JNK/SAPK activation by TRAF2. Mol Cell 2:389–395

Nolan B, Kim R, Duffy A, Sheth K, De M, Miller C, Chari R, Bankey P (2000) Inhibited neutrophil apoptosis: proteasome dependent NF-kappaB translocation is required for TRAF-1 synthesis. Shock 14(3):290–294

Pan G, Ni J, Wei YF, Yu G, Gentz R, Dixit VM (1997) An antagonist decoy receptor and a death domain-containing receptor for TRAIL. Science 277:815–818

Park SM, Kim HS, Choe J, Lee TH (1999) Differential induction of cytokine genes and activation of mitogen-activated protein kinase family by soluble CD40 ligand and TNF in a human follicular dendritic cell line. J Immunol 163(2):631–638

Pfeffer K (2003) Biological functions of tumor necrosis factor cytokines and their receptors. Cytokine Growth Factor Rev 14:185–191

Pradet-Balade B, Medema JP, Lopez-Fraga M, Lozano JC, Kolfschoten GM, Picard A, Martinez-A C, Garcia-Sanz JA, Hahne M (2002) An endogenous hybrid mRNA encodes TWE-PRIL, a functional cell surface TWEAK-APRIL fusion protein. EMBO J 21(21):5711–5720

Pullen SS, Miller HG, Everdeen DS, Dang TT, Crute JJ, Kehry MR (1998) CD40-tumor necrosis factor receptor-associated factor (TRAF) interactions: regulation of CD40 signaling through multiple TRAF binding sites and TRAF hetero-oligomerization. Biochemistry 37(34):11836–11845

Rabizadeh S, Bredesen DE (2003) Ten years on: mediation of cell death by the common neurotrophin receptor p75(NTR). Cytokine Growth Factor Rev 14(3–4):225–239

Ren CL, Morio T, Fu SM, Geha RS (1994) Signal transduction via CD40 involves activation of lyn kinase and phosphatidylinositol-3-kinase, and phosphorylation of phospholipase Cγ2. J Exp Med 179:673–680

Rennert PD, James D, Mackay F, Browning JL, Hochman PS (1998) Lymph node genesis is induced by signaling through the lymphotoxin beta receptor. Immunity 9(1):71–79

Rothe J, Mackay F, Bluethmann H, Zinkernagel R, Lesslauer W (1994) Phenotypic analysis of TNFR1-deficient mice and characterization of TNFR1-deficient fibroblasts in vitro. Circ Shock 44(2):51–56

Saito K, Ohara N, Hotokezaka H, Fukumoto S, Yuasa K, Naito M, Fujiwara T, Nakayama K (2004) Infection-induced up-regulation of the costimulatory molecule 4-1BB in osteoblastic cells and its inhibitory effect on M-CSF/RANKL-induced in vitro osteoclastogenesis. J Biol Chem 279(14):13555–13563

Sanchez I, Hughes RT, Mayer BJ, Yee K, Woodgett JR, Avruch J, Kyriakis JM, Zon LI (1994) Role of SAPK/ERK kinase-1 in the stress-activated pathway regulating transcription factor c-Jun. Nature 372:794–798

Scaffidi C, Krammer PH, Peter ME (1999) Isolation and analysis of components of CD95 (APO-1/Fas) death-inducing signaling complex. Methods 17(4):287–291

Scallon B, Cai A, Solowski N, Rosenberg A, Song XY, Shealy D, Wagner C (2002) Binding and functional comparisons of two types of tumor necrosis factor antagonists. J Pharmacol Exp Ther 301(2):418–426

Schwenzer R, Siemienski K, Liptai S, Schubert G, Peters N, Scheurich P, Schmid RM, Wajant H (1999) The human tumor necrosis factor (TNF) receptor-associated factor 1 gene (TRAF1) is up-regulated by cytokines of the TNF ligand family and modulates TNF-induced activation of NF-kappaB and c-Jun N-terminal kinase. J Biol Chem 274(27):19368–19374

Seitz C, Muller P, Krieg RC, Männel DN, Hehlgans T (2001) A novel p75TNF receptor isoform mediating NFkappa B activation. J Biol Chem 276(22):19390–19395

Senftleben U, Cao Y, Xiao G, Greten FR, Krahn G, Bonizzi G, Chen Y, Hu Y, Fong A, Sun SC, Karin M (2001) Activation by IKKalpha of a second, evolutionary conserved, NF-kappa B signaling pathway, Science 293(5534):1495–1499

Shanahan JC, St Clair W (2002) Tumor necrosis factor-alpha blockade: a novel therapy for rheumatic disease. Clin Immunol 103(3 Part 1):231–242

Shi C, Leonardi A, Kyriakis J, Siebenlist U, Kehrl JH (1999) TNF-mediated activation of the SAPK pathway: TRAF2 recruits and activates germinal center kinase related. J Immunol 163: 3279–3285

Shiels H, Li X, Schumacker PT, Maltepe E, Padrid PA, Sperling A, Thompson CB, Lindsten T (2000) TRAF4 deficiency leads to tracheal malformation with resulting alterations in air flow to the lungs. Am J Pathol 157(2):679–688

Shin HH, Kim SJ, Lee HS, Choi HS (2004) The soluble glucocorticoid-induced tumor necrosis factor receptor causes cell cycle arrest and apoptosis in murine macrophages. Biochem Biophys Res Commun 316(1):24–32

Shiozaki EN, Chai J, Shi Y (2002) Oligomerization and activation of caspase-9, induced by Apaf-1 CARD. Proc Natl Acad Sci USA 99:4197–4202

Siegel RM, Frederiksen JK, Zacharias DA, Chan FK, Johnson M, Lynch D, Tsien RY, Lenardo MJ (2000) Fas preassociation required for apoptosis signaling and dominant inhibition by pathogenic mutations. Science 288(5475):2354–2357

Sil AK, Maeda S, Sano Y, Roop DR, Karin M (2004) IkappaB kinase-alpha acts in the epidermis to control skeletal and craniofacial morphogenesis. Nature 428(6983):660–664

Song HY, Régnier CH, Kirschning CJ, Goeddel DV, Rothe M (1997) TNF-mediated kinase cascades: bifurcation of NF-kappaB and JNK/SAPK pathways at TRAF2. Proc Natl Acad Sci USA 94:9792–9796

Soong L, Xu JC, Grewal IS, Kima P, Sun J, Longley BJ Jr, Ruddle NH, McMahon-Pratt D, Flavell RA (1996) Disruption of CD40–CD40L interactions results in an enhanced susceptibility to *Leishmania amazonesis* infection. Immunity 4:263–273

Speiser DE, Lee SY, Wong B, Arron J, Santana A, Kong YY, Ohashi PS, Choi Y (1997) A regulatory role for TRAF1 in antigen-induced apoptosis of T cells. J Exp Med 185(10):1777–1783

Spinicelli S, Nocentini G, Ronchetti S, Krausz LT, Bianchini R, Riccardi C (2002) GITR interacts with the pro-apoptotic protein Siva and induces apoptosis. Cell Death Differ 9(12):1382–1384

Stanger BZ, Leder P, Lee TH, Kim E, Seed B (1995) RIP: a novel protein containing a death domain that interacts with Fas/APO-1 (CD95) in yeast and causes cell death. Cell 81:513–523

Stout RD, Suttles J, Xu J, Grewal IS, Flavell RA (1996) Impaired T cell-mediated macrophage activation in CD40 ligand-deficient mice. J Immunol 156:8–11

Sutherland CL, Heath AW, Pelech SL, Young PR, Gold MR (1996) Differential activation of the ERK, JNK, and p38 MAP kinases by CD40 and the B cell antigen receptor. J Immunol 157:3381–3390

Suzuki I, Fink PJ (1998) Maximal proliferation of cytotoxic T lymphocytes requires reverse signaling through Fas ligand. J Exp Med 187(1):123–128

Suzuki Y, Nakabayashi Y, Takahashi R (2001) Ubiquitin–protein ligase activity of X-linked inhibitor of apoptosis protein promotes proteosomal degradation of caspase-3 and enhances its anti-apoptotic effect in Fas-induced cell death. Proc Natl Acad Sci USA 98: 8662–8667

Takeda K, Takeuchi O, Tsujimura T, Itami S, Adachi O, Kawai T, Sanjo H, Yoshikawa K, Terada N, Akira S (1999) Limb and skin abnormalities in mice lacking IKKa. Science 284:313–316

Takeuchi M, Rothe M, Goeddel D (1996) Anatomy of TRAF2. Distinct domains for nuclear factor-κB activation and association with tumor necrosis factor signaling proteins. J Biol Chem 271:19935–19942

Tanaka M, Fuentes ME, Yamaguchi K, Durnin MH, Dalrymple SA, Hardy KL, Goeddel DV (1999) Embryonic lethality, liver degeneration, and impaired NF-κB activation in IKKB-deficient mice. Immunity 10:421–429

Tang G, Yang J, Minemoto Y, Lin A (2001) Blocking caspase-3-mediated proteolysis of IKKbeta suppresses TNF-alpha-induced apoptosis. Mol Cell 8:1005–1016

Tartaglia LA, Ayers TM, Wong GH, Goeddel DV (1993) A novel domain within the 55 kD TNF receptor signals cell death. Cell 74(5):845–853

ten Hove T, van Montfrans C, Peppelenbosch MP, van Deventer SJ (2002) Infliximab treatment induces apoptosis of lamina propria T lymphocytes in Crohn's disease. Gut 50(2):206–211

Tumanov AV, Kuprash DV, Nedospasov SA (2003) The role of lymphotoxin in development and maintenance of secondary lymphoid tissues. Cytokine Growth Factor Rev 14:275–288

Uren AG, Pakusch M, Hawkins CJ, Puls KL, Vaux DL (1996) Cloning and expression of apoptosis inhibitory protein homologs that function to inhibit apoptosis and/or bind tumor necrosis factor receptor associated factors. Proc Natl Acad Sci USA 93:4974–4978

Vandenabeele P, Declercq W, Beyaert R, Fiers W (1995) Two tumour necrosis factor receptors: structure and function. Trends Cell Biol. 5(10):392–393

van Deventer SJ (2001) Transmembrane TNF-alpha, induction of apoptosis, and the efficacy of TNF-targeting therapies in Crohn's disease. Gastroenterology 121(5):1242–1246

Vandevoorde V, Haegeman G, Fiers W (1997) Induced expression of trimerized intracellular domains of the human tumor necrosis factor (TNF) p55 receptor elicits TNF effects. J Cell Biol 137(7):1627–1638

van Essen D, Kikutani H, Gray D (1995) CD40 ligand-transduced co-stimulation of T cells in the development of helper function. Nature 378:620–623

von Tresckow B, Kallen KJ, von Strandmann EP, Borchmann P, Lange H, Engert A, Hansen HP (2004) Depletion of cellular cholesterol and lipid rafts increases shedding of CD30. J Immunol 172(7):4324–4331

Wang C, Deng L, Hong M, Akkaraju GR, Inoue J, Chen ZJ (2001) TAK1 is a ubiquitin-dependent kinase of MKK and IKK. Nature 412:346–351

Wang CY, Mayo MW, Korneluk RG, Goeddel DV, Baldwin AS Jr (1998) NF-kappaB antiapoptosis: induction of TRAF1 and TRAF2 and c-IAP1 and c-IAP2 to suppress caspase-8 activation. Science 281(5383):1680–1683

Wang J, Lobito AA, Shen F, Hornung F, Winoto A, Lenardo MJ (2000) Inhibition of Fas-mediated apoptosis by the B cell antigen receptor through c-FLIP. Eur J Immunol 30:155–163

Wei MC, Lindsten T, Mootha VK, Weiler S, Gross A, Ashiya M, Thompson CB, Korsmeyer SJ (2000) tBID, a membrane-targeted death ligand, oligomerizes BAK to release cytochrome *c*. Genes Dev 14:2060–2071

Wheller K, Pound JD, Gordon J, Jefferis R (1993) Engagement of CD40 lowers the threshold for activation of resting B cells via antigen receptor. Eur J Immunol 23:1165–1168

Wong BR, Besser D, Kim N, Arron JR, Vologodskaia M, Hanafusa H, Choi Y (1999) TRANCE, a TNF family member, activates Akt/PKB through a signaling complex involving TRAF6 and c-Src. Mol Cell 4(6):1041–1049

Yamamoto H, Kishimoto T, Minamoto S (1998) NF-kappaB activation in CD27 signaling: involvement of TNF receptor-associated factors in its signaling and identification of functional region of CD27. J Immunol 161(9):4753–4759

Yamamoto Y, Verma UN, Prajapati S, Kwak YT, Gaynor RB (2003) Histone H3 phosphorylation by IKK-α is critical for cytokine-induced gene expression. Nature 423:655–659

Yan M, Dai T, Deak JC, Kyriakis JM, Zon LI, Woodgett JR, Templeton DJ (1994) Activation of SAPK by MEKK1 phosphorylation of its activator SEK1. Nature 372:798–800

Yang X, Chang HY, Baltimore D (1998) Autoproteolytic activation of pro-caspases by oligomerization. Mol Cell 1:319–325

Ye X, Mehlen P, Rabizadeh S, VanArsdale T, Zhang H, Shin H, Wang JJ, Leo E, Zapata J, Hauser CA, Reed JC, Bredesen DE (1999) TRAF family proteins interact with the common neurotrophin receptor and modulate apoptosis induction. J Biol Chem 274(42):30202–30208

Yeh WC, Shahinian A, Speiser D, Kraunus J, Billia F, Wakeham A, de la Pompa JL, Ferrick D, Hum B, Iscove N, Ohashi P, Rothe M, Goeddel DV, Mak TW (1997) Early lethality, functional NF-kappaB activation, and increased sensitivity to TNF-induced cell death in TRAF2-deficient mice. Immunity 7:715–725

Yin L, Wu L, Wesche H, Arthur CD, White JM, Goeddel DV, Schreiber RD (2001) Defective LT-α receptor-induced NF-κB transcriptional activity in NIK-deficient mice. Science 291:2162–2165

Yuasa T, Ohno S, Kehrl JH, Kyriakis JM (1998) TNF signaling to SAPK/JNK and p38. Germinal center kinase couples TRAF2 to MAPK/ERK kinase kinase-1 and SAPK while RIP associates with a MAPKKK upstream of MKK6 and p38. J Biol Chem 273:22681–22692

Yujiri T, Ware M, Widmann C, Oyer R, Russell D, Chan E, Zaitsu Y, Clarke P, Tyler K, Oka Y, Fanger GR, Henson P, Johnson GL (2000) MEK kinase 1 gene disruption alters cell migration and c-Jun NH2-terminal kinase regulation but does not cause a measurable defect in NF-kappa B activation. Proc Natl Acad Sci USA 97(13):7272–7277

Zhu N, Ramirez LM, Lee RL, Magnuson NS, Bishop GA, Gold MR (2002) CD40 signaling in B cells regulates the expression of the Pim-1 kinase via the NF-κB pathway. J Immunol 168:744–754

Zou H, Henzel WJ, Liu X, Lutschg A, Wang X (1997) Apaf-1, a human protein homologous to *C. elegans* CED-4, participates in cytochrome *c*-dependent activation of caspase-3. Cell 90: 405–413

Regulation of Immunoreceptor Activities: The Paradigm of the Type I Fcε Receptor

JAKUB ABRAMSON, ISRAEL PECHT

8.1 Introduction

Antigen-specific activation of immunocytes is initiated upon clustering their respective multichain immune recognition receptors (MIRR) (e.g. TCR, BCR, Fc receptors; Keegan and Paul 1992). All members of the MIRR family consist of membranal subunits responsible for antigen recognition, the majority of which are non-covalently associated with one or more transmembrane subunits, containing conserved 18-amino-acid-long sequences (E/DxxYxxLxxxxxxxYxxL) (where x denotes any amino acid) within their cytoplasmic tails. These motifs were first identified and described by Michael Reth in 1989 (Reth 1989), and were later named immunoreceptor tyrosine-based activation motifs (ITAMs) (Cambier 1995), as they are responsible for coupling the respective MIRRs to their cellular signaling cascades. The key event required for initiating the signaling machinery is phosphorylation of the ITAMs' canonical tyrosines by protein tyrosine kinases (PTKs) which become proximal to the receptors upon their aggregation (Paolini et al. 1991). Signals generated thereafter involve many of the same coupling molecules, which become activated and subsequently induce events such as the transient rise in free cytosolic calcium ion concentrations, gene transcription and exo- or endocytosis, etc.

Here, we illustrate a receptor's stimulus-response coupling cascade using type I Fcε receptor (FcεRI) as a model. In the first part of this chapter (Sect. 8.2), we discuss in detail the current understanding of the molecular mechanisms that couple the FcεRI stimulus to the secretion of proinflammatory mediators. Our focus will then be on distinct regulatory mechanisms, primarily those exerting inhibition of the above cascade (Sect. 8.3).

8.2 Type I Fcε Receptor (FcεRI)-Mediated Cell Responses

8.2.1 FcεRI Structure, Expression and Function

FcεRI is a tetrameric molecule composed of three polypeptide subunits: α-, β- and a disulfide-linked γ-chain homodimer (Metzger 1992). The extracellular part of the α-subunit contains two extracellular Ig-like domains (Garman et al. 1998),

Springer Series in Biophysics
S. Damjanovich (Ed.)
Biophysical Aspects of Transmembrane Signaling

which bind with high affinity (Kd=10^{-9}–10^{-10} M) to the Fc part of monomeric IgE class antibodies (Helm et al. 1988). The single transmembrane stretch of the α-subunit contains several charged amino acid residues, which together with hydrophobic interactions mediate the association of the α-subunit with the ITAM containing γ- and β-chains (Zidovetzki et al. 2003). While the γ-subunit is essential for the FcεRI-induced signal transduction (Letourneur and Klausner 1991), the β-subunit has been proposed to function as an amplifier (Lin et al. 1996) and/or as a possible suppressor, i.e., a modulator (Erdei et al. 1995, 1998; Wilson et al. 1995; Furumoto et al. 2004), of the γ-chain-mediated signaling events.

While the FcεRI of rodents has an obligatory $\alpha\beta\gamma_2$ tetrameric structure, human FcεRI is expressed as both trimeric ($\alpha\gamma_2$) and tetrameric ($\alpha\beta\gamma_2$) structures (Kinet 1999). The cellular distribution of human and rodent FcεRI is also different. Rodent FcεRI is only expressed on mast cells (MCs) and basophils, whereas expression of human FcεRI extends to monocytes (Maurer et al. 1993), eosinophils (Gounni et al. 1994), platelets (Joseph et al. 1997), Langerhans cells (Bieber et al. 1992; Wang et al. 1992) and dendritic cells (DC) (Maurer et al. 1996). Interestingly, normal human monocytes express 10- to 100-fold less Fcε receptors than circulating basophils (Maurer et al. 1993; Kinet 1999).

Clustering of the FcεRI on mast cells and basophils triggers a biochemical cascade that culminates in secretion of granule-stored mediators (Barsumian et al. 1981), and the de novo synthesis and secretion of arachidonic acid metabolites (Lewis et al. 1982; Razin et al. 1983; Siraganian et al. 1983) and several cytokines (Wodnar-Filipowicz et al. 1989; Gordon et al. 1990; Galli et al. 1991; Gordon and Galli 1991; Sayama et al. 2002). Upon secretion, these mediators affect many other cells: in particular, they mobilize immune system cells (e.g., neutrophils) and thus initiate an early immune response to pathogens such as parasites and bacteria (reviewed in Malaviya and Abraham 2001). On the other hand, secretion of these mediators is also often associated with the induction of type I IgE-dependent allergic reactions (Kinet 1999).

The relevance of FcεRI expression on the antigen-presenting cells (APC) (such as monocytes and DC/Langerhans cells) is not yet fully understood. However, several studies suggest that FcεRI functions as an allergen-focusing molecule on these cells (Maurer et al. 1995). It has been shown that allergens are more efficiently taken up, processed and presented to T cells following targeting to APC via FcεRI compared to allergen binding to APC in the conventional manner in vivo (Stingl and Maurer 1997; Kraft et al. 2001). This seems to contribute to IgE-mediated delayed-type hypersensitivity reactions that are thought to play a pivotal role in atopic diseases. Indeed, it has been reported that Langerhans cells of atopic individuals display elevated levels of FcεRI (particularly in atopic dermatitis skin). In addition, recent studies of FcεRI signal transduction and function on APC suggest additional mechanisms by which FcεRI engagement on APC could affect inflammatory reactions. Specifically, engagement of FcεRI on APC has been shown to induce major signaling events such as activation of protein tyrosine kinases and increased activation of transcription factors, eventually resulting in secretion of several cytokines (TNF-α) and chemokines (IL-8, MCP-1) (Kraft et al. 2002a,b).

8.2.2
FcεRI Stimulus-Response Coupling Cascade on Mast Cells

Our understanding of stimulus-response coupling networks in cells of the immune system has undergone dramatic advancement during the past decade, together with the field of molecular signaling, which is one of the fastest evolving fields in the life sciences. It is constantly changing, making it difficult to present a final paradigm. To illustrate this, we first briefly present the "original" model of the FcεRI stimulus-response coupling cascade in mast cells, and then follow this with detailed updates summarizing current understanding.

8.2.2.1
The "Original" Model

In unperturbed mast cells, a fraction of the PTK Lyn molecules is constitutively associated with the FcεRI β-chain (probably via its SH3/SH4 domains) (Vonakis et al. 1997, 2001). Upon FcεRI clustering, the adjacent Lyn transphosphorylates the ITAMs present on both β- and γ-chains of the receptor complex (Eiseman and Bolen 1992; Pribluda et al. 1994). While the phosphorylated β-chain functions as a binding site for the SH2 domain of activated Lyn (and hence the β-chain may serve as an amplifier), the phosphorylated γ-chain dimer serves as a docking site for SH2 domains of another PTK–Syk, which is then activated by proximal Lyn (Kihara and Siraganian 1994; Scharenberg et al. 1995). These two activated PTKs then phosphorylate a set of several signal-relay or adaptor molecules such as the linker for activation of T cells (LAT) (Saitoh et al. 2000) and enzymes including phosphatidylinositol 3-kinase (PI3 K) and Bruton's tyrosine kinase (Btk) (Kawakami et al. 1994), which in turn contribute to the activation of phospholipase C-γ (PLC-γ) (Nadler et al. 2000). Activated PLC-γ subsequently hydrolyzes the membranal phosphatidylinositol-3,4-bisphosphate [PI(3,4)P2] into the two second messengers, inositol-1,4,5-trisphosphate (IP3) and diacylglycerol (DAG). IP3 then binds to specific receptors on the endoplasmic reticulum (ER), causing release of Ca^{2+} ions from ER internal stores. Depletion of the latter triggers influx of extracellular calcium and results in the prolongation of the transient elevation of free cytosolic Ca^{2+} ion concentration, which is required for the FcεRI-mediated responses. DAG is essential for the activation of certain isoforms of protein kinase C (PKC), which are required for the secretory response as well as coupling of the stimulus to the nucleus (Zhang et al. 1997b).

However, during the last 5 years, some novel insights have been obtained concerning the above stimulus-response cascade. Specifically: (1) binding of monomeric IgE molecules to the FcεRI has been reported to initiate several important cellular responses, suggesting that it does not only play a passive role; (2) the initiation of FcεRI signaling has been suggested to depend on the involvement of plasma membrane microdomains known as lipid rafts; (3) PTK Lyn, which was considered to be the key kinase of the FcεRI coupling cascade, was found to be dispensable for the secretory response, implying that an alternative PTK must be involved; (4) a novel PTK required for coupling the receptor stimulus to the secretory response has recently been identified as Fyn; (5) the FcεRI-mediated PLC-γ activation, which was originally thought to be dependent on Btk and PI3 K activity, is apparently controlled by in different mechanisms; (6) regulation of FcεRI-induced gene transcription and subsequent protein de novo synthesis are currently better understood.

All these findings are presented in detail below.

8.2.2.2 The Function of Monomeric IgE (mIgE)

The long-standing paradigm of the FcεRI stimulus-response coupling cascade states that IgE binding to FcεRI endows it with specific antigen recognition capacity and that only upon FcεRI clustering (e.g., with multivalent antigen or anti-IgE Abs) does triggering of the cascade take place (Beaven and Metzger 1993; Turner and Kinet 1999). Recently, however, several reports have suggested that monomeric IgE (mIgE) binding to FcεRI initiates multiple signaling pathways, which result in several important cellular responses. These range from enhanced cell adhesion to secretion of stored and/or de novo synthesized mediators (Kalesnikoff et al. 2001; Oka et al. 2004). Unfortunately, upon critical examination of these reports, considerable discrepancies are observed in experimental protocols, results and conclusions. Hence, the above observation requires a more elaborate and critical discussion.

Probably the first observation suggesting that FcεRI binding of mIgE is capable of inducing a biological response was reported by Hsu and Yamaguchi, who showed that exposure of (human and rodent) mast cells to IgE results in a striking (up to 32-fold) upregulation of surface expression of FcεRI, both in vitro or in vivo (Hsu and MacGlashan 1996; Yamaguchi et al. 1997a,b). Further studies have established that this enhanced expression is in fact caused by the increased half-life of the surface-resident FcεRI rather than by its increased synthesis and/or transport to the plasma membrane (Borkowski et al. 2001; Kubo et al. 2001).

More recently, two independent studies have reported that mIgE binding leads to the prolonged survival of mouse mast cells under growth-factor-limiting conditions (Asai et al. 2001; Kalesnikoff et al. 2001). However, these two studies differed in the proposed mechanisms: Kalesnikoff et al. found that mIgE binding to FcεRI induces tyrosine phosphorylation of several intracellular components (e.g., the receptor's β- and γ-chains, Shc and SHIP), as well as activation of PI3 K, Akt and MAP kinases (Erks, JNK, p38). This response was followed by increased de-novo cytokine synthesis which was suggested to be responsible for the observed cell survival in an autocrine-dependent fashion (Kalesnikoff et al. 2001). In contrast, Asai et al. did not detect any signaling events, or significant cytokine secretion by the mIgE-treated cells. Further confusion has been caused by several more recent papers demonstrating that, in contrast to the studies of Kalesnikoff and Asai, mIgEs are capable of inducing secretion of granule-stored mediators by the rat mucosal mast cells of the RBL-2H3 line (Oka et al. 2004; Pandey et al. 2004). These marked differences in cellular responses naturally raised several key questions: namely, what is the nature of the assumed stimulus produced by mIgE binding and what mechanism couples it to the biological response? Does mIgE binding to FcεRI induce in it a conformational change that couples it to the biological response or does it rather mediate aggregation of the FcεRI resident molecules?

The simplest rational could be that the supposed mIgE still contains traces of IgE oligomers, which upon FcεR binding cause its clustering. However, in most of the studies, considerable efforts were made in mIgE purification and in ascertaining its monomeric nature. Hence one may exclude this explanation.

Further insights into the events induced upon mIgE binding have been provided by a recent study by Kitaura et al. (2003). They found that mIgE binding does not induce any biological responses in mast cells ablated of Syk, implying that the coupling does involve tyrosine phosphorylation (although not always detectable) mediated by this PTK. More importantly, however, their (and other) studies demonstrated that addition of monovalent haptens for which the IgE is specific abrogates the responses (Kalesnikoff et al. 2001; Kitaura et al. 2003; Pandey et al. 2004). This raised the possibility that FcεRI resident IgEs may be clustered via binding of their antigen recognition site to an epitope on their own structure or to another cell-surface component. Such an interaction may result from artificial experimental conditions employed in all these studies, where the rather high cell-surface expression levels of FcεRI (10- to 100-fold higher than those of TCR) are all occupied by one and the same monoclonal IgE. Moreover, one is dealing with the two-dimensional concentration of the FcεRI–IgE complexes, which is therefore markedly higher (Schweitzer-Stenner and Pecht 2005). Thus, it is reasonable to assume that under such non-physiological conditions, even a low affinity of the IgE will still lead to either homo- or heterotypic aggregation and induce clustering of the receptors. In line with this rational is the observation that different monoclonal IgEs were found to have distinct capacities to induce FcεRI aggregation and the subsequent biological responses (Kitaura et al. 2003). While IgE-class mAbs such as SPE7 induce significant cytokine production with strong antiapoptotic effects, others (such as H1) display a lower (or undetectable) capacity to induce cytokine secretion and exert less robust survival effects. Significantly, recent studies of mAb SPE7 have demonstrated that its binding site may exist in at least two distinct (unbound) conformations, which markedly differ in their epitope binding specificities (James and Tawfik 2003; James et al. 2003). The first conformation has a relatively flat binding site, whereas the other isomer exhibits a deep hole typical of hapten binding antibodies. As expected, screening studies have demonstrated that SPE7 binds to several structurally unrelated compounds with a broad range of affinities (James et al. 2003).

8.2.2.3 Lipid Rafts and FcεRI Activation

The current model for the initiation of FcεRI signaling proposes the existence of plasma membrane microdomains known as lipid rafts (Baird et al. 1999). These detergent-insoluble structures (enriched in sphingolipids and cholesterol) are thought to control the spatial organization of key coupling molecules, including receptors, non-receptor PTKs and phosphatidylinositol-associating proteins.

According to this model, FcεRI (which is excluded from rafts in unperturbed cells) translocates upon its aggregation into these microdomains (Field et al. 1995, 1997), where it interacts or becomes proximal to Src-family kinases. These enzymes are constitutively associated with lipid rafts (Honda et al. 2000), as they are myristoylated/palmitoylated on their N-terminus (saturated fatty acid preferentially associates with cholesterol and hence with rafts). This interaction and proximity are suggested to initiate their trans-activation, and thereby the subsequent signaling events which eventually result in a biological response.

In order to maintain an efficient response, it is important that Src family members are strictly controlled: Their negative regulation is carried out by the C-terminal Src kinase (Csk), which is recruited to the lipid rafts upon FcεRI clustering by a transmembrane adaptor named Cas-binding protein (CBP) (see Sect. 8.3.2). There, Csk phosphorylates these kinases at an autoregulatory C-terminal tyrosine (Honda et al. 1997) and keeps their activity in check (see Sect. 8.3.1). In contrast, the positive regulation involves CD45, a transmembrane protein tyrosine phosphatase (PTP), which dephosphorylates the regulatory tyrosine, and thereby activates the PTK and potentiates the FcεRI-induced secretory response (Berger et al. 1994; Adamczewski et al. 1995; Murakami et al. 2000; Hamaguchi et al. 2001).

8.2.2.4
Lyn Activation and Its Significance

The above-described original linear cascade model (8.2.2.1), where Lyn functions as the exclusive PTK initiating FcεRI signaling, has served as a paradigm for a long time (Nadler et al. 2000). However, more recent studies have suggested that Lyn may not be the sole Src family member required for FcεRI activation and proposed the activity of another PTK (Parravicini et al. 2002). The first evidence supporting this notion mainly came from studies of bone marrow mast cells (BMMC) derived from Lyn-deficient mice (Nishizumi and Yamamoto 1997; Kawakami et al. 2000b). It was reasoned that if Lyn is the only PTK required for initiating the FcεRI signal transduction cascade, then Lyn ablation should abrogate all FcεRI-mediated downstream events (in a manner analogous to the Syk knock-out; Costello et al. 1996; see below). Indeed, defects in FcεRI-mediated calcium mobilization and tyrosyl phosphorylation were observed in Lyn-deficient mast cells. Surprisingly, however, FcεRI-induced degranulation, MAP kinase activation and cytokine production were found operative in these lyn–/– mast cells (Nishizumi and Yamamoto 1997; Kawakami et al. 2000b). Recent studies by Rivera and colleagues have demonstrated that lyn–/– mast cells display even higher secretory response to FcεRI clustering than wild-type (WT) cells (Parravicini et al. 2002). Furthermore, the activation of Akt, PKCδ and MAP kinases (e.g., JNK and Erk), as well as the subsequent production and secretion of cytokines such as IL-2 and TNF-α, are markedly enhanced (and/or prolonged) in the Lyn-deficient mast cells (Kawakami et al. 2000b). These data therefore imply that Lyn is dispensable for the FcεRI-induced secretory response and that an alternative mechanism(s) must be involved (see Sect. 8.2.2). Moreover, these results demonstrate that, in addition to its activating capacity, Lyn is also involved in regulatory functions (Harder et al. 2001), which modulate the FcεRI-induced responses. Such a "dual" role for Lyn has previously been observed in B cells derived from Lyn-ablated mice, which are hyper-responsive to membranal IgM clustering-induced proliferation, while they are impaired in that induced by membranal IgD aggregation (Wang et al. 1996; Horikawa et al. 1999). Additional support for a regulatory function of Lyn stems from studies showing that Lyn is responsible for phosphorylating inhibitory receptors, such as FcγRIIB (Malbec et al. 1998), gp49B1, PIR-B (Ho et al. 1999) and MAFA (Xu et al. 2001). These receptors are involved in the negative regulation of the FcεRI-mediated response, for example, discussed in more detail in Section 8.3 of this chapter.

Finally, it should be stressed that none of the studies involving Lyn-deficient mast cells (MC) monitored the tyrosine phosphorylation of γ-subunits [despite the fact that several reviews (Turner and Kinet 1999; Nadler et al. 2000) report that it is, like β-chain, abolished in lyn–/– cells]. Since the tyrosine phosphorylation of γ-ITAM was shown to be essential for initiating the FcεRI-mediated secretory response, logically an additional PTK must be involved in its phosphorylation. Honda et al. (2000) provided indirect evidence that PTK Fyn can also induce tyrosine phosphorylation of both β- and γ-receptor subunits (Honda et al. 2000). To provide a complete answer to this crucial question, additional studies (e. g., using fyn–/– lyn–/– double knockout mice) are necessary.

8.2.2.5 Fyn–Gab-2–PI3 K-Mediated Signaling

As mentioned above, Lyn's exclusive role has been challenged, and another FcεRI proximal PTK, the Src family member Fyn (feline yes-related protein), was clearly shown to be an essential element of the coupling cascade. Evidence that Fyn is responsible for inducing signaling events crucial for mast cell degranulation was provided by Parravicini et al. (2002), who showed that fyn–/– BMMCs are (unlike lyn–/– BMMC) impaired (close to 90%) in their FcεRI-induced secretory response. The same group also identified a downstream effector of Fyn activity – the adaptor protein Gab2 – which was found to exhibit markedly reduced tyrosine phosphorylation in fyn–/– but not in lyn–/– mast cells upon FcεRI clustering (Parravicini et al. 2002). Furthermore, mast cells derived from gab2–/– and fyn–/– mice display similar phenotype (e. g., defective activation of PI3 K, and secretory response) (Gu et al. 2001), suggesting that Gab2 may be a key coupling element between Fyn and PI3 K activation.

PI3 K is assumed to play a central role in linking the FcεRI stimulus with the mast cell secretory response, as its specific inhibition by wortmannin results in a dramatic suppression of FcεRI-induced degranulation (and the de novo synthesis of several cytokines) (Barker et al. 1995). The products of PI3 K activity are: phosphatidylinositol-3,4-bisphosphate [PI(3,4)P2] and phosphatidylinositol-3,4,5-trisphosphate [PI(3,4,5)P3], which can bind to pleckstrin homology (PH) domains of several signaling proteins such as Btk, PLC-γ, Akt and PDK1 (see Fig. 8.1). This interaction facilitates recruitment of the above proteins to the plasma

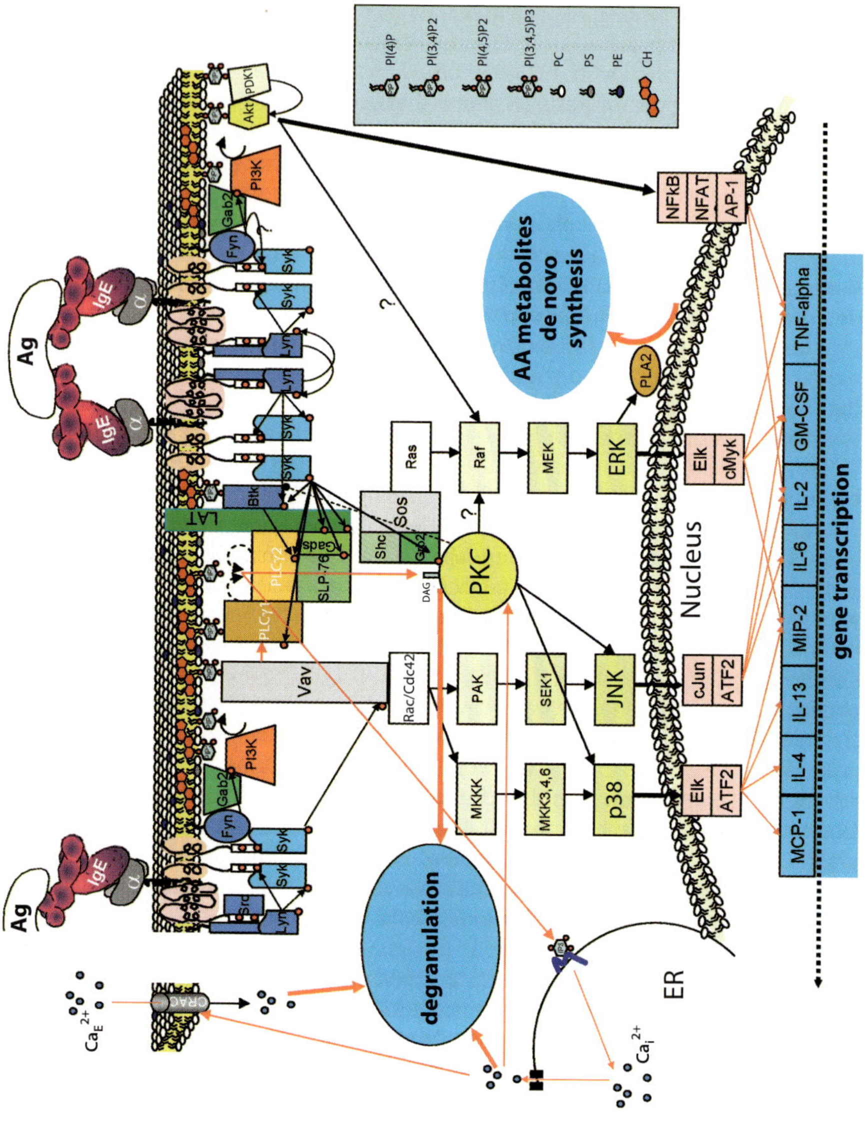
Ag
IgE
α
Src
Lyn
Syk
Fyn
Gab2
PI3K
Akt
PDK1
Btk
LAT
PLCγ2
Gads
SLP-76
PLCγ1
Vav
Shc
Grb2
Sos
Ras
Raf
MEK
ERK
PKC
DAG
Rac/Cdc42
PAK
SEK1
JNK
MKKK
MKK3,4,6
p38
PLA2
AA metabolites de novo synthesis
degranulation
CRAC
Nucleus
ER
NFkB
NFAT
AP-1
Elk
cMyk
cJun
ATF2
MCP-1
IL-4
IL-13
MIP-2
IL-6
IL-2
GM-CSF
TNF-alpha
gene transcription
PI(4)P
PI(3,4)P2
PI(4,5)P2
PI(3,4,5)P3
PC
PS
PE
CH

Fig. 8.1. A scheme illustrating the network coupling the type I Fc receptor for IgE (FcεRI) to mast cell secretory responses: aggregation of IgE-residing in FcεRI leads to translocation of receptors into lipid rafts, where they interact with raft-associated Src family kinases such as Fyn and Lyn. This interaction is a prerequisite for efficient phosphorylation of the ITAMs present on both β- and γ-chains of the FcεRI complex. The phosphorylated γ-chain dimer then serves as a docking site for PTK–Syk, which becomes activated mostly by proximal Lyn associated with β-chain. Activated Syk then phosphorylates membrane-associated protein LAT which becomes a docking site for a number of cytosolic proteins such as SLP-76, Gads and PLC-γ. This multiprotein complex is then phosphorylated by Syk and/or Lyn. In parallel, activated Fyn phosphorylates the adaptor protein Gab2, which then associates with PI3 K and thereby enables its recruitment to the plasma membrane and subsequent activation. Active PI3 K catalyzes the production of PI(3,4)P2 and PI(3,4,5)P3 which bind to pleckstrin homology (*PH*) domains of several signaling proteins such as Btk, PLC-γ, Vav, Akt and PDK1. This interaction also promotes translocation of the above proteins to the plasma membrane, where they become activated. Tyrosine-phosphorylated PLC-γ catalyzes hydrolysis of PIP_2 in the membrane, forming IP_3 and 1,2-diacylglycerol – second messengers that release Ca^{2+} from internal stores and activate PKC, respectively. Binding of IP_3 to specific receptors in the endoplasmic reticulum results in a depletion of Ca^{2+} stores, which activates store-operated Ca^{2+} entry (I_{CRAC}) from the extracellular medium. Btk, SLP-76, LAT and PLC-γ are all essential for generating signals for sustained Ca^{2+} influx. Both elevated Ca^{2+} levels and activated PKC eventually induce secretion of granule-stored mediators. Tyrosine phosphorylation and activation of other enzymes and adaptors, including Vav, Shc, Grb2 and Sos, stimulate small GTPases such as Rac, Cdc42 and Ras. These pathways lead to activation of the ERK, JNK and p38 MAP kinases, phosphorylation of transcription factors that induce the de novo synthesis and secretion of cytokines, and activation of cytoplasmic phospholipase A_2 ($cPLA_2$) to release arachidonic acid (*AA*) metabolites. Adaptor proteins are illustrated in *green*; PTKs in *blue*; S/T kinases in *yellow*; lipases in *brown*; GTPases in *white*; GDP/GTP exchange factors in *grey*; transcription factors in *pink*. *Black arrows* indicate (de)phosphorylation; *red arrows* binding/activation

membrane, where they may become activated. This step is assumed to be essential for coupling the activating stimulus to the respective biological response. Taken together, the Fyn–Gab2 complex therefore appears to be an important link between FcεRI clustering and PI3 K activation, and thus is essential for coupling the FcεRI stimulus to the secretory response (Fig. 8.1).

8.2.2.6 The Syk–LAT–SLP-76 Signaling Pathway

The spleen tyrosine kinase (Syk) is expressed by hematopoietic, epithelial and endothelial cells (Taniguchi et al. 1991). In mast cells, Syk plays an essential role in coupling the FcεRI-mediated tyrosine phosphorylation to the secretion of granules stored and de novo synthesized inflammatory mediators (reviewed in Siraganian et al. 2002). One of the first indications of the critical role of Syk in the FcεRI signaling pathway came from the early studies of Siraganian and colleagues, who identified a variant of the RBL-2H3 cell line with no detectable Syk expression and activity (Zhang et al. 1996). In these Syk-deficient TB1A2 cells, FcεRI aggregation leads to tyrosine phosphorylation of the receptor's β- and γ-subunits, while it did not induce any detectable tyrosine phosphorylation of other cellular proteins and any secretory response. Reconstitution of Syk expression by transfection restored the protein tyrosine phosphorylation as well as the increase in intracellular free Ca^{2+} and histamine release in response to the FcεRI aggregation (Zhang et al. 1996). These findings were then confirmed by the studies of Costello et al. (1996), who used mast cells derived from Syk-deficient mice. These cells also failed to degranulate, as well as to synthesize and secrete leukotrienes or cytokines upon FcεRI clustering.

Activated Syk has been shown to phosphorylate a number of cytosolic substrates (see Fig. 8.1), such as Shc (Jabril-Cuenod et al. 1996), Cbl (Ota et al. 1996) and SLP-76 (SH2 domain-containing leukocyte protein of 76 kDa) (Hendricks-Taylor et al. 1997), some of which were also shown to play an essential role in coupling the upstream signals to the secretory response. One of the most important ones, however, seems to be LAT. This small transmembrane adaptor protein couples the activated PTKs to their downstream substrates, as, upon its tyrosine phosphorylation, LAT recruits them to the plasma membrane, where they may become activated (see Fig. 8.1). LAT-deficient BMMCs display an intact tyrosine phosphorylation of FcεRI β- and γ-chains, of Syk, and Vav following FcεRI aggregation. However, the tyrosine phosphorylation of the downstream elements including SLP-76, PLC-γ1 and PLC-γ2, as well as Ca^{2+} mobilization are dramatically reduced. Hence, the FcεRI-induced degranulation, MAPK activation and subsequent cytokine production are disrupted in LAT-deficient BMMCs (Saitoh et al. 2000). Consistent with these results, BMMCs from SLP76(–/–) mice fail to degranulate and to secrete IL-6 upon FcεRI clustering, suggesting that the Syk-dependent LAT–SLP-76 complex is essential for FcεRI-induced secretory response (Pivniouk et al. 1999).

Taken together, the current model of FcεRI-induced activation of mast cells involves two independent coupling pathways (Fig. 8.1): the first involving (Lyn)–Syk–LAT, and the second that of Fyn–Gab2–PI3 K signaling. Both pathways are activated in parallel, and complement each other's induction of the secretory response.

However, even this updated model does not provide a complete understanding of the FcεRI coupling cascade. In particular, it is still unclear why Lyn, which is thought to operate upstream of Syk (reviewed in Turner and Kinet 1999), is dispensable (Nishizumi and Yamamoto 1997), while Syk is essential for mediating secretory responses (Costello et al. 1996). The simple explanation could be that, in addition to Lyn, another PTK may also activate Syk. Indeed, experiments using Lyn-deficient mast cells demonstrated that FcεRI-induced tyrosine phosphorylation of Syk is delayed and dramatically reduced, yet is still clearly detectable (Parravicini et al. 2002). Furthermore, in vitro kinase assays using GST-HS1 as substrate demonstrated that Syk is significantly activated even in Lyn–/– mast cells in response to FcεRI clustering (Kawakami et al. 2000b). However, it is still unclear which PTK contributes to Syk activation. Therefore additional studies using, e. g., double knockout mice are needed for its identification.

More recently, Reth et al. came up with a novel concept of Syk's activation and involvement in BCR signaling. They found that Syk can interact with BCR independent of Lyn activity, and, unlike Lyn, can phosphorylate both tyrosines of the BCR ITAM motif (Rolli et al. 2002). These data therefore suggest that Syk may not necessarily be controlled by Lyn's activity and can actually initiate the BCR signaling cascade by itself. Whether this model can also be applied to other activating receptors, such as FcεRI, is still unknown.

8.2.2.7 The (Btk)–PLC-γ–PKC Coupling Cascade

In addition to the PTK family members discussed above, FcεRI clustering was also found to activate Bruton's tyrosine kinase (Btk) (Kawakami et al. 1994). Btk is a member of the Btk/Tec family of PTKs, which are distinguished by the presence of a large N-terminal region containing a PH domain, a single SH3 and SH2 domains, and by the lack of an autoregulatory C-terminal tyrosine (Kawakami et al. 1996, 1999). The importance of Btk in the FcεRI coupling cascade has been demonstrated in studies using mast cells derived from both btk–/– and x-linked immunodeficiency (Xid) mice (which express a mutated Btk) (Hata et al. 1998a; Kawakami et al. 2000b). These mast cells display a reduction in the transient rise of free cytosolic Ca^{2+} ions, degranulation and, particularly, the production and secretion of certain cytokines. In contrast, tyrosine phosphorylation levels of several proteins such as SLP-76, LAT and Vav are normal in these cells (Hata et al. 1998a; Kawakami et al. 2000b).

According to an accepted model, the PH domain of Btk is required for its recruitment to the plasma membrane (PM), where it binds to PI(3,4,5)P3, a product of PI3 K activation (Salim et al. 1996). There, Btk becomes an accessible substrate for Syk and Lyn, which activate it by tyrosine phosphorylation (Rawlings et al. 1996; Kawakami et al. 2000a). Btk then activates both PLC-γ1 and PLCγ2 isoforms (Schneider et al. 1992; Wen et al. 2002), which have also been recruited to the PM by their PH domains (i.e., in a PI3 K-dependent manner) upon FcεRI clustering. However, more recent findings contradict this model and call for its reconsideration.

First, Btk was recently found to be dispensable for PLC-γ1/2 activation, as Btk-deficient BMMCs demonstrate no or minimal changes in PLCγ1 and PLCγ2 phosphorylation respectively (Kawakami et al. 2000b). Second, the two PLCγ isoforms seem to differ in their intracellular localization and mechanisms of activation (particularly regarding species differences). While PLC-γ1 is recruited from the cytosol to the plasma membrane upon FcεRI clustering, PLC-γ2 was shown to be plasma membrane associated even in unperturbed RBL-2H3 cells (Barker et al. 1998). Finally, FcεRI-induced activation of PLC-γ2 is not blocked by wortmannin (PI3 K inhibitor) in RBL-2H3 cells, suggesting that it is independent of PI3 K activity (Barker et al. 1998) [PLC-γ1 activation is independent of PI3 K in human mast cells (Tkaczyk et al. 2003)].

The PI3 K-independent mechanism of PLC-γ1/2 activation seems to involve a multiple signaling complex assembled on the transmembrane adaptor LAT with molecules such as SLP-76 and Gads (Pivniouk et al. 1999; Saitoh et al. 2000; Fig. 8.1). This complex serves as a scaffold, which keeps PLC-γ in proximity to the plasma membrane, so that it can be activated by proximal PTKs such as Lyn and Syk. Mast cells deficient in LAT or SLP-76 exhibit significantly reduced PLC-γ1 and completely abolished PLC-γ2 activation upon FcεRI clustering. Furthermore, the guanine nucleotide exchange factor Vav was also shown to regulate PLC-γ activity, yet by a mechanism that is still not fully understood (Manetz et al. 2001).

When activated, both PLC-γ isoforms hydrolyze membrane-associated phosphatidylinositol 4,5-bishosphate [PI(4,5)P2] into the second messengers, i.e., (1) inositol 1,4,5-triphosphate (IP3) and () diacylglycerol (DAG). While the former is required for calcium release from internal stores and subsequent activation of the Ca^{2+}-dependent isoforms (i.e., α- and β) of protein kinase C (PKC), the latter activates Ca^{2+}-independent PKC isoforms (δ-, ε- and ζ) (Ozawa et al. 1993; Chang et al. 1997). It has been shown in several studies that activated PKC is essential (and sufficient) for FcεRI-induced MC degranulation (Sagi-Eisenberg et al. 1985). Furthermore, PKC was shown to directly activate several MAP kinases (e.g., p38 and JNK) and thus to control the de novo cytokine production (Razin et al. 1994; Zhang et al. 1997b).

Taken together, recent data suggest that both PLCγ isoforms are regulated by two distinct mechanisms and that Btk is dispensable for their activation. How then does Btk regulate calcium mobilization (which should reflect PLC-γ activity)? A possible explanation may come from recent studies of BCR signaling. It has been shown that BCR clustering induces Btk association with phosphatidylinositol 4-phosphate 5-kinases (PIP5 K), the enzymes that synthesize PI(4,5)P2 (Saito et al. 2003). This enzyme–enzyme interaction provides a shuttling mechanism that allows Btk to stimulate the production of PI(4,5)P2, which is then hydrolyzed by PLC-γ into the second messengers. Whether this mechanism is also involved in the FcεRI coupling cascade is yet to be shown.

8.2.2.8
The Role of MAP Kinases in FcεRI-Induced Gene Transcription

Concomitantly with the secretion of granule-stored mediators (first phase response), FcεRI aggregation initiates a network of biochemical processes, which couple the stimulus to nuclear responses such as gene transcription and subsequent de novo protein synthesis (see Fig. 8.1).

DNA microarray analysis of FcεRI-induced gene expression has recently demonstrated that after 1–2 h of FcεRI stimulation, more than 2,400 genes (half of them of unknown function) exhibit 2- to 200-fold changes in their mRNA levels (Sayama et al. 2002). The FcεRI-induced transcriptional processes include changes in the expression of a dozen different cytokines (IL-1–6, IL-9–14, IL-16, IL-18, MIF, TNF-α, GM-SCF, LIF, CSF-1, etc.) and chemokines (IL-8, MCP-1–4, MIP-1, MIP-3, RANTES, etc.) as well as that of numerous adhesion molecules (α- and β-integrins, ICAM-1, SLAM, CD40L, CD82, CD83, etc.) (Galli et al. 1991; Gonzalez-Espinosa et al. 2003). These newly synthesized proteins are then either secreted by the cell (cytokines and chemokines) and may thus contribute to sustained inflammatory responses (second phase response), or expressed on the MC surface (e. g., adhesion molecules) in order to mediate/modulate interactions with other immunocytes or epithelial cells.

FcεRI-induced transcription of the above genes is regulated by a variety of transcription factors (e. g., Elk, NFAT, AP-1), which, upon their activation, translocate into the nucleus and bind to specific gene promoters (Turner and Cantrell 1997; Kitaura et al. 2000). These factors are activated by multiple coupling pathways, which particularly involve the mitogen-activated protein kinases (MAPK) (Fukamachi et al. 1993; Kimata et al. 2000) and the Ser/Thr protein kinase Akt (PKB) (Kitaura et al. 2000). It has been shown that FcεRI aggregation activates three distinct members of the MAPK family, namely Erk (extracellular signal-regulated kinases) (Fukamachi et al. 1993), JNK (c-Jun NH2-terminal kinase) (Ishizuka et al. 1996) and p38 MAPK (Zhang et al. 1997a). In general, these kinases are 60–70% identical to each other and differ mainly in the sequence and size of their activation loop. Each MAPK subfamily activates different transcription factors regulating distinct genes (Johnson and Lapadat 2002). In contrast, one specific gene can also be regulated by distinct transcription factors and thus by different MAPK pathways (e. g., FcεRI-induced GM-CSF production may be driven by both the ERK and JNK pathways) (Kimata et al. 2000).

It has been shown that the intensity of extracellular stimulus exerted by FcεRI (i. e., Ag concentration, FcεRI occupancy by specific IgE) controls which MAPK pathway will be preferentially activated. While a weak stimulus favors the p38 MAPK pathway, leading to the production of lymphokines such as IL-4, IL-13 and MCP-1, a stronger stimulus favors the Erk coupling pathway (Gonzalez-Espinosa et al. 2003), causing the production and secretion of, e. g., TNF-α and metabolites derived from free arachidonic acid (AA) (Zhang et al. 1997a; Miura et al. 1999). Unlike cytokines, the latter class of mediators, which includes prostaglandins (PGD) and leukotrienes (LT), is synthesized and secreted in a very different time frame following FcεRI clustering; While secretion of most cytokines usually requires at least 2–6 h for completion, secretion of newly synthesized fatty acid metabolites is essentially complete after 15 min and thus immediately follows the secretion of granule-stored mediators. All MAP kinases are activated by the dual phosphorylation of their Thr and Tyr residues in a "TXY" activation motif by the dual-specificity MAPKK. The MAPKK, in turn, is phosphorylated by a serine/threonine kinase termed MAPKKK (Dong et al. 2002; Johnson and Lapadat 2002).

Below we discuss the role of the three MAPK family members and of Akt in FcεRI-induced de novo synthesis.

Extracellular Signal-Regulated Kinase (Erk)

In general, the Erk signaling pathway is involved in the regulation of diverse biological responses such as cell growth and differentiation. It is activated in response to a wide range of extracellular stimuli, including growth factors, cytokines, virus infection, GPCR ligands, carcinogens, transforming agents and antigens. When activated, cytosolic Erk translocates to the nucleus and activates transcription by phosphorylation of kinases such as p90RSK and MSK-1 and transcription factors such as Elk-1, c-Myc and Stat3.

Activation of Erk1 and 2 occurs upon double phosphorylation of their Thr-Glu-Tyr motif by an upstream MAP kinase kinase (MKK1/2) (reviewed in Dong et al. 2002; Johnson and Lapadat 2002). MKK1/2 is activated by a serine/threonine protein kinase Raf-1 which exists in three different isoforms (i.e., c-Raf, B-Raf, A-Raf). The most common activator of Raf-1 is the proto-oncogene Ras. Active, GTP-bound Ras recruits Raf-1 to the plasma membrane, where Raf-1 is activated by phosphorylation. One of the Raf-1 isoforms (B-Raf) can also be activated by another small GTPase – Rap – which has an identical effector domain to that of Ras. Interestingly, Rap was shown to suppress activation of c-Raf, and therefore serves as both the activator as well as the inhibitor of Erk signaling. Further known activators of Raf include PTK Src, PKC and Akt (reviewed in Chong et al. 2003).

In FcεRI-activated mast cells, the Raf–MKK–Erk pathway is one of the most important coupling elements involved in induced gene transcription. FcεRI clustering was shown to activate all Erk pathway components, including Ras (Prieschl et al. 1995), Raf-1, MKK, Erk1/2 itself (Fukamachi et al. 1993) and the transcription factor Elk (which is downstream of Erk) (Turner and Cantrell 1997; Fig. 8.1). On the other hand, Erk seems also to be activated by a Ras-independent mechanism, as inhibitors of Ras (lovastatin and the farnesylation-specific inhibitor BZA-5B) do not affect FcεRI-induced Erk activation (Graham et al. 1998). Furthermore, unlike other MAPK pathways, FcεRI-mediated Erk activation seems to be independent of PI3 K activity, as PI3 K-specific inhibitors wortmannin or LY294002 have only marginal or no effect on its activation (Ishizuka et al. 1999; Miura and MacGlashan 2000).

FcεRI-induced Erk activation initiates transcription of specific cytokine genes such as TNF-α (Zhang et al. 1997a; Csonga et al. 1998; Miura et al. 1999; Gonzalez-Espinosa et al. 2003), GM-CSF (Kimata et al. 2000) and MIP-2 (Miura et al. 1999; Numahata et al. 2003), while it has no or little effect on, e.g., IL-4 (Ishizuka et al. 1999) or IL-13 (Gibbs and Grabbe 1999; Miura et al. 1999) production. Interestingly, in contrast to previous data, Gelfand et al. reported that FcεRI-induced TNF-α synthesis is not regulated by the Erk pathway in mouse BMMC (Ishizuka et al. 1999).

In addition, several studies have shown that Erk1/2 is also involved in the regulation of AA metabolism and leukotriene de novo synthesis. A selective inhibitor of Erk activity (PD098059) was found to dramatically inhibit FcεRI-induced $cPLA_2$ activation and subsequent LTC_4 and PGD_2 secretion (Gibbs and Grabbe 1999; Miura et al. 1999; Kimata et al. 2000).

c-Jun Amino-Terminal Kinase (JNK)

Another member of the mitogen-activated protein kinase superfamily is the c-Jun amino-terminal kinase (JNK). JNK is activated following its dual phosphorylation at a Thr-Pro-Tyr motif mostly in response to stress conditions (e.g., heat shock, UV, proinflammatory cytokines or antigens). When activated, JNK phosphorylates c-Jun, a component of the AP-1 transcription factor, which in turn increases its transcriptional activity (Franklin et al. 1992; Pulverer et al. 1991; Smeal et al. 1991). Moreover, JNK was shown to also activate other transcription factors such as ATF2 (Gupta et al. 1995; Livingstone et al. 1995), Elk-1 (Whitmarsh et al. 1995) and Sap-1a (Janknecht and Hunter 1997).

More recently, FcεRI aggregation has been shown to activate JNK as well as protein kinases upstream of JNK, such as MEK kinase 1 (MEKK1) and JNK kinase (JNKK) (Ishizuka et al. 1996). Further studies have suggested that FcεRI-induced JNK activation is mostly dependent on PI3 K and its downstream effector Btk. Kawakami et al. (1997) showed that JNK activity is compromised in btk-null (or Xid) mast cells. These findings are in line with observations demonstrating that the PI3 K inhibitor wortmannin significantly suppresses JNK activation and subsequent TNF-α production following FcεRI clustering (Ishizuka et al. 1996, 1999)

In addition, FcεRI-mediated JNK activation is also dependent on the small GTPase Rac1, which is activated by the specific guanine nucleotide exchange factor (GEF) Vav (Teramoto et al. 1997; Arudchandran et al. 2000; Fig. 8.2).

FcεRI-induced JNK signaling was shown to activate several transcription factors, such as c-Jun, Elk-1 and Sap-1a, which, in turn, initiate transcription of cytokine genes, such as IL-2, IL-6 (Song et al. 1999) and TNF-α (Hata et al. 1998b). Production of GM-CSF in response to FcεRI aggregation (as well as IL-10 and IL-13 in mast cells activated by LPS) (Masuda et al. 2002) also depends on JNK.

MAP Kinase of 38 kDa (p38)

Over the years, it has been established that, like JNK, p38 is activated mostly in response to stress conditions such as heat shock, UV light, changes in osmotic pressure, pro-inflammatory cytokines, growth factors and upon clustering of a MIRR such as FcεRI.

There are four known p38 MAP kinase isoforms (α-δ), and all are activated upon phosphorylation of their Thr-Gly-Tyr motif. At least three different MAPK kinases (MKK3, MKK4 and MKK6) can activate p38. These are regulated by other upstream serine-threonine kinases (MAPKK kinases, e. g., Tak1, Ask1 and Mlk3), which are controlled by GEF-proteins such as Rac1 and Cdc42 and probably by some other, still poorly understood, mechanisms (see Fig. 8.2).

Studies using wortmannin, or of mast cells derived from Btk-null mice, have demonstrated that FcεRI-mediated p38 activation is significantly (but not fully) inhibited. This suggests that the FcεRI-mediated activation of p38 is, similarly to JNK, dependent on Btk and PI3 K activity (Kawakami et al. 1997).

The FcεRI-induced p38 pathway was shown to regulate expression of genes for several cytokines, such as IL-4 (Hirasawa et al. 2000), IL-13 (Gibbs et al. 2002), MCP-1 (Gonzalez-Espinosa et al. 2003), MIP-2 (Numahata et al. 2003) and, only marginally, TNF-α (Gonzalez-Espinosa et al. 2003).

Protein Kinase B (PKB/Akt)

Protein kinase B (PKB), better known as Akt, is a serine/threonine protein kinase belonging to the "AGC" kinase superfamily (reviewed in Scheid and Woodgett 2003). Interest in Akt has intensified upon recognition of its role in promoting cell survival signals, which mostly lead to the inactivation of a series of pro-apoptotic proteins. Akt sequence is composed of three functionally distinct regions: an N-terminal pleckstrin homology (PH) domain, a central catalytic domain and C-terminal hydrophobic motif (HM). The PH domain directs Akt to PI3K-generated membranal phosphoinositides PI(3,4,5)P3 and PI(3,4)P2. At the plasma membrane, proximal Akt molecules become activated by autophosphory-

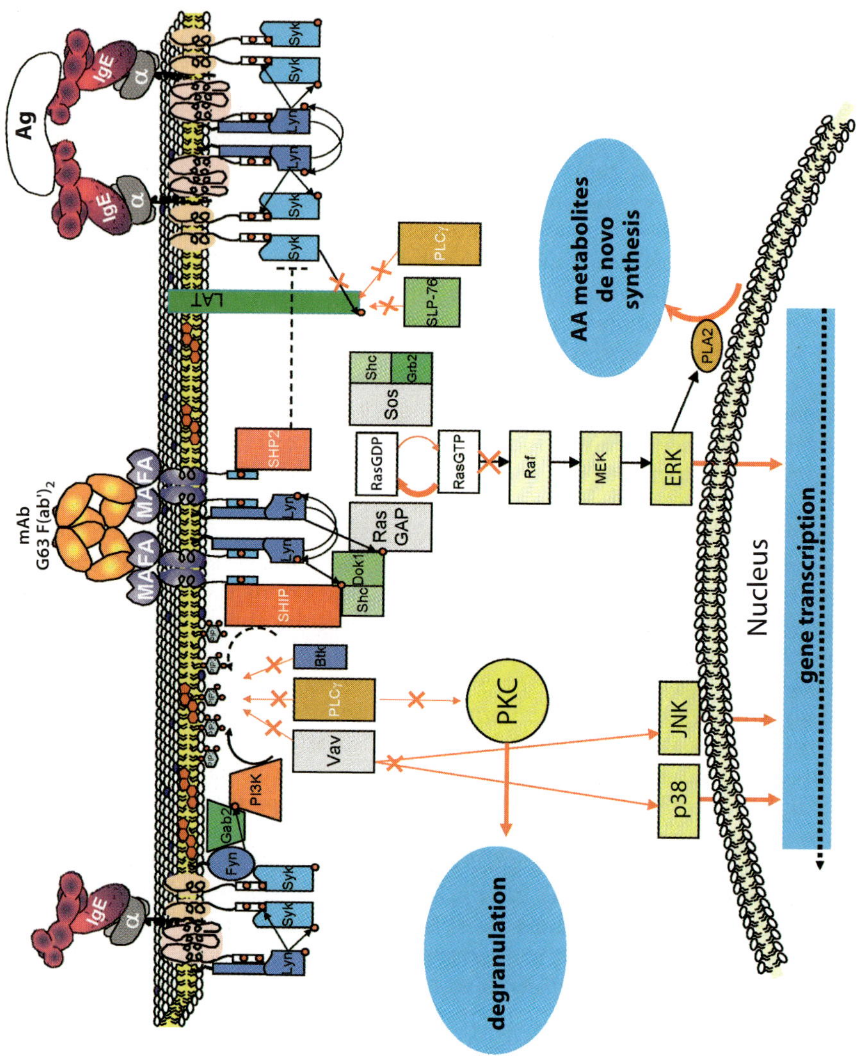
Ag
IgE
α
Syk
Lyn
LAT
PLCγ
SLP-76
mAb
G63 F(ab')2
MAFA
SHP2
Shc
Grb2
Sos
RasGDP
RasGTP
Ras GAP
Dok1
SHIP
Raf
MEK
ERK
PLA2
AA metabolites de novo synthesis
Btk
Vav
PI3K
Gab2
Fyn
PKC
JNK
p38
Nucleus
gene transcription
degranulation

Fig. 8.2. A scheme illustrating the mechanism of MAFA-mediated inhibition of response to FcεRI stimulus: MAFA clustering leads to a rapid tyrosine phosphorylation of its ITIM sequence by PTK Lyn which is associated with MAFA even in unperturbed cells. Once phosphorylated, ITIM binds directly to several SH2 domain-containing molecules, such as Lyn, SHP-2 and SHIP. At the plasma membrane, SHP-2 dephosphorylates Syk (and thereby suppresses generation of the LAT signaling complex), while SHIP interferes with PI3 K signaling. Both are thereby suppressing FcεRI-induced recruitment and activation of PLCγ. This leads to inhibition of downstream coupling events of the FcεRI stimulus, culminating in suppression of degranulation and gene expression. In parallel, SHIP becomes phosphorylated and recruits the adaptor molecule Dok-1, which then binds the Ras GTPase-activating protein (RasGAP). RasGAP is a negative regulator of the Ras signaling pathway and its recruitment to the plasma membrane inhibits activation of Erk. This eventually results in suppression of de novo synthesis of proteins and AA metabolites. *Black arrows* indicate phosphorylation; *black striped arrows* dephosphorylation; *red arrows* binding/activation; *red strikes* suppression; *green* adaptor proteins; *blue* PTKs; *red* PTPs; *yellow* S/T kinases; *brown* lipases; *white* GTPases; *gray* GDP/GTP exchange factors

lation (i.e., by catalytically competent Akt) at the Ser-473 and by phosphoinositide-dependent kinase-1 (PDK-1) which phosphorylates its Thr-308 (Toker and Newton 2000). Akt phosphorylated at both sites becomes active and phosphorylates target proteins, most of which are involved in the regulation of gene transcription.

Akt has recently been shown to become active in response to FcεRI stimulation (Kitaura et al. 2000), and its activation was shown to depend on Fyn (Parravicini et al. 2002), Btk and Syk (Kitaura et al. 2000). Mast cells deficient in either Fyn, Btk or Syk all exhibit a reduced activity of Akt compared to WT cells. Significantly, *lyn–/–* cells showed a several-fold more robust and prolonged Akt activation than WT cells, suggesting that Lyn acts as a negative regulator of the Akt signaling cascade (Kitaura et al. 2000; Parravicini et al. 2002).

Furthermore, the results of Kawakami and colleagues demonstrated that Akt regulates activity of several transcription factors, such as NF-κB, NF-AT and AP-1, which are critical for the expression of the IL-2 and TNF-α genes. In their experiments, the overexpression of WT Akt in BMMC significantly affected the FcεRI-induced activities of NF-AT and AP-1 and concomitantly led to an increased secretion of IL-2 and TNF-α compared to vector-transfected cells or transfectants overexpressing a mutated form of Akt (Kitaura et al. 2000).

8.2.2.9 Overview of the Current Model of the FcεRI Stimulus-Response Coupling Network

According to the current model, FcεRI translocates upon its aggregation into lipid rafts (Field et al. 1995, 1997), where it becomes proximal to raft-associated src kinases such as Lyn or Fyn. This interaction is thought to initiate tyrosine phosphorylation of the ITAMs present on both β- and γ-chains of the FcεRI complex (Eiseman and Bolen 1992; Pribluda et al. 1994). The phosphorylated γ-chain dimer recruits the PTK Syk, which is then activated mostly by proximal Lyn (associated with phosphorylated ITAM of β-chain). Nevertheless, current data suggest that Lyn is not the sole PTK that phosphorylates ITAMs and activates Syk, and thus is dispensable for FcεRI-induced secretory responses (Nishizumi and Yamamoto 1997; Kawakami et al. 2000b; Parravicini et al. 2002). Activated Syk then phosphorylates the membrane-associated protein LAT, which becomes a docking site for a number of cytosolic proteins such as SLP-76, Gads and PLC-γ2. Recruited proteins are sequentially phosphorylated by proximal PTKs (Syk, Lyn, etc.), which leads to creation of the multiprotein complex and eventually to PLCγ activation.

In parallel with Syk activity, activated Fyn phosphorylates the adaptor protein Gab2, which then associates with PI3 K and thereby enables its recruitment to the plasma membrane and subsequent activation. Active PI3 K catalyzes the production of PI(3,4)P2 and PI(3,4,5)P3, which bind to the PH domains of several signaling proteins such as Btk, PLC-γ, Vav, Akt or PDK1. This interaction also promotes translocation of the above proteins to the plasma membrane, where they become activated. It is noteworthy that both PLCγ isoforms seem to be regulated by two distinct mechanisms involving either PI3 K activity and/or formation of the LAT-centered multiprotein complex. Therefore, an efficient initiation of FcεRI stimulus coupling cascade involves two independent signaling paths: one involving Syk activity and the other involving activity of Fyn. Both paths are activated in parallel, and are complementing each other's induction of the secretory response.

When activated, both PLC-γ isoforms hydrolyze membrane-associated PI(4,5)P2 into the second messengers, i.e., (1) IP3 and (2) DAG. IP3 binds to specific receptors in the endoplasmic reticulum, which then triggers release and later depletion of Ca^{2+} stores, leading to activation of store-operated Ca^{2+} entry (I_{CRAC}) from the extracellular medium. The continuation of the transient elevation of free cytosolic Ca^{2+} ion concentration eventually results in activation of the Ca^{2+}-dependent isoforms (i.e., α- and β) of protein kinase C (PKC). DAG activates Ca^{2+}-independent PKC isoforms (δ-, ε- and ζ) (Ozawa et al. 1993; Chang et al. 1997). It has been shown in several studies that activated PKC is essential (and sufficient) for FcεRI-induced MC degranulation (Sagi-Eisenberg et al. 1985). Furthermore, PKC was shown to directly activate several MAP kinases (e.g., p38 and JNK) and thus to control de novo cytokine production (Razin et al. 1994; Zhang et al. 1997b).

Tyrosine phosphorylation and activation of other enzymes and adaptors, including Vav, Shc, Grb2 and Sos, stimulate small GTPases such as Rac, Cdc42 and Ras. These pathways lead to activation of the ERK, JNK and p38 MAP kinases and the subsequent phosphorylation of transcription factors that induce the de novo synthesis and secretion of cytokines.In parallel, the ERK pathway activates cytoplasmic phospholipase A_2 ($cPLA_2$) and thereby the de novo synthesis and release of AA derivatives.

8.3 Mechanisms Regulating Response to the FcεRI Stimulus

In order to avoid hyper-responsiveness and maintain an effective immune response, activation signaling has to be controlled by regulatory mechanisms that can either terminate it or attenuate it.

The simplest mechanism would probably involve a loss of activating stimuli (i.e. antigen dissociation from the receptor), thereby resulting in receptor disengagement. However, the dissociation rate of multivalent antigens from FcεRI is usually very slow and therefore cannot be a cause of effective signal termination in vivo. The modulation of FcεRI signaling is therefore based on a wide range of complex mechanisms that respond to specific stimuli. Since phosphorylation of both proteins and lipid-based molecules was shown to be a crucial element in coupling FcεRI to the secretory response, dephosphorylation by specific phosphatases seems to be the most effective mode of inhibition. Indeed, phosphatases have been shown to be an important part of a built-in autoregulatory apparatus, which is activated in parallel with or following the activating network in order to keep it in check. This may be illustrated by the tyrosine phosphorylated ITAM of the FcεRI β-chain, which is known to bind to the protein tyrosine phosphatases SHP-1 and SHP-2, as well as the phosphatidylinositol 5′-phosphatase SHIP. These enzymes, when recruited to this receptor component, counteract the PTK-based activation signal and thereby downregulate the cellular response.

Other mechanisms involve regulation of the cell-surface FcεRI expression levels. These can be mediated either by the internalization and degradation of receptors (see below: Cbl family) or by downregulation of FcεRI synthesis and expression (see below: β_T).

Finally, negative regulation of FcεRI-induced responses is mediated by several inhibitory receptors expressed on the cell surface by mechanisms which are discussed in more detail below (see Sect. 8.3.3).

8.3.1 Regulatory Cytosolic Molecules

8.3.1.1 Phosphatases

SH2 Domain-Containing Inositol 5′ Phosphatase (SHIP): The "Gatekeeper" of Mast Cell Degranulation

As discussed in Section 8.2.2, one of the most important pathways coupling the stimulus produced by aggregated FcεRI to the secretory response of cells involves PI3 K (Yano et al. 1993; Barker et al. 1995). Its activity's main product is the plasma membrane-associated second messenger PI(3,4,5)P_3, which is the site of recruitment of several proteins (such as Btk, PLC-γ, Akt and PDK1) to the plasma membrane, where they become activated and couple the upstream signals to the secretory response.

To ascertain that this pathway is accurately controlled, several regulatory mechanisms have evolved. One of them involves a tumor suppressor, phosphatase and tensin homologue (PTEN, also known as MMAC1/TEP1), which hydrolyses the newly synthesized PI(3,4,5)P_3 back to PI(4,5)P_2 (Maehama and Dixon 1998). Another mechanism is carried out by two different SH2-containing inositol 5′-phosphatases – SHIP1 (Ono et al. 1996) and its more ubiquitous homologue SHIP2 (Pesesse et al. 1997). Both enzymes, when recruited to the plasma membrane, dephosphorylate PI(3,4,5)P_3 to PI(3,4)P_2 (Pesesse et al. 1998). This reduces the pool of PI(3,4,5)P_3 as well as the recruitment of certain PH-containing proteins to the plasma membrane, resulting in downregulation of the cellular response (Ono et al. 1996; Malbec et al. 1998). In immunocytes, SHIP was shown to serve as one of the most important molecules in downregulation of the activating signals produced by several different ITAM-containing receptors, including BCR (Liu et al. 1998), TCR (Jensen et al. 2001) and FcεRI (Ono et al. 1996; Xu et al. 2001). SHIP's inhibitory action is mostly dependent on ITIM-containing receptors (see Sect. 8.3.3), which upon phosphorylation of their ITIM tyrosines become binding sites for SHIP's SH2 domain. It is interesting to note, however, that also several ITAM-containing molecules, including TCR (Osborne et al. 1996) and FcεRI (Osborne et al. 1996; Kimura et al. 1997a; Muraille et al. 2000), may recruit SHIP to the plasma membrane (via their phosphorylated ITAMs) upon their engagement.

The importance of SHIP in FcεRI signaling became apparent from studies using BMMCs derived from SHIP–/– mice. These showed that SHIP–/– BMMCs are far more prone to FcεRI-mediated cell degranulation as well as inflammatory cytokine production and secretion (Kalesnikoff et al. 2002). Furthermore, unlike WT cells, the former degranulated vigorously even in response to binding mIgE (see Sect. 8.2.2) alone (Huber et al. 1998). These findings led Krystal and colleagues to propose that SHIP functions as a "gatekeeper" of mast cell response to FcεRI by keeping the PI3 K-generated PI-3,4,5-P3 levels in check, thereby regulating the subsequent secretory response (Huber et al. 1998).

SH2 Domain-Containing Protein Tyrosine Phosphatase-1 (SHP-1)

SHP-1 (also known as SHPTP-1, SHP, HCP, PTPN6, PTPH6 and PTP1C) is a 67-kDa cytosolic enzyme expressed mainly in hematopoietic cells (Matthews et al. 1992). It contains two adjacent amino-terminal SH2 domains, followed by a phosphatase domain and a carboxy-terminal region containing two conserved tyrosine phosphorylation sites (Martin et al. 1999).

As a tyrosine phosphatase, SHP-1 counteracts the action of protein tyrosine kinases, which are mostly activated upon engagement of ITAM-containing receptors.

In mast cells, overexpression of WT SHP-1 was shown to suppress the FcεRI-induced tyrosine phosphorylation of several intracellular components, including the FcεRI β- and γ-subunits and PTK Syk. In contrast, FcεRI-mediated histamine release was unaffected, suggesting that the extent of the receptor's tyrosine phosphorylation is not the sole determinant of the FcεRI-induced secretory response (Kimura et al. 1997b). Moreover, SHP-1 appears to exert a dual effect on the FcεRI-mediated signaling network, as it was also shown to enhance phosphorylation of JNK and subsequent TNF-α production (Xie et al. 2000). Similar results were observed in Lyn-deficient mast cells, which are severely impaired in their FcεRI-induced cellular protein tyrosine phosphorylation pattern (including that of Syk and β-chain), and yet are normal or augmented in their secretory response (see Sect. 8.2.2).

Taken together, SHP-1 does not appear to be the key phosphatase keeping the secretory response to FcεRI activation in check.

SH2 Domain-Containing Protein Tyrosine Phosphatase-2 (SHP-2)

SHP-2 (also known as Syp, PTP1D, PTP2C and SHPTP-2) is a widely expressed cytosolic enzyme of 72 kDa. Upon FcεRI aggregation, SHP-2 was shown to bind directly via its SH2 domains to the tyrosine-phosphorylated β-subunit of FcεRI (Kimura et al. 1997b). There it becomes activated by tyrosine phosphorylation from a so far unknown PTK (Kimura et al. 1997b). In vitro experiments demonstrated that both β- and γ-subunits of FcεRI can be dephosphorylated by recombinant SHP-2, suggesting the existence of a negative feedback loop which regulates the extent of this receptor's tyrosine phosphorylation. SHP-2 is therefore assumed to be part of a built-in autoregulatory apparatus of FcεRI-mediated signaling. Moreover, SHP-2 is found to bind directly to the tyrosine-phosphorylated ITIM sequences of several inhibitory receptors, such as MAFA (Xu et al. 2001), PECAM-1 (Sagawa et al. 1997a), LMIR (Kumagai et al. 2003) and FcγRIIB (Fong et al. 1996), and thus it serves as the mediator of their regulatory actions. When recruited to the plasma membrane, SHP-2 dephosphorylates numerous substrates, including Syk, and thus counteracts the activating signaling mediated by protein tyrosine kinases (Xu and Pecht 2001).

Megakaryocytic Protein Tyrosine Phosphatase-2 (PTP-MEG2)

PTP-MEG2 is a 68-kDa non-receptor PTP originally cloned from human endothelial cells and megakaryocyte lines which has recently been discovered in several cell lines of hematopoietic lineage (Gu et al. 1992). The C-terminal region of MEG2 contains a conserved PTP catalytic domain, whereas the N-terminal domain has significant homology to yeast SEC14p (Bankaitis et al. 1989), and was recently shown to bind membranal PI(3,4,5)P3 and phosphatidylserine (Huynh et al. 2003; Zhao et al. 2003).

In mast cells, PTP-MEG2 was found on the secretory granules, probably as a result of binding of the N-terminal Sec14p homology domain to PI(3,4,5)P3 on secretory vesicle membranes. Overexpression of PTP-MEG2 in RBL-2H3 cells leads to a dramatic enlargement of these intracellular granules, suggesting that it regulates their formation (Wang et al. 2002). Therefore PTP-MEG2 may likely be involved in the modulation of the mast cell secretory response, though probably in an indirect manner.

Haematopoietic Protein Tyrosine Phosphatase (HePTP)

HePTP (also known as PTPN7, LC-PTP and PTPNI) is a 40-kDa protein tyrosine phosphatase expressed in mast cells, T cells and B cells. It has an ~55% sequence homology with CD45, yet unlike CD45 it is expressed in the cells' cytoplasm (Adachi et al. 1994). Primary lymphocytes from HePTP–/– mice show enhanced activation of ERK upon TCR aggregation, suggesting that HePTP serves as a physiological regulator of ERK (Gronda et al. 2001). Although the involvement of HePTP in mast cell function is not completely understood, it has been shown that FcεRI aggregation induces its tyrosine phosphorylation in RBL-2H3 cells (Swieter et al. 1995). Further experiments showed that HePTP does not immunoprecipitate with molecules such as Lyn, Syk and FcεRI subunits, suggesting that they are not physiological substrates of HePTP. Taken together, HePTP, as a physiological regulator of ERK, may control the later phases of mast cell response to FcεRI at an unidentified site in the coupling cascade.

8.3.1.2
Protein Kinases

C-Terminal Src Kinase (Csk) – A PTK with an Intrinsic Inhibitory Function

C-terminal Src kinase (Csk) is a cytoplasmic PTK expressed ubiquitously, albeit in larger amounts in hematopoietic cells (reviewed in Latour and Veillette 2001). Recent studies have shown that Csk is a potent negative regulator of MIRR signaling in activated T cells (Chow et al. 1993), B cells (Hata et al. 1994) and mast cells (Honda et al. 1997). This inhibitory activity correlates with the exquisite ability of Csk to phosphorylate the inhibitory carboxy-terminal tyrosine of the Src family kinases such as Lyn, Lck and Fyn. Once this tyrosyl residue is phosphorylated, it creates a binding site for the kinases' own SH2 domain. Such an interaction results in a "closed" conformation of Src kinases, where the catalytic domain is blocked and thus inactive (Cooper and Howell 1993). Disruption of this intramolecular interaction (e.g., through dephosphorylation of the C-terminal tyrosine by CD45) results in kinase activity ("open" conformation).

Csk shares structural similarities with Src family PTKs but lacks the C-terminal regulatory tyrosine and the N-terminal myristylation signal required for membrane localization (Nada et al. 1991). Structural analyses indicate that, in addition to its catalytic domain, the Src homology 3 (SH3) and 2 (SH2) regions of Csk are essential for its capacity to inhibit cell signaling. The SH3 domain of Csk was found to bind constitutively to the proline-enriched protein tyrosine phosphatase (PEP) and PTP-PEST [for Pro(P), Glu(E), Ser(S) and Thr(T)-rich domains], two cytosolic PTPs abundantly expressed in hemopoietic cells (Cloutier and Veillette 1996; Davidson et al. 1997). This suggests that in addition to the direct suppression of Src kinases, Csk also recruits PTPs, which may further contribute to the inhibition of phosphotyrosine-dependent signaling.

The involvement of Csk in the negative regulation of FcεRI signaling has been demonstrated by Honda et al. (1997), who generated RBL-2H3 sublines overexpressing different Csk variants. Results of these studies showed that several cellular events, including Lyn activation, tyrosyl phosphorylation of Syk, elevation of free intracellular calcium concentration and histamine release, are delayed in Csk overexpressing RBL-2H3 cells. On the other hand, the amplitude of histamine release is rather augmented (Honda et al. 1997), similarly to that observed upon FcεRI clustering in lyn–/– mast cells (Nishizumi and Yamamoto 1997; Kawakami et al. 2000b).

8.3.1.3 Adaptor Proteins

Downstream of Tyrosine Kinases (Dok) Proteins – Modulators of FcεRI-Induced De Novo Cytokine Synthesis

Dok proteins (Dok-1–Dok-5) are members of a recently discovered family of adaptors tyrosyl-phosphorylated by a range of PTKs. Dok-1, which is a prototype member of the family, was originally identified as a tyrosyl-phosphorylated molecule of ~62 kDa, associated with the RasGAP protein – a negative regulator of Ras (Yamanashi and Baltimore 1997). Dok-1 possesses several characteristic domains and motifs: at its N-terminus it contains a PH domain that is followed by a phosphotyrosine-binding (PTB) domain. Its C-terminal tail harbors ten potential tyrosyl-phosphorylation sites and seven PxxP motifs, which may serve as potential docking sites for SH2 and SH3 domain-containing proteins, respectively. Indeed, phosphorylation of the multiple tyrosine residues at the C-terminus has been shown to create docking sites for proteins containing SH2 domains, including RasGAP, Nck, SHIP, CrkL and Csk (Lock et al. 1999; Dunant et al. 2000; Martelli et al. 2001).

Both Dok-1 and Dok-2 have been recently shown to be involved in the FcεRI stimulus-response network. FcεRI clustering induces marked tyrosine phosphorylation of the above two proteins, with a concomitant increase in their binding to RasGAP (and to several other adaptors). This suggests that Dok proteins negatively regulate the FcεRI-induced Ras/Raf1/Erk signaling cascade. Indeed, both Ras and Erk activation as well as the subsequent de novo TNF-α synthesis are markedly reduced in RBL-2H3 cells overexpressing Dok-1 (Abramson et al. 2003). In contrast, the FcεRI-induced release of preformed mediators is apparently unaffected by Dok-1 overexpression, suggesting that Dok-1 selectively regulates the late phase of mast cells response.

The regulatory function of Dok-1 in mast cells is further supported by findings demonstrating that it is also involved in coupling the inhibitory action of ITIM-bearing receptors such as MAFA (Abramson and Pecht 2002) and FcγRIIB (Ott et al. 2002).

Signal Transducing Adaptor Protein (STAP-2): Suppressor of PLCγ Activity

STAP-2 (also known as breast tumor kinase substrate – BKS) is a recently discovered adaptor molecule containing PH- and SH2-like domains, which has been identified as a c-fms-interacting protein by yeast two-hybrid screening.

Overexpression of STAP-2/BSK in RBL-2H3 cells results in suppression of FcεRI-mediated tyrosine phosphorylation of PLC-γ, calcium mobilization and degranulation. Furthermore, STAP-2/BSK was found to associate with PLC-γ in vivo, suggesting that it negatively controls FcεRI-mediated calcium mobilization and degranulation by direct modulation of PLC-γ activity (Yamamoto et al. 2003).

Grb2-Associated Binder 2 (Gab2): One Adaptor with Two Opposing Roles

Gab2 is a recently identified 97-kDa molecule that is a member of the Dos/Gab adaptor molecules family, which also includes mammalian Gab1, IRS and the *Drosophila* daughter of sevenless Dos. Gab2 contains an N-terminal PH domain followed by a long region with multiple tyrosine-containing motifs and two proline-rich domains (Gu et al. 1998; Zhao et al. 1999). The tyrosine-containing motifs, once phosphorylated, provide binding sites for SH2 domain-containing signaling molecules including the p85 subunit of PI3-kinase, PLCγ and SHP-2 (Zhao et al. 1999). The proline-rich regions of Gab2 are potential binding sites for SH3-domain-containing proteins such as Src-family protein tyrosine kinases.

As already mentioned, Gab2 has been proposed to function as the missing link between FcεRI stimulus and PI3 K activation, and thus to play an essential role in the evoked secretory response (Parravicini et al. 2002; see Sect. 8.2.2). Indeed, Gab2–/– mast cells display defective degranulation and cytokine gene expression in response to FcεRI clustering (Gu et al. 2001). However, in addition to its activating role, Gab2 also seems to function as an inhibitory molecule, which can modulate MIRR-induced signaling. It has been previously shown that T cells derived from transgenic mice expressing WT Gab2 exhibit impaired T-cell responses, while T cells from Gab2-deficient mice show enhanced proliferation upon TCR stimulation (Pratt et al. 2000; Yamasaki et al. 2001).

Similarly, Siraganian et al. found that RBL-2H3 cells overexpressing Gab2 exhibit a lower FcεRI-induced tyrosine phosphorylation (e.g., of Syk, FcεRI β- and γ-subunits), degranulation and cytokine production (e.g., TNF-α and IL-6) (Xie et al. 2002). These results therefore suggest that Gab2 expression levels may regulate the amplitude of the FcεRI-induced secretory response. The activating potential of Gab2 seems to be mediated by PI3 K, while the inhibitory capacity is likely due to SHP-2 which associates with tyrosine-phosphorylated Gab2.

8.3.1.4 E3 Ubiquitin Ligases – Negative Regulation by Protein Endocytosis and Degradation

MIRR signaling has also been found to be regulated through selective protein endocytosis and degradation. The mechanism of such a process generally involves multiple steps, including covalent addition of ubiquitin (Ub) molecules to an internal lysine side chain of a target protein, followed by proteolysis of the ubiquitinated protein in the proteasome – a multisubunit protease complex (reviewed in Marmor and Yarden 2004).

The process of ubiquitination is carried out by a cascade that involves several enzymes of the ubiquitin–protein isopeptide ligase family (E3s). These provide specificity to the ubiquitin system because they are responsible for substrate recognition and for promoting Ub ligation to the target protein. The Cbl-related protein family members belong to this class of E3 ubiquitin ligases (Joazeiro et al. 1999). In immunocytes these are employed in the above mechanisms of modulating the stimuli induced by a variety of activating receptors, notably the MIRRs.

The Cbl family (reviewed by Rao et al. 2002) comprises three members – c-Cbl, Cbl-b and Cbl-3 (Cbl-3 is not expressed in lymphocytes). Their structure is highly conserved and contains the following characteristic motifs: (1) an amino-terminal variant SH2 domain, (2) a RING finger with E3 ubiquitin ligase activity,

(3) multiple proline-rich regions and sites of tyrosine phosphorylation, and (4) a leucine zipper-like sequence.

c-Cbl – A Suppressor of PTK and MIRR Expression

The first evidence suggesting that c-Cbl is involved in the negative regulation of immunoreceptor signaling stems from the studies of Samelson et al. They reported that overexpression of c-Cbl in RBL-2H3 cells leads to inhibition of the FcεRI-induced Syk tyrosine phosphorylation (and its activity), with a concomitant suppression of serotonin release (Ota and Samelson 1997). Further studies revealed that c-Cbl, as a ubiquitin ligase, mediates ubiquitination of Syk and Src and their subsequent targeting to the proteasome, where they are degraded. More recent studies have demonstrated that c-Cbl is also responsible for the ubiquitination of the FcεRI β- and γ-subunits and the receptor's subsequent internalization and degradation (Paolini et al. 2002). This is induced by a Syk-dependent phosphorylation of c-Cbl with a concomitant increase in their binding. This Syk–Cbl complex is then recruited to the cell membrane, where the activated c-Cbl ubiquitinates both receptor β- and γ-subunits as well as Lyn (Kyo et al. 2003) and Syk (Ota et al. 1996). This implies the existence of a negative feedback loop, where Syk activity is required for its own degradation and that of both FcεRI β- and γ-chains. The eventual proteasome-dependent degradation of ubiquitinated receptor complexes and activated Syk plays a crucial and general role in the downregulation of the coupling networks of all MIRR family members (TCR, BCR, FcεRI, etc.).

Cbl-b

Cbl-b is another member of the Cbl-family of E3 ubiquitin–protein ligases (Keane et al. 1995). While c-Cbl mostly catalyzes ubiquitination of the MIRR components, recent studies have indicated that Cbl-b rather promotes ubiquitination of downstream signaling molecules such as PTKs (e.g. Lyn, Syk) (Sohn et al. 2003), PLC-γ1, PKC (Heissmeyer et al. 2004), the CrkL–C3G complex (Zhang et al. 2003b), PI3 K (Fang et al. 2001; Zhang et al. 2003a) and Gab-2 (Qu et al. 2004).

In mast cells, Cbl-b overexpression leads to a strong inhibition of FcεRI-mediated tyrosine phosphorylation of the β- and γ-receptor subunits of Gab2 and of Syk. Furthermore, Gab2 expression levels are also reduced upon Cbl-b overexpression, suggesting that Cbl-b induces its degradation. As expected, Cbl-b was shown to suppress FcεRI-induced cell degranulation, MAPK activation and subsequent cytokine gene transcription (Qu et al. 2004). These data therefore suggest that Cbl-b also plays an important autoregulatory role in FcεRI-induced secretory responses.

8.3.2 Transmembrane Adaptors Modulating FcεRI Signaling

8.3.2.1 Adaptor Proteins

Csk-Binding Protein (Cbp): A Transmembrane Adaptor Protein with an Inhibitory Function

Cbp [also known as phosphoprotein associated with GEMs (PAG)] is a recently discovered 75- to 85-kDa membrane protein that regulates the activity of Src family kinases (Brdicka et al. 2000; Kawabuchi et al. 2000). Cbp, like LAT, possesses a short extracellular domain, a transmembrane region and a long cytoplasmic domain, which contain several potential sites of tyrosine phosphorylation. It is palmitoylated, probably at two cysteines positioned in the juxtamembrane portion of its cytoplasmic domain. This palmitoylation is presumed to be responsible for targeting Cbp to lipid rafts (Brdickova et al. 2001). When phosphorylated on a specific tyrosine residue, Cbp becomes a docking site for the SH2 domain of Csk (the negative regulator of Src family kinases; see Sect. 8.3.1).

Recently, Cbp has also been shown to negatively regulate the FcεRI-mediated response (Ohtake et al. 2002). FcεRI aggregation leads to a very rapid (a matter of seconds) tyrosyl-phosphorylation of Cbp, with its concomitant increased binding to Csk. Overexpression of Cbp in RBL-2H3 cells was shown to inhibit FcεRI-mediated cell degranulation. In addition, elevated levels of Cbp expression significantly decreased tyrosine phosphorylation of both FcεRI subunits, suggesting that Cbp/PAG is regulating the earliest events induced upon this receptor aggregation. Thus, Cbp/PAG is likely to function as a negative feedback regulator of mast cell response to the FcεRI stimulus.

8.3.2.2 Tetraspanins

β_T: A Suppressor of FcεRI Expression and Maturation

Recently, an alternative spliced product of the β-gene has been identified (Donnadieu et al. 2003). The corresponding protein, designated β_T, is a truncated form of the FcεRI β-subunit and migrates on SDS-PAGE as a 21.4-kDa molecule. While the β-subunit has been proposed to serve as an amplifier of FcεRI expression (Donnadieu et al. 2000), the β_T isoform has been shown to interfere with α-chain maturation and therefore it negatively regulates FcεRI expression (Donnadieu et al. 2003). The importance of β-expression on rodent mast cells was previously demonstrated in studies using mast cells derived from mice deficient in the β-subunit. These cells express no FcεRI on their surface and demonstrate no IgE-mediated passive cutaneous anaphylaxis (PCA) (Hiraoka et al. 1999). The results therefore suggest that β_T is competing with β for control of FcεRI expression levels at the cell surface, and thus helps to regulate a proper biological response. The molecular mechanisms determining which isoform should be expressed still need to be resolved.

CD81: A Tetraspanin Modulating FcεRI-Induced Degranulation
CD81 (TAPA-1) is a 26-kDa membrane protein containing four transmembrane (TM) stretches and two extracellular (EC) loops which classify it as a member of the tetraspanin (TM4) protein superfamily.

In mast cells of the RBL-2H3 line, CD81 was shown to act as a negative regulator of FcεRI-induced degranulation. Binding of two different CD81-specific monoclonal antibodies (5D1 and 1A12) was found to downregulate the FcεRI-induced mast cell secretory response as well as the IgE-dependent passive cutaneous anaphylaxis reactions (Fleming et al. 1997).

Unlike typical inhibitory receptors (see Sect. 8.3.3), CD81 lacks an immunoreceptor tyrosine-based inhibitory motif (ITIM) within its cytoplasmic domains and fails to inhibit FcεRI-induced tyrosine phosphorylation, the $[Ca^{2+}]i$ signal, and leukotriene C_4 production. Molecular mechanisms underlying CD81-mediated inhibition remain elusive, but they clearly differ from those employed by ITIM-containing receptors.

More recently, another similar tetraspanin molecule – CD63 – has also been shown to suppress the FcεRI-induced secretory response when clustered by its specific antibodies (Kraft et al. 2005).

8.3.3 Modulation of the FcεRI Signaling Cascade by ITIM-Containing Receptors

While this chapter has so far dealt with attenuation or termination of immune responses by a relatively wide range of diverse regulatory mechanisms, probably the most important one involves the family of inhibitory receptors, which regulates the activating cascades triggered primarily by the ITAM-containing receptor family.

Lymphocyte receptors that display inhibitory functions now number about 30. These are transmembrane glycoproteins and can be grouped into two major classes on the basis of their extracellular domains, which classifies them as members of either the Ig-like or the C-type lectin superfamilies respectively (Gergely et al. 1999). Their intracellular domains contain one or more copies of ITIMs which share a consensus amino acid sequence – (I/V/L/S)xYxx(L/V) – where x denotes any amino acid (Bolland and Ravetch 1999). Practically all these receptors, upon coclustering with an ITAM-containing receptor, undergo phosphorylation on their tyrosine residue in the ITIM sequence by a Src family PTK. The phosphotyrosines then become binding sites for SH2 domain-containing molecules that carry out the inhibitory actions. So far, these have been found to be protein tyrosine- and/or phosphatidylinositol-phosphatases, which, upon recruitment to the ITIMs at the plasma membrane and following subsequent activation, dephosphorylate their respective cellular substrates which would otherwise mediate activation. As a result, the coclustering of inhibitory and activating receptors eventually leads to a reduction in or ablation of the cellular response. It is, however, noteworthy that not all ITIM-bearing receptors require coclustering with their activating counterparts. Several receptors such as MAFA and PECAM were shown to modulate FcεRI-induced responses even without the coclustering. Furthermore, inhibitory receptors bind to their specific extracellular ligands (see Table 8.1), which may also induce their inhibitory action independent of their coclustering with ITAM-containing receptors. Interestingly, in some cases the activating (PIR-A) and inhibitory (PIR-B) receptors may bind to the same molecule (e.g., HLA-I molecules), suggesting that one and the same ligand may exert two opposing responses and can thus either up- or downregulate the biological response. It also implies that an effective response is achieved not only by the presence/intensity of extracellular stimuli, but also by the expression levels of respective (i.e., activating/inhibitory) receptors.

In this section we review the current knowledge on those ITIM-bearing receptors that have been shown to interfere with the FcεRI-induced stimuli.

Table 8.1. ITIM-containing receptors expressed on mast cells

Names	Ligands	ITIM sequences	In vivo (in vitro) recruited
FcγRIIB (CD32)	IgG (Fc part)	ITYSLL	SHIP (SHP-1, SHP-2)
LIRs (ILT, CD85)	HLA class I	VTYAEV SIYATL VTYAQL	SHP-1
PIR-B	HLA class I	SLYASV VTYAQL SVYATL	SHP-1 (SHP-2)
gp49B1	$\alpha_V\beta_3$ (integrin)	IVYAQV	SHP-1 (SHP-2)
		VTYAQL	
PECAM (CD31)	$\alpha_V\beta_3$ (integrin) PECAM (Ig-like receptor) CD38	VQYTEV TVYSEV	SHP-2 SHP-1 (weak)
SIRPα (SHPS-1)	CD47 (Ig-like receptor associated with $\alpha_V\beta_3$)	ITYADL	SHP-1; SHP-2
Siglec-3	Sialic acid residues	LHYASL TEYSEV	SHP-1
LMIR-1 (MAIR-1)	?	VEYSTL LHYSSV	SHP-1, SHP-2, SHIP
MAFA (KLRG1)	Mannose binding capacity	SIYSTL	SHIP, SHP-2

8.3.3.1
ITIM-Bearing Immunoglobulin-like Receptors

Type IIB Fcγ Receptor (FcγRIIB)

FcγRIIB (CD32) is the first identified member of the inhibitory receptor superfamily. It is expressed throughout the hematopoietic system in three alternatively spliced isoforms (FcγRIIB1, FcγRIIB1′ and FcγRIIB2). While FcγRIIB1 and FcγRIIB2 are expressed in both humans and rodents, the FcγRIIB1′ isoform is found only in mice. All isoforms share the same extracellular region, with two C2-type Ig-like domains (reviewed in Daeron 1997), yet differ in their intracellular parts, containing a conserved ITIM sequence.

The inhibitory role of FcγRIIB was first described in B cells, where it functions as a negative regulator of BCR-mediated cell proliferation (Phillips and Parker 1984). Under physiological conditions, FcγRIIB (because of its relatively low affinity) only binds to the Fc part of IgG molecules which are in the immune complex with a specific antigen. Therefore, if this antigen is also recognized by a specific BCR (or FcεRI-bound IgE on mast cells) the immune complex will mediate coclustering of both FcγRIIB and BCR (or FcεRI-IgE). This will then trigger an inhibitory cascade, abrogating the activating machinery induced by the clustering of the respective ITAM-containing receptor.

The importance of FcγRIIB in regulating immune responses was particularly evident in FcγRIIB-null mice, which exhibit enhanced antibody production, anaphylactic responses and autoimmune symptoms (Takai et al. 1996; Yuasa et al. 1999).

Coclustering of FcγRIIB with FcεRI on RBL-2H3 cells was shown to inhibit FcεRI-mediated degranulation and TNF-α secretion (Daeron et al. 1995). When coclustered with FcεRI, FcγRIIB undergoes a rapid tyrosine phosphorylation within its ITIM sequence by the FcεRI-associated Lyn (Malbec et al. 1998). While synthetic peptides with sequences corresponding to the tyrosine phosphorylated FcγRIIB ITIM bind SHP-1, SHP-2 and SHIP from mast cell lysates, neither SHP-1 nor SHP-2 were found to associate with FcγRIIB in vivo (Fong et al. 1996). Furthermore, FcγRIIB-mediated inhibition is unimpaired in mast cells derived from SHP-1 null mice (Fong et al. 1996), supporting the predominant role of SHIP in mediating the FcγRIIB inhibitory activity in these mast cells. As detailed in Section 8.3.1, SHIP contains a catalytic domain that hydrolyses the 5′-phosphate of PI(3,4,5)P3 and PI(1,3,4,5)P4. Thus its activity interferes with FcεRI-induced recruitment and activation of Btk and phospholipase Cγ1 (PLCγ1). This leads to inhibition of downstream coupling events of the FcεRI stimulus, culminating in suppression of degranulation and gene expression (Ono et al. 1996).

More recently, an additional inhibitory mechanism of FcγRIIB in B cells and mast cells has been described (Tamir et al. 2000; Ott et al. 2002). When FcγRIIB binds SHIP, the enzyme becomes phosphorylated and recruits the adapter molecule Dok-1, which then binds the Ras GTPase-activating protein (RasGAP). RasGAP is a negative regulator of the Ras signaling pathway and its recruitment to the plasma membrane inhibits activation of Erk and subsequent B-cell proliferation. It is noteworthy, however, that inhibition of FcεRI-induced activation of MAP kinases by FcγRIIB in Dok-1-deficient mast cells was found to be similar to that of WT cells (Ott et al. 2002). These results may imply the existence of an additional/complementary inhibitory mechanism that operates in the absence of Dok-1 (such as Dok-2).

Significantly, FcγRIIB was also found to inhibit c-kit-induced mast cell proliferation, demonstrating that FcγRIIB inhibitory activity is not limited to the regulation of ITAM-mediated signals (Malbec et al. 1999).

Regulation mechanism(s) of the expression of ITIM-bearing receptors such as FcγRIIB are, so far, largely unknown. However, a recent study showed that in vitro, IL-4 decreases the B-cell-surface levels of FcγRIIB (Rudge et al. 2002). IL-4 had similar effects on three other ITIM-containing receptors, namely, paired Ig-like receptor [(PIR)-B; see below], Siglec-2 (CD22; see below) and CD72. Moreover, the decrease in FcγRIIB and Siglec-2 expression were, as expected, accompanied by a decline in inhibitory function of these receptors (Rudge et al. 2002). These findings have important implications for the induction of allergic diseases, where Th2 polarization favors the production of IL-4 leading to the generation of IgE (which is known to upregulate FcεRI expression; see Sect. 8.2.2).

gp49B1

Both gp49B1and gp49A murine proteins are close homologues of the human killer cell Ig-like receptors (KIRs) family. Their extracellular parts contain two C2-type Ig-like domains (Arm et al. 1991; Castells et al. 1994). The cytoplasmic domain of gp49B1 contains two different ITIMs, whereas that of gp49A does not (McCormick et al. 1999).

gp49B1 is expressed on the surface of mouse mast cells (Katz et al. 1996), macrophages, natural killer (NK) cells (Wang et al. 1997, 2000), activated T cells (Gu et al. 2003) and neutrophils (Zhou et al. 2003). When coclustered with respective activating receptors or upon binding to its ligand – $\alpha_v\beta_3$ integrin – gp49B1 was found to modulate biological responses of the above cells, i. e., to suppress NK cell-mediated target cell lysis (Rojo et al. 1997), to prevent neutrophil-dependent vascular injury in response to lipopolysaccharide (LPS) (Zhou et al. 2003) or to downregulate IgG-induced TNF-α production in macrophages (Matsumoto et al. 2001).

Coclustering of gp49B1 with FcεRI on mast cells inhibits the induced secretion of stored pro-inflammatory mediators, as well as of newly generated lipid mediators (e. g., leukotriene C4) (Katz et al. 1996). The molecular mechanism(s) by which gp49B1 inhibits FcεRI-induced mast cell activation involve SHP-1 rather than SHP-2 or the inositol phosphatase SHIP (Lu-Kuo et al. 1998; Lu-Kuo et al. 1999). In-vivo experiments demonstrated no association of gp49B1 with SHIP and/or SHP-2, suggesting that SHP-1 is the major phosphatase responsible for gp49B1 inhibitory action. Still, gp49B1 was found to inhibit the FcεRI-induced secretory response of SHP-1-deficient BMMCs only partially, suggesting that SHP-1 may not be the exclusive mediator of gp49B1 inhibitory activity (Lu-Kuo et al. 1999). It is also noteworthy that under in-vitro conditions, SHP-2 binds to phosphorylated tyrosine-containing peptides corresponding to both gp49B ITIM sequences, while SHIP was found to bind only to its second ITIM (ETQD**VTYAQL**CIR) (Kuroiwa et al. 1998). It is therefore still unclear whether SHP-2, SHIP and/or other, still unknown, molecules are also involved in the gp49B1 inhibitory action. Interestingly, mutations of both ITIM tyrosines completely abolish the gp49B1 inhibitory activity, while mutation of either one reduces it only partially (Lu-Kuo et al. 1999), suggesting that both ITIMs are required for attaining full inhibitory function.

More recently, Castells et al. (2001) reported that the integrin $\alpha_v\beta_3$ binds to gp49B1 in solution as well as in cell–cell binding studies, suggesting that $\alpha_v\beta_3$ is its natural ligand. Moreover, soluble $\alpha_v\beta_3$ binding to gp49B1 on BMMCs was shown to inhibit the FcεRI-mediated secretory response. Since $\alpha_v\beta_3$ is expressed on a wide variety of cells, including those of the vasculature, the above results suggest that gp49B1 can constitutively downregulate mast cell activation. Indeed, experiments using gp49B1 null mice demonstrated an augmented mast cell degranulation and the resultant early tissue inflammation in IgE-dependent passive cutaneous anaphylaxis (PCA) (Daheshia et al. 2001). Furthermore, gp49B1-deficient mice died more frequently from anaphylactic shock compared to normal mice in response to a systemic antigen challenge.

Paired Ig-like Receptors of Inhibition (PIR-B)

Both paired Ig-like receptors of activation (PIR-A) and inhibition (PIR-B) were originally identified as murine protein molecules with a high sequence similarity to the KIR, LIR, murine gp49B and human Fc receptors for IgA (Hayami et al. 1997). PIRs are also members of the immunoglobulin (Ig) superfamily, as their extracellular part consists of six Ig-like domains. Both types of PIR are expressed on various hematopoietic cell lineages, such as B cells, mast cells, macrophages, granulocytes and dendritic cells (DCs), but not on T and NK cells (Kubagawa et al. 1997; Chen et al. 1998; Kubagawa et al. 1999; Roach et al. 1999).

Coclustering PIR-B with FcεRI on mouse mast cells was shown to inhibit the FcεRI-mediated secretory response (Uehara et al. 2001). This inhibitory capacity is presumably mediated by its cytoplasmic tail, which contains three potential ITIMs (see Table 8.1). Two of these motifs, when phosphorylated on their tyrosines, were shown to recruit SHP-1 (Blery et al. 1998; Maeda et al. 1998), which was thought to downregulate B-cell and mast cell activation (Maeda et al. 1999). More recently, however, SHP-1 has been shown to be entirely dispensable for PIR-B inhibitory activity, since PIR-B-mediated inhibition of FcεRI-induced secretion was unimpaired in SHP-1-deficient mast cells (Uehara et al. 2001). These results suggest that some other molecule(s) are likely to be involved in the inhibitory activity of PIR-B.

Efforts to identify a specific ligand for PIR-B have so far been unsuccessful. Still, several observations suggest that it interacts with class I MHC-encoded molecules. It has recently been reported that HLA-G (MHC class Ib molecule) tetramers bind to cells transfected by PIR-B cDNA, suggesting a direct interaction between HLA-G and PIR-B molecules. Furthermore, analysis of PIR-B expression and its tyrosine phosphorylation status in DC from HLA-G-transgenic mice showed that it is highly phosphorylated (Liang et al. 2002).

Signal Regulatory Protein-α (SIRPα)

SIRP molecules [also known as: SHPS-1 (src homology 2 domain-containing phosphatase substrate-1), BIT (brain immunoglobulin-like molecule with a tyrosine-based activation motif), P84 and MFR (macrophage fusion receptor)] all belong to a large family of membrane proteins which consists of at least 15 related SIRP variants. This large family can be divided into two subtypes based on the presence (in SIRPα) or absence (in SIRPβ) of an intracellular domain-containing ITIM sequence (Kharitonenkov et al. 1997). Both SIRPα and SIRPβ molecules are encoded by separate genes clustered in the same locus (human chromosome 20p12.2–13 and mouse chromosome 2). The presence of three Ig-like domains in the extracellular part classifies them as members of the immunoglobulin superfamily. However, unlike other ITIM-bearing receptors of the Ig-superfamily, expression of which is usually restricted to immunocytes, SIRPs are expressed by virtually all human tissues, including the brain, lung, liver and kidney (SIRP is actually the first ITIM-containing receptor found to be expressed by non-hemopoietic cells) (Oshima et al. 2002). More recently, however, different SIRP variants have been also discovered in hematopoietic and to lesser extent in myeloid cells (Adams et al. 1998; Veillette et al. 1998; Ghannadan et al. 2002).

As expected from their broad tissue distribution, SIRPs were shown to be involved in diverse biological responses. These include suppression of anchorage-independent cell growth, negative regulation of immune cells, self-recognition of red blood cells, mediation of macrophage multinucleation, skeletal muscle differentiation, entrainment of circadian clock, neuronal survival and synaptogenesis (reviewed in Oshima et al. 2002).

When artificially expressed in RBL-2H3 cells (as a chimeric molecule containing the extracellular domain of murine FcγRIIB and the TM and IC domains of human SIRPα), and coclustered with FcεRI, it inhibits Ca^{2+} mobilization, ERK activation and the subsequent release of secretory granule-stored mediators and cytokine synthesis. This suppression involves recruitment of SHP-1 and SHP-2 to the cytoplasmic tail, which contains two ITIMs and a proline-rich region (Lienard et al. 1999).

Recently, CD47 (also known as integrin-associated protein – IAP), another Ig-superfamily member involved in cell adhesion processes, has been identified as a SIRPα ligand (Jiang et al. 1999; Seiffert et al. 1999; Vernon-Wilson et al. 2000). Interestingly, CD47 was first identified as a protein coisolated with the $\alpha_v\beta_3$ integrin, which is a ligand for other inhibitory receptors – gp49B1 and PECAM (see Table 8.1). CD47–SIRPα binding was reported to prevent the phenotypic and functional maturation of human immature DCs and to suppress interleukin-12 (IL-12) production by mature DCs (Latour et al. 2001).

Leukocyte Ig-Like Receptors (LIR)

The human inhibitory LIRs (also known as Ig-like transcript, ILT, or CD85) contain one, two or four Ig-like domains in their extracellular parts and two to four ITIMs in their cytosolic domains. This inhibitory family comprises so far of six members (LIR-1, -2, -3, -5, -8, LAIR-1), where LIR-5 represents the closest human ortholog of the mouse inhibitory receptor gp49B1. LIRs are mostly expressed on human mast cells, basophils, eosinophils, B cells, mononuclear phagocytes, neutrophils and NK cells (Arm et al. 1997; Colonna et al. 1997; Tedla et al. 2001). It has been shown that lung mast cells express mRNA encoding LIR-3 (Arm et al. 1997), and basophils and eosinophils each express LIR-2 and LIR-3 proteins (Tedla et al. 2001). More recently, it has been shown that coligation of LIR3 to FcεRI by means of a second monoclonal antibody significantly inhibits histamine release, leukotriene production and IL-4 secretion from human basophils (Sloane et al. 2004). Similarly, serotonin secretion is suppressed upon coclustering of FcεRI with ILT2 (LIR-1) artificially expressed in RBL cells (Colonna et al. 1997). The inhibitory mechanisms involve protein tyrosine phosphatase 1 (SHP-1), which has recently been found to be recruited by tyrosine-phosphorylated ITIMs of ILT2 upon pervanadate treatment of RBL-2H3 cells (Bellon et al. 2002). LIR-1 and LIR-2 bind classical and non-classical MHC-class I molecules (Navarro et al. 1999), which may therefore provide a constitutive down-regulation of cell activation, because of the widespread expression of MHC class I molecules.

Leukocyte Mono-Ig-Like Receptor 1 (LMIR1)

LMIR1 (also known as myeloid-associated Ig-like receptor, MAIR; Yotsumoto et al. 2003) and LMIR2 were originally discovered on FcεRI-stimulated BMMCs, in an attempt to identify additional regulatory molecules (Kumagai et al. 2003). Their extracellular domains contain a single Ig-like domain, hence its classification. While LMIR1 contains two ITIM sequences in its cytoplasmic domain, the LMIR2 lacks both and has only a short cytoplasmic domain. However, LMIR2 associates (via its TM domain) with several ITAM-containing adaptors, such as DAP10, DAP12 and γ-chain, and thus serves as an activating receptor. Besides its abundant expression in mouse mast cells, the mRNA of LMIR molecules is also found at high levels in B cells and myeloid cells. Nonetheless, LMIR1/2 is not expressed in T cell lines and no ligand(s) for LMIR molecules have so far been identified.

LMIR1 capacity to counteract activating stimuli in immune cells was first reported for B cells, where coclustering of LMIR1 with the BCR caused a rapid tyrosyl phosphorylation of its intracellular ITIMs, which, in turn, recruited the tyrosine phosphatases SHP-1 and SHP-2 as well as the inositol polyphosphate 5-phosphatase, SHIP (Kumagai et al. 2003). Coclustering of LMIR1 with FcεRI on mouse mast cells was found to interfere with the signaling of the latter receptor (Yotsumoto et al. 2003). Furthermore, its high expression levels in mast cells suggest that LMIR1 plays an important role in the modulation of FcεRI-mediated responses.

Platelet Endothelial Cell Adhesion Molecule-1 (PECAM-1)

PECAM-1 (also known as CD31) is a newly discovered member of the Ig superfamily containing six extracellular Ig domains and two cytosolic ITIMs (interestingly, the cytosolic tail also contains a deduced ITAM-like sequence) (Lu et al. 1997). PECAM-1 was found to be expressed in humans, mice and rats, mostly at the lateral junctions of endothelial cells, and at lower levels on the surfaces of neutrophils, monocytes, platelets, NK cells, T and B cell subsets, and on mast cells. PECAM-1 was found to be involved in a number of biological processes, including leukocyte transmigration, cell migration, angiogenesis and cell adhesion as well as modulation of intracellular signaling (reviewed in Jackson 2003).

In the RBL-2H3 line, PECAM-1 was shown to undergo rapid tyrosine phosphorylation on its ITIMs in response to FcεRI clustering alone (without any need for the coclustering of receptors) (Sagawa et al. 1997b). Studies based on plasmon resonance measurements of peptides corresponding to its ITIM sequence showed that, once phosphorylated, it binds to SHP-2 and SHP-1, yet to the former with considerably higher affinity (Hua et al. 1998). SHP-2 therefore may be the major mediator of its inhibitory activity (Sagawa et al. 1997a). Recent studies using PECAM-1-deficient mice have further demonstrated that PECAM-1 plays an important role in modulation of FcεRI-induced responses. PECAM-1–/– mice exhibited an enhanced IgE-mediated systemic anaphylaxis and an increased sensitivity to local cutaneous IgE-dependent anaphylaxis compared to PECAM-1+/+ mice. Interestingly, in vitro analyses demonstrated that PECAM-1-deficient BMMCs also exhibit enhanced serotonin release in response to FcεRI stimulation compared to WT BMMCs (Wong et al. 2002). This phenomenon was not observed (in vitro) in other inhibitory receptor knock-outs, including FcγRIIB–/– and gp49B1–/– BMMCs. A possible explanation may be that FcεRI induces rapid tyrosine phosphorylation of PECAM-1 on its ITIMs, while FcγRIIB and gp49b1 require coclustering with the FcεRI in order to mediate inhibitory signaling in vitro. The mechanism whereby FcεRI clustering induces PECAM-1 phosphorylation on its ITIMs is not clear, but may involve PECAM's homophilic interactions (i.e., PECAM-1 on one cell serves as a ligand for PECAM-1 on another cell), which have been reported recently (Sun et al. 1996). Other physiological ligands of PECAM-1 include $\alpha_v\beta_3$ integrin (Piali et al. 1995) and CD38 (Deaglio et al. 1998). It is noteworthy that $\alpha_v\beta_3$ integrin also serves as a ligand for gp49B1, another ITIM-bearing Ig-like receptor that modulates FcεRI signaling (see above). This further suggests that $\alpha_v\beta_3$, being an adhesion molecule, plays an important role in the negative regulation of FcεRI-mediated secretory responses in vivo.

Sialic Acid Binding Ig-like Lectins (Siglec)

Human mast cells and basophils were shown to express several members of the Siglec family. So far, at least 11 different Siglec molecules have been identified in humans and all but Siglec-1 possess ITIM or ITIM-like sequences (reviewed in Crocker and Varki 2001; Crocker 2002). The Ig-like domains present in their extracellular regions bind sialic acid residues, which are typically found attached to extracellular domains of certain glycoproteins present in the PM.

Siglecs are mostly expressed on monocytes, macrophages, neutrophils, NK cells and B cells in both humans and mice. Human mast cells and basophils have long been known to express Siglec-3, also known as CD33 (Valent et al. 1989). More recently, Siglec-5 and Siglec-8 have been discovered on human mast cells and basophils (Kikly et al. 2000; Ghannadan et al. 2002; Nutku et al. 2003). Still, the question whether these Siglecs have the capacity to inhibit the FcεRI-induced secretory response has not yet been addressed. Additional studies are therefore required to better understand their role in the above cells.

8.3.3.2 ITIM-Bearing C-Type Lectins

Mast Cell Function-Associated Antigen (MAFA)

MAFA (its mouse homologue renamed as KLRG1) is a type II membranal glycoprotein originally discovered on the surface of the rat mucosal-type mast cells of the RBL-2H3 line. This was in the frame of a search for membrane molecules capable of modulating the FcεRI secretory response (Ortega and Pecht 1988). The monoclonal antibody (mAb) G63, which was shown to bind it specifically, efficiently inhibited the FcεRI-induced secretory response of these cells yet required MAFA clustering. More recently, the genes encoding the human and mouse MAFA homologues were cloned and sequenced (Blaser et al. 1998; Butcher et al. 1998). In humans, MAFA was found to be expressed not only on basophils but also on peripheral NK cells. Significantly, the mouse MAFA homologue was not detected on mice mast cells, and was therefore renamed as killer cell lectin-like receptor G1 (KLRG1), as it is expressed only by lymphokine-activated NK cells and by effector CD8 T cells. The inhibitory function of KLRG1 on these cells has been shown in transfection studies, which demonstrated that it inhibits both cytokine production and NK cell-mediated cytotoxicity (Robbins et al. 2002). In addition, MAFA (KLRG1) has been discovered on a subset of human NK cells and antigen-experienced T cells that lack proliferative capacity (Voehringer et al. 2001, 2002).

MAFA's extracellular domain shares a marked homology with the carbohydrate recognition domain (CRD) of several members of the calcium-dependent (C-type) lectins (sugar-binding proteins) superfamily (Guthmann et al. 1995), such as the type II FcεR (CD 23), CD 69, CD 72 and several killer cell inhibitory receptors (KIR). So far, terminal mannose binding capacity has been established for rat MAFA (Binsack and Pecht 1997). This might suggest MAFA's potential involvement in cell–cell or cell–matrix interactions. Nevertheless, the sugar binding capacity of MAFA does not exclude the possibility of binding to a protein epitope of a ligand, as has been found, e. g., for CD23.

The intracellular domain of MAFA contains a single ITIM sequence (SIYSTL), which classifies MAFA as part of the inhibitory receptors family (Guthmann et al. 1995). However, unlike other ITIM-containing receptors, it inhibits the FcεRI stimulus without the requirement for their coclustering, i. e., MAFA clustering alone causes inhibition. Still, its coclustering with FcεRI was shown to markedly increase the inhibition (Licht et al. 2005).

MAFA clustering leads to a rapid tyrosine phosphorylation on its ITIM sequence by Lyn (serine residues also undergo phosphorylation). Lyn is associated with MAFA even in unperturbed cells, and in vitro measurements established that this interaction involves Lyn's SH3 domain and a PAAP sequence on MAFA's intracellular tail. This association may be important for Lyn's trans-activation and subsequent phosphorylation of MAFA ITIM tyrosine. Once phosphorylated, the ITIM binds directly to several SH2 domain-containing molecules, such as Lyn, SHP-2 and SHIP (see Fig. 8.2), but not SHP-1 or Syk (Xu et al. 2001). Moreover, SHIP binding exhibits considerably higher affinity than SHP-2 binding, suggesting that SHIP is the major phosphatase that is recruited upon clustering MAFA. Indeed, WT SHIP overexpression in RBL-2H3 cells causes, upon MAFA clustering, a markedly stronger inhibition of the FcεRI-induced secretory response than in control cells (Xu et al. 2001). In addition, SHIP undergoes considerable phosphorylation on its tyrosines upon MAFA clustering, providing a binding site for SH2 domain-containing molecules, which are thereby recruited to the plasma membrane (Abramson and Pecht 2002). These molecules include, for example, two adaptor proteins of the Dok family (Dok-1 and Dok-2), which in turn also undergo tyrosine phosphorylation. Phosphorylated Dok-1 then binds the Ras GTPase-activating protein (RasGAP) that ultimately inhibits activation of the Ras-ERK signaling pathway (see Fig. 8.2).

Indeed, MAFA has recently been shown to interfere with FcεRI-induced Erk-1/2 and p38 MAPK activation and thereby with the late phase of secretory response (unpubl. data). The FcεRI-induced gene transcription and de novo synthesis of cytokines including IL-1, IL-4, IL-6, IL-10, IL-12, TNF-α and IFN-γ are selectively suppressed upon MAFA clustering, while that of IL-3, IL-5 and IL-16 are unaffected (unpubl. results).

8.4 Epilogue

Crucial life-maintaining biological activities like those performed by the immune system obviously require close and strict regulation. In this chapter, we have reviewed primarily the different currently identified cell-membrane components involved in negative regulation of the cellular response to one member of the MIRR family, the type I Fcε receptor. As it may become evident from this discussion, the field is still developing and at a fast rate. New proteins endowed with regulatory function(s) are still being discovered. Moreover, our current understanding of the complex networks by which they exert regulation is still quite limited. Also, among other issues, we hardly addressed here the exceptionally important regulatory elements operating at the level of the different key players in these networks, namely, enzymes and adaptor proteins. One particular example illustrating this topic is the protein tyrosine kinases family, where intriguing control elements are built into their structure. This topic is of course too broad and out of the scope of this chapter but would complement it. As an evolving field in life sciences, our review is bound to become outdated soon, but we have tried to provide a snapshot of it at this point in time.

Acknowledgements. The research done in I. Pecht's laboratory is supported by a grant from the Israel National Science Foundation, founded by the Israel Academy of Science and Humanities and by the E. and B. Shoon Foundation.

References

Abramson J, Pecht I (2002) Clustering the mast cell function-associated antigen (MAFA) leads to tyrosine phosphorylation of p62(Dok) and SHIP and affects RBL-2H3 cell cycle. Immunol Lett 82:23–28

Abramson J, Rozenblum G, Pecht I (2003) Dok protein family members are involved in signaling mediated by the type 1 Fc epsilon receptor. Eur J Immunol 33:85–91

Adachi M, Miyachi T, Sekiya M, Hinoda Y, Yachi A, Imai K (1994) Structure of the human Lc-Ptp (Heptp) gene – similarity in genomic organization within protein-tyrosine-phosphatase genes. Oncogene 9:3031–3035

Adamczewski M, Numerof RP, Koretzky GA, Kinet JP (1995) Regulation by CD45 of the tyrosine phosphorylation of high affinity IgE receptor beta- and gamma-chains. J Immunol 154:3047–3055

Adams S, van der Laan LJW, Vernon-Wilson E, de Lavalette CR, Dopp EA, Dijkstra CD, Simmons DL, van den Berg TK (1998) Signal-regulatory protein is selectively expressed by myeloid and neuronal cells. J Immunol 161:1853–1859

Arm JP, Gurish MF, Reynolds DS, Scott HC, Gartner CS, Austen KF, Katz HR (1991) Molecular-cloning of Gp49, a cell-surface antigen that is preferentially expressed by mouse mast-cell progenitors and is a new member of the immunoglobulin superfamily. J Biol Chem 266:15966–15973

Arm JP, Nwankwo C, Austen KF (1997) Molecular identification of a novel family of human Ig superfamily members that possess immunoreceptor tyrosine-based inhibition motifs and homology to the mouse gp49B1 inhibitory receptor. J Immunol 159:2342–2349

Arudchandran R, Brown MJ, Peirce MJ, Song JS, Zhang J, Siraganian RP, Blank U, Rivera J (2000) The Src homology 2 domain of Vav is required for its compartmentation to the plasma membrane and activation of c-Jun NH(2)-terminal kinase 1. J Exp Med 191:47–60

Asai K, Kitaura J, Kawakami Y, Yamagata N, Tsai M, Carbone DP, Liu FT, Galli SJ, Kawakami T (2001) Regulation of mast cell survival by IgE. Immunity 14:791–800

Baird B, Sheets ED, Holowka D (1999) How does the plasma membrane participate in cellular signaling by receptors for immunoglobulin E? Biophys Chem 82:109–119

Bankaitis VA, Malehorn DE, Emr SD, Greene R (1989) The *Saccharomyces-cerevisiae* Sec14 gene encodes a cytosolic factor that is required for transport of secretory proteins from the yeast Golgi-complex. J Cell Biol 108:1271–1281

Barker SA, Caldwell KK, Hall A, Martinez AM, Pfeiffer JR, Oliver JM, Wilson BS (1995) Wortmannin blocks lipid and protein kinase activities associated with PI 3-kinase and inhibits a subset of responses induced by Fc epsilon R1 cross-linking. Mol Biol Cell 6:1145–1158

Barker SA, Caldwell KK, Pfeiffer JR, Wilson BS (1998) Wortmannin-sensitive phosphorylation, translocation, and activation of PLCgamma1, but not PLCgamma2, in antigen-stimulated RBL-2H3 mast cells. Mol Biol Cell 9:483–496

Barsumian EL, Isersky C, Petrino MG, Siraganian RP (1981) IgE-induced histamine release from rat basophilic leukemia cell lines: isolation of releasing and nonreleasing clones. Eur J Immunol 11:317–323

Beaven MA, Metzger H (1993) Signal-transduction by Fc-receptors – the Fc-epsilon-RI case. Immunol Today 14:222–226

Bellon T, Kitzig F, Sayos J, Lopez-Botet M (2002) Mutational analysis of immunoreceptor tyrosine-based inhibition motifs of the Ig-like transcript 2 (CD85j) leukocyte receptor. J Immunol 168:3351–3359

Berger SA, Mak TW, Paige CJ (1994) Leukocyte common antigen (CD45) is required for immunoglobulin E-mediated degranulation of mast cells. J Exp Med 180:471–476

Bieber T, Delasalle H, Wollenberg A, Hakimi J, Chizzonite R, Ring J, Hanau D, Delasalle C (1992) Human epidermal Langerhans cells express the high-affinity receptor for immunoglobulin-e (Fc-epsilon-RI). J Exp Med 175:1285–1290

Binsack R, Pecht I (1997) The mast cell function-associated antigen exhibits saccharide binding capacity. Eur J Immunol 27:2557–2561

Blaser C, Kaufmann M, Pircher H (1998) Virus-activated CD8 T cells and lymphokine-activated NK cells express the mast cell function-associated antigen, an inhibitory C-type lectin. J Immunol 161:6451–6454

Blery M, Kubagawa H, Chen CC, Vely F, Cooper MD, Vivier E (1998) The paired Ig-like receptor PIR-B is an inhibitory receptor that recruits the protein-tyrosine phosphatase SHP-1. Proc Natl Acad Sci USA 95:2446–2451

Bolland S, Ravetch JV (1999) Inhibitory pathways triggered by ITIM-containing receptors. Adv Immunol 72:149–177

Borkowski TA, Jouvin MH, Lin SY, Kinet JP (2001) Minimal requirements for IgE-mediated regulation of surface Fc epsilon RI. J Immunol 167:1290–1296

Brdicka T, Pavilstova D, Leo A, Bruyns E, Korinek V, Angelisova P, Scherer J, Shevchenko A, Shevchenko A, Hilgert I, Cerny J, Drbal K, Kuramitsu Y, Kornacker B, Horejsi V, Schraven B (2000) Phosphoprotein associated with glycosphingolipid-enriched microdomains (PAG), a novel ubiquitously expressed transmembrane adaptor protein, binds the protein tyrosine kinase Csk and is involved in regulation of T cell activation. J Exp Med 191: 1591–1604

Brdickova N, Brdicka T, Andera L, Spicka J, Angelisova P, Milgram SL, Horejsi V (2001) Interaction between two adapter proteins, PAG and EBP50: a possible link between membrane rafts and actin cytoskeleton. FEBS Lett 507:133–136

Butcher S, Arney KL, Cook GP (1998) MAFA-L, an ITIM-containing receptor encoded by the human NK cell gene complex and expressed by basophils and NK cells. Eur J Immunol 28:3755–3762

Cambier JC (1995) New nomenclature for the Reth motif (or ARH1/TAM/ARAM/YXXL). Immunol Today 16:110

Castells MC, Wu XY, Arm JP, Austen KF, Katz HR (1994) Cloning of the Gp49B gene of the immunoglobulin superfamily and demonstration that one of its 2 products is an early-expressed mast-cell surface protein originally described as Gp49. J Biol Chem 269:8393–8401

Castells MC, Klickstein LB, Hassani K, Cumplido JA, Lacouture ME, Austen KF, Katz HR (2001) gp49B1-alpha(v)beta3 interaction inhibits antigen-induced mast cell activation. Nat Immunol 2:436–442

Chang EY, Szallasi Z, Acs P, Raizada V, Wolfe PC, Fewtrell C, Blumberg PM, Rivera J (1997) Functional effects of overexpression of protein kinase C-alpha, -beta, -delta, -epsilon, and -eta in the mast cell line RBL-2H3. J Immunol 159:2624–2632

Chen CC, Gartland GL, Kubagawa H, Cooper MD (1998) Cellular distribution and function of paired immunoglobulin-like receptors, PIR-A and PIR-B. FASEB J 12:A917

Chong H, Vikis HG, Guan KL (2003) Mechanisms of regulating the Raf kinase family. Cell Signal 15:463–469

Chow LML, Fournel M, Davidson D, Veillette A (1993) Negative regulation of T-cell receptor signaling by tyrosine protein-kinase p50(csk). Nature 365:156–160

Cloutier JF, Veillette A (1996) Association of inhibitory tyrosine protein kinase p50(csk) with protein tyrosine phosphatase PEP in T cells and other hemopoietic cells. EMBO J 15:4909–4918

Colonna M, Navarro F, Bellon T, Llano M, Garcia P, Samaridis J, Angman L, Cella M, LopezBotet M (1997) A common inhibitory receptor for major histocompatibility complex class I molecules on human lymphoid and myelomonocytic cells. J Exp Med 186:1809–1818

Cooper JA, Howell B (1993) The when and how of Src regulation. Cell 73:1051–1054

Costello PS, Turner M, Walters AE, Cunningham CN, Bauer PH, Downward J, Tybulewicz VL (1996) Critical role for the tyrosine kinase Syk in signalling through the high affinity IgE receptor of mast cells. Oncogene 13:2595–2605

Crocker PR (2002) Siglecs: sialic-acid-binding immunoglobulin-like lectins in cell–cell interactions and signalling. Curr Opin Struct Biol 12:609–615

Crocker PR, Varki A (2001) Siglecs, sialic acids and innate immunity. Trends Immunol 22:337–342

Csonga R, Prieschl EE, Jaksche D, Novotny V, Baumruker T (1998) Common and distinct signaling pathways mediate the induction of TNF-alpha and IL-5 in IgE plus antigen-stimulated mast cells. J Immunol 160:273–283

Daeron M (1997) Fc receptor biology. Annu Rev Immunol 15:203–234

Daeron M, Latour S, Malbec O, Espinosa E, Pina P, Pasmans S, Fridman WH (1995) The same tyrosine-based inhibition motif, in the intracytoplasmic domain of Fc gamma RIIB, regulates negatively BCR-, TCR-, and FcR-dependent cell activation. Immunity 3:635–646

Daheshia M, Friend DS, Grusby MJ, Austen KF, Katz HR (2001) Increased severity of local and systemic anaphylactic reactions in gp49B1-deficient mice. J Exp Med 194:227–234

Davidson D, Cloutier JF, Gregorieff A, Veillette A (1997) Inhibitory tyrosine protein kinase p50(csk) is associated with protein-tyrosine phosphatase PTP-PEST in hemopoietic and non-hemopoietic cells. J Biol Chem 272:23455–23462

Deaglio S, Morra M, Mallone R, Ausiello CM, Prager E, Garbarino G, Dianzani U, Stockinger H, Malavasi F (1998) Human CD38 (ADP-ribosyl cyclase) is a counter-receptor of CD31, an Ig superfamily member. J Immunol 160:395–402

Dong C, Davis RJ, Flavell RA (2002) MAP kinases in the immune response. Annu Rev Immunol 20:55–72

Donnadieu E, Jouvin MH, Kinet JP (2000) A second amplifier function for the allergy-associated Fc(epsilon)RI-beta subunit. Immunity 12:515–523

Donnadieu E, Jouvin MH, Rana S, Moffatt MF, Mockford EH, Cookson WO, Kinet JP (2003) Competing functions encoded in the allergy-associated F(c)epsilonRIbeta gene. Immunity 18:665–674

Dunant NM, Wisniewski D, Strife A, Clarkson B, Resh MD (2000) The phosphatidylinositol polyphosphate 5-phosphatase SHIP1 associates with the dok1 phosphoprotein in bcr-Abl transformed cells. Cell Signal 12:317–326

Eiseman E, Bolen JB (1992) Engagement of the high-affinity IgE receptor activates src protein-related tyrosine kinases. Nature 355:78–80

Erdei A, Andreev S, Pecht I (1995) Complement peptide C3a inhibits IgE-mediated triggering of rat mucosal mast-cells. Int Immunol 7:1433–1439

Erdei A, Rong X, Ischenko A, Pecht I (1998) Inhibition of IgE-mediated triggering of mast cells by C3a; association of the complement-peptide with the beta-chain of Fc epsilon RI. Mol Immunol 35:413

Fang D, Wang HY, Fang N, Altman Y, Elly C, Liu YC (2001) Cbl-b, a RING-type E3 ubiquitin ligase, targets phosphatidylinositol 3-kinase for ubiquitination in T cells. J Biol Chem 276:4872–4878

Field KA, Holowka D, Baird B (1995) Fc epsilon RI-mediated recruitment of p53/56lyn to detergent-resistant membrane domains accompanies cellular signaling. Proc Natl Acad Sci USA 92:9201–9205

Field KA, Holowka D, Baird B (1997) Compartmentalized activation of the high affinity immunoglobulin E receptor within membrane domains. J Biol Chem 272:4276–4280

Fleming TJ, Donnadieu E, Song CH, Laethem FV, Galli SJ, Kinet JP (1997) Negative regulation of Fc epsilon RI-mediated degranulation by CD81. J Exp Med 186:1307–1314

Fong DC, Malbec O, Arock M, Cambier JC, Fridman WH, Daeron M (1996) Selective in vivo recruitment of the phosphatidylinositol phosphatase SHIP by phosphorylated Fc gammaRIIB during negative regulation of IgE-dependent mouse mast cell activation. Immunol Lett 54:83–91

Franklin CC, Sanchez V, Wagner F, Woodgett JR, Kraft AS (1992) Phorbol ester-induced amino-terminal phosphorylation of human JUN but not JUNB regulates transcriptional activation. Proc Natl Acad Sci USA 89:7247–7251

Fukamachi H, Takei M, Kawakami T (1993) Activation of multiple protein kinases including a MAP kinase upon Fc epsilon RI cross-linking. Int Arch Allergy Immunol 102:15–25

Furumoto Y, Nunomura S, Terada T, Rivera J, Ra C (2004) The FcepsilonRIbeta immunoreceptor tyrosine-based activation motif exerts inhibitory control on MAPK and IkappaB kinase phosphorylation and mast cell cytokine production. J Biol Chem 279(47):49177–87

Galli SJ, Gordon JR, Wershil BK (1991) Cytokine production by mast-cells and basophils. Curr Opin Immunol 3:865–873

Garman SC, Kinet JP, Jardetzky TS (1998) Crystal structure of the human high-affinity IgE receptor. Cell 95:951–961

Gergely J, Pecht I, Sarmay G (1999) Immunoreceptor tyrosine-based inhibition motif-bearing receptors regulate the immunoreceptor tyrosine-based activation motif-induced activation of immune competent cells. Immunol Lett 68:3–15

Ghannadan M, Hauswirth AW, Schernthaner GH, Muller MR, Klepetko W, Schatzl G, Sperr WR, Buhring HJ, Valent P (2002) Detection of novel CD antigens on the surface of human mast cells and basophils. Int Arch Allergy Immunol 127:299–307

Gibbs BF, Grabbe J (1999) Inhibitors of PI 3-kinase and MEK kinase differentially affect mediator secretion from immunologically activated human basophils. J Leuk Biol 65:883–890

Gibbs BF, Plath KE, Wolff HH, Grabbe J (2002) Regulation of mediator secretion in human basophils by p38 mitogen-activated protein kinase: phosphorylation is sensitive to the effects of phosphatidylinositol 3-kinase inhibitors and calcium mobilization. J Leukoc Biol 72:391–400

Gonzalez-Espinosa C, Odom S, Olivera A, Hobson JP, Martinez ME, Oliveira-Dos-Santos A, Barra L, Spiegel S, Penninger JM, Rivera J (2003) Preferential signaling and induction of allergy-promoting lymphokines upon weak stimulation of the high affinity IgE receptor on mast cells. J Exp Med 197:1453–1465

Gordon JR, Galli SJ (1991) Release of both preformed and newly synthesized tumor-necrosis-factor-alpha (TNF-alpha)/cachectin by mouse mast-cells stimulated via the Fc-epsilon-RI – a mechanism for the sustained action of mast-cell derived TNF-alpha during IgE-dependent biological responses. J Exp Med 174:103–107

Gordon JR, Burd PR, Galli SJ (1990) Mast-cells as a source of multifunctional cytokines. Immunol Today 11:458–464

Gounni AS, Lamkhioued B, Kinet JP, Capron A, Capron M (1994) Expression of the high-affinity receptor for IgE (Fc-epsilon-RI) by human eosinophils. FASEB J 8:A979

Graham TE, Pfeiffer JR, Lee RJ, Kusewitt DF, Martinez AM, Foutz T, Wilson BS, Oliver JM (1998) MEK and ERK activation in ras-disabled RBL-2H3 mast cells and novel roles for geranylgeranylated and farnesylated proteins in Fc epsilonRI-mediated signaling. J Immunol 161: 6733–6744

Gronda M, Arab S, Iafrate B, Suzuki H, Zanke BW (2001) Hematopoietic protein tyrosine phosphatase suppresses extracellular stimulus-regulated kinase activation. Mol Cell Biol 21:6851–6858

Gu H, Saito K, Klaman LD, Shen J, Fleming T, Wang Y, Pratt JC, Lin G, Lim B, Kinet JP, Neel BG (2001) Essential role for Gab2 in the allergic response. Nature 412:186–190

Gu HH, Pratt JC, Burakoff SJ, Neel BG (1998) Cloning of p97/Gab2, the major SHP2-binding protein in hematopoietic cells, reveals a novel pathway for cytokine-induced gene activation. Mol Cell 2:729–740

Gu MX, Warshawsky I, Majerus PW (1992) Cloning and expression of a cytosolic megakaryocyte protein-tyrosine-phosphatase with sequence homology to retinaldehyde-binding protein and yeast Sec14P. Proc Natl Acad Sci USA 89:2980–2984

Gu XG, Laouar A, Wan JM, Daheshia M, Lieberman J, Yokoyama WM, Katz HR, Manjunath N (2003) The gp49B1 inhibitory receptor regulates the IFN-gamma responses of T cells and NK cells. J Immunol 170:4095–4101

Gupta S, Campbell D, Derijard B, Davis RJ (1995) Transcription factor ATF2 regulation by the JNK signal transduction pathway. Science 267:389–393

Guthmann MD, Tal M, Pecht I (1995) A secretion inhibitory signal transduction molecule on mast cells is another C-type lectin. Proc Natl Acad Sci USA 92:9397–9401

Hamaguchi T, Takahashi A, Manaka A, Sato M, Osada H (2001) TU-572, a potent and selective CD45 inhibitor, suppresses IgE-mediated anaphylaxis and murine contact hypersensitivity reactions. Int Arch Allergy Immunol 126:318–324

Harder KW, Parsons LM, Armes J, Evans N, Kountouri N, Clark R, Quilici C, Grail D, Hodgson GS, Dunn AR, Hibbs ML (2001) Gain- and loss-of-function Lyn mutant mice define a critical inhibitory role of Lyn in the myeloid lineage. Immunity 15:603–615

Hata A, Sabe H, Kurosaki T, Takata M, Hanafusa H (1994) Functional-analysis of Csk in signal-transduction through the B-cell antigen receptor. Mol Cell Biol 14:7306–7313

Hata D, Kawakami Y, Inagaki N, Lantz CS, Kitamura T, Khan WN, Maeda-Yamamoto M, Miura T, Han W, Hartman SE, Yao L, Nagai H, Goldfeld AE, Alt FW, Galli SJ, Witte ON, Kawakami T (1998a) Involvement of Bruton's tyrosine kinase in Fc epsilon RI-dependent mast cell degranulation and cytokine production. J Exp Med 187:1235–1247

Hata D, Kitaura J, Hartman SE, Kawakami Y, Yokota T, Kawakami T (1998b) Bruton's tyrosine kinase-mediated interleukin-2 gene activation in mast cells – dependence on the c-Jun N-terminal kinase activation pathway. J Biol Chem 273:10979–10987

Hayami K, Fukuta D, Nishikawa Y, Yamashita Y, Inui M, Ohyama Y, Hikida M, Ohmori H, Takai T (1997) Molecular cloning of a novel murine cell-surface glycoprotein homologous to killer cell inhibitory receptors. J Biol Chem 272:7320–7327

Heissmeyer V, Macian F, Im SH, Varma R, Feske S, Venuprasad K, Gu H, Liu YC, Dustin ML, Rao A (2004) Calcineurin imposes T cell unresponsiveness through targeted proteolysis of signaling proteins. Nature Immunol 5:255–265

Helm B, Marsh P, Vercelli D, Padlan E, Gould H, Geha R (1988) The mast cell binding site on human immunoglobulin E. Nature 331:180–183

Hendricks-Taylor LR, Motto DG, Zhang J, Siraganian RP, Koretzky GA (1997) SLP-76 is a substrate of the high affinity IgE receptor-stimulated protein tyrosine kinases in rat basophilic leukemia cells. J Biol Chem 272:1363–1367

Hiraoka S, Furumoto Y, Koseki H, Takagaki Y, Taniguchi M, Okumura K, Ra C (1999) Fc receptor beta subunit is required for full activation of mast cells through Fc receptor engagement. Int Immunol 11:199–207

Hirasawa N, Sato Y, Fujita Y, Ohuchi K (2000) Involvement of a phosphatidylinositol 3-kinase-p38 mitogen activated protein kinase pathway in antigen-induced IL-4 production in mast cells. Biochim Biophys Acta 1456:45–55

Ho LH, Uehara T, Chen CC, Kubagawa H, Cooper MD (1999) Constitutive tyrosine phosphorylation of the inhibitory paired Ig-like receptor PIR-B. Proc Natl Acad Sci USA 96:15086–15090

Honda Z, Suzuki T, Hirose N, Aihara M, Shimizu T, Nada S, Okada M, Ra C, Morita Y, Ito K (1997) Roles of C-terminal Src kinase in the initiation and the termination of the high affinity IgE receptor-mediated signaling. J Biol Chem 272:25753–25760

Honda Z, Suzuki T, Kono H, Okada M, Yamamoto T, Ra C, Morita Y, Yamamoto K (2000) Sequential requirements of the N-terminal palmitoylation site and SH2 domain of Src family kinases in the initiation and progression of FcepsilonRI signaling. Mol Cell Biol 20:1759–1771

Horikawa K, Nishizumi H, Umemori H, Aizawa S, Takatsu K, Yamamoto T (1999) Distinctive roles of Fyn and Lyn in IgD- and IgM-mediated signaling. Int Immunol 11:1441–1449

Hsu C, MacGlashan D (1996) IgE antibody up-regulates high affinity IgE binding on murine bone marrow-derived mast cells. Immunol Lett 52:129–134

Hua CT, Gamble JR, Vadas MA, Jackson DE (1998) Recruitment and activation of SHP-1 protein-tyrosine phosphatase by human platelet endothelial cell adhesion molecule-1 (PECAM-1) – identification of immunoreceptor tyrosine-based inhibitory motif-like binding motifs and substrates. J Biol Chem 273:28332–28340

Huber M, Helgason CD, Damen JE, Liu L, Humphries RK, Krystal G (1998) The src homology 2-containing inositol phosphatase (SHIP) is the gatekeeper of mast cell degranulation. Proc Natl Acad Sci USA 95:11330–11335

Huynh H, Wang XD, Li WH, Bottini N, Williams S, Nika K, Ishihara H, Godzik A, Mustelin T (2003) Homotypic secretory vesicle fusion induced by the protein tyrosine phosphatase MEG2 depends on polyphosphoinositides in T cells. J Immunol 171:6661–6671

Ishizuka T, Oshiba A, Sakata N, Terada N, Johnson GL, Gelfand EW (1996) Aggregation of the FcepsilonRI on mast cells stimulates c-Jun amino-terminal kinase activity. A response inhibited by wortmannin. J Biol Chem 271:12762–12766

Ishizuka T, Chayama K, Takeda K, Hamelmann E, Terada N, Keller GM, Johnson GL, Gelfand EW (1999) Mitogen-activated protein kinase activation through Fc epsilon receptor I and stem cell factor receptor is differentially regulated by phosphatidylinositol 3-kinase and calcineurin in mouse bone marrow-derived mast cells. J Immunol 162:2087–2094

Jabril-Cuenod B, Zhang C, Scharenberg AM, Paolini R, Numerof R, Beaven MA, Kinet JP (1996) Syk-dependent phosphorylation of Shc. A potential link between FcepsilonRI and the Ras/mitogen-activated protein kinase signaling pathway through SOS and Grb2. J Biol Chem 271:16268–16272

Jackson DE (2003) The unfolding tale of PECAM-1. FEBS Lett 540:7–14

James LC, Tawfik DS (2003) The specificity of cross-reactivity: promiscuous antibody binding involves specific hydrogen bonds rather than nonspecific hydrophobic stickiness. Protein Sci 12:2183–2193

James LC, Roversi P, Tawfik DS (2003) Antibody multispecificity mediated by conformational diversity. Science 299:1362–1367

Janknecht R, Hunter T (1997) Activation of the Sap-1a transcription factor by the c-Jun N-terminal kinase (JNK) mitogen-activated protein kinase. J Biol Chem 272:4219–4224

Jensen WA, Marschner S, Ott VL, Cambier JC (2001) FcgammaRIIB-mediated inhibition of T-cell receptor signal transduction involves the phosphorylation of SH2-containing inositol 5-phosphatase (SHIP), dephosphorylation of the linker of activated T-cells (LAT) and inhibition of calcium mobilization. Biochem Soc Trans 29:840–846

Jiang PH, Lagenaur CF, Narayanan V (1999) Integrin-associated protein is a ligand for the P84 neural adhesion molecule. J Biol Chem 274:559–562

Joazeiro CAP, Wing SS, Huang HK, Leverson JD, Hunter T, Liu YC (1999) The tyrosine kinase negative regulator c-Cbl as a RING-type, E2-dependent ubiquitin–protein ligase. Science 286: 309–312

Johnson GL, Lapadat R (2002) Mitogen-activated protein kinase pathways mediated by ERK, JNK, and p38 protein kinases. Science 298:1911–1912

Joseph M, Gounni AS, Kusnierz JP, Vorng H, Sarfati M, Kinet JP, Tonnel AB, Capron A, Capron M (1997) Expression and functions of the high-affinity IgE receptor on human platelets and megakaryocyte precursors. Eur J Immunol 27:2212–2218

Kalesnikoff J, Huber M, Lam V, Damen JE, Zhang J, Siraganian RP, Krystal G (2001) Monomeric IgE stimulates signaling pathways in mast cells that lead to cytokine production and cell survival. Immunity 14:801–811

Kalesnikoff J, Baur N, Leitges M, Hughes MR, Damen JE, Huber M, Krystal G (2002) SHIP negatively regulates IgE + antigen-induced IL-6 production in mast cells by inhibiting NF-kappa B activity. J Immunol 168:4737–4746

Katz HR, Vivier E, Castells MC, McCormick MJ, Chambers JM, Austen KF (1996) Mouse mast cell gp49B1 contains two immunoreceptor tyrosine-based inhibition motifs and suppresses mast cell activation when coligated with the high-affinity Fc receptor for IgE. Proc Natl Acad Sci USA 93:10809–10814

Kawabuchi M, Satomi Y, Takao T, Shimonishi Y, Nada S, Nagai K, Tarakhovsky A, Okada M (2000) Transmembrane phosphoprotein Cbp regulates the activities of Src-family tyrosine kinases. Nature 404:999–1003

Kawakami Y, Yao L, Miura T, Tsukada S, Witte ON, Kawakami T (1994) Tyrosine phosphorylation and activation of Bruton tyrosine kinase upon Fc epsilon RI cross-linking. Mol Cell Biol 14: 5108–5113

Kawakami Y, Yao L, Han W, Kawakami T (1996) Tec family protein-tyrosine kinases and pleckstrin homology domains in mast cells. Immunol Lett 54:113–7

Kawakami Y, Miura T, Bissonnette R, Hata D, Khan WN, Kitamura T, Maeda-Yamamoto M, Hartman SE, Yao L, Alt FW, Kawakami T (1997) Bruton's tyrosine kinase regulates apoptosis and JNK/SAPK kinase activity. Proc Natl Acad Sci USA 94:3938–3942

Kawakami Y, Kitaura J, Hata D, Yao LB, Kawakami T (1999) Functions of Bruton's tyrosine kinase in mast and B cells. J Leukoc Biol 65:286–290

Kawakami Y, Kitaura J, Hartman SE, Lowell CA, Siraganian RP, Kawakami T (2000a) Regulation of protein kinase Cbetal by two protein-tyrosine kinases, Btk and Syk. Proc Natl Acad Sci USA 97:7423–7428

Kawakami Y, Kitaura J, Satterthwaite AB, Kato RM, Asai K, Hartman SE, Maeda-Yamamoto M, Lowell CA, Rawlings DJ, Witte ON, Kawakami T (2000b) Redundant and opposing functions of two tyrosine kinases, Btk and Lyn, in mast cell activation. J Immunol 165: 1210–1219

Keane MM, Riverolezcano OM, Mitchell JA, Robbins KC, Lipkowitz S (1995) Cloning and characterization of Cbl-B – a Sh3 binding-protein with homology to the c-Cbl protooncogene. Oncogene 10:2367–2377

Keegan AD, Paul WE (1992) Multichain immune recognition receptors: similarities in structure and signaling pathways. Immunol Today 13:63–68

Kharitonenkov A, Chen ZJ, Sures I, Wang HY, Schilling J, Ullrich A (1997) A family of proteins that inhibit signalling through tyrosine kinase receptors. Nature 386:181–186

Kihara H, Siraganian RP (1994) Src homology 2 domains of Syk and Lyn bind to tyrosine-phosphorylated subunits of the high affinity IgE receptor. J Biol Chem 269:22427–22432

Kikly KK, Bochner BS, Freeman SD, Tan KB, Gallagher KT, D'alessio KJ, Holmes SD, Abrahamson JA, Erickson-Miller CL, Murdock PR, Tachimoto H, Schleimer RP, White JR (2000) Identification of SAF-2, a novel Siglec expressed on eosinophils, mast cells, and basophils. J Allergy Clin Immunol 105:1093–1100

Kimata M, Inagaki N, Kato T, Miura T, Serizawa I, Nagai H (2000) Roles of mitogen-activated protein kinase pathways for mediator release from human cultured mast cells. Biochem Pharmacol 60:589–594

Kimura T, Sakamoto H, Appella E, Siraganian RP (1997a) The negative signaling molecule SH2 domain-containing inositol-polyphosphate 5-phosphatase (SHIP) binds to the tyrosine-phosphorylated beta subunit of the high affinity IgE receptor. J Biol Chem 272:13991–13996

Kimura T, Zhang J, Sagawa K, Sakaguchi K, Appella E, Siraganian RP (1997b) Syk-independent tyrosine phosphorylation and association of the protein tyrosine phosphatases SHP-1 and SHP-2 with the high affinity IgE receptor. J Immunol 159:4426–4434

Kinet JP (1999) The high-affinity IgE receptor (Fc epsilon RI): from physiology to pathology. Annu Rev Immunol 17:931–972

Kitaura J, Asai K, Maeda-Yamamoto M, Kawakami Y, Kikkawa U, Kawakami T (2000) Akt-dependent cytokine production in mast cells. J Exp Med 192:729–740

Kitaura J, Song J, Tsai M, Asai K, Maeda-Yamamoto M, Mocsai A, Kawakami Y, Liu FT, Lowell CA, Barisas BG, Galli SJ, Kawakami T (2003) Evidence that IgE molecules mediate a spectrum of effects on mast cell survival and activation via aggregation of the FcepsilonRI. Proc Natl Acad Sci USA 100:12911–12916

Kraft S, Katoh N, Novak N, Koch S, Bieber T (2001) Unexpected functions of FcepsilonRI on antigen-presenting cells. Int Arch Allergy Immunol 124:35–37

Kraft S, Novak N, Katoh N, Bieber T, Rupec RA (2002a) Aggregation of the high-affinity IgE receptor Fc epsilon RI on human monocytes and dendritic cells induces NF-kappa B activation. J Invest Dermatol 118:830–837

Kraft S, Wessendorf JHM, Haberstok J, Novak N, Wollenberg A, Bieber T (2002b) Enhanced expression and activity of protein-tyrosine kinases establishes a functional signaling pathway only in Fc epsilon RIhigh Langerhans cells from atopic individuals. J Invest Dermatol 119:804–811

Kraft S, Fleming T, Billingsley JM, Lin SY, Jouvin MH, Storz O, Kinet JP (2005) Anti-CD63 antibodies suppress IgE-dependent allergic reactions in vitro and in vivo. J Exp Med 201(3): 385–396

Kubagawa H, Burrows PD, Coopers MD (1997) A novel pair of immunoglobulin-like receptors expressed by B cells and myeloid cells. Proc Natl Acad Sci USA 94:5261–5266

Kubagawa H, Chen CC, Ho LH, Shimada T, Gartland L, Mashburn C, Uehara T, Ravetch JV, Cooper MD (1999) Biochemical nature and cellular distribution of the paired immunoglobulin-like receptors, PIR-A and PIR-B. J Exp Med 189:309–317

Kubo S, Matsuoka K, Taya C, Kitamura F, Takai T, Yonekawa H, Karasuyama H (2001) Drastic up-regulation of Fc epsilon RI on mast cells is induced by IgE binding through stabilization and accumulation of Fc epsilon RI on the cell surface. J Immunol 167:3427–3434

Kumagai H, Oki T, Tamitsu K, Feng SZ, Ono M, Nakajima H, Bao YC, Kawakami Y, Nagayoshi K, Copeland NG, Gilbert DJ, Jenkins NA, Kawakami T, Kitamura T (2003) Identification and characterization of a new pair of immunoglobulin-like receptors LMIR1 and 2 derived from murine bone marrow-derived mast cells. Biochem Biophys Res Commun 307: 719–729

Kuroiwa A, Yamashita Y, Inui M, Yuasa T, Ono M, Nagabukuro A, Matsuda Y, Takai T (1998) Association of tyrosine phosphatases SHP-1 and SHP-2, inositol 5-phosphatase SHIP with gp49B1, and chromosomal assignment of the gene. J Biol Chem 273:1070–1074

Kyo S, Sada K, Qu X, Maeno K, Miah SM, Kawauchi-Kamata K, Yamamura H (2003) Negative regulation of Lyn protein-tyrosine kinase by c-Cbl ubiquitin-protein ligase in Fc aepsilon RI-mediated mast cell activation. Genes Cells 8:825–836

Latour S, Veillette A (2001) Proximal protein tyrosine kinases in immunoreceptor signaling. Curr Opin Immunol 13:299–306

Latour S, Tanaka H, Demeure C, Mateo V, Rubio M, Brown EJ, Maliszewski C, Lindberg FP, Oldenborg A, Ullrich A, Delespesse G, Sarfati M (2001) Bidirectional negative regulation of human T and dendritic cells by CD47 and its cognate receptor signal-regulator protein-alpha: down-regulation of IL-12 responsiveness and inhibition of dendritic cell activation. J Immunol 167:2547–2554

Letourneur F, Klausner RD (1991) T-cell and basophil activation through the cytoplasmic tail of T-cell-receptor zeta-family proteins. Proc Natl Acad Sci USA 88:8905–8909

Lewis RA, Soter NA, Diamond PT, Austen KF, Oates JA, Roberts LJ (1982) Prostaglandin D2 generation after activation of rat and human mast cells with anti-IgE. J Immunol. 129:1627–1631

Liang SY, Baibakov B, Horuzsko A (2002) HLA-G inhibits the functions of murine dendritic cells via the PIR-B immune inhibitory receptor. Eur J Immunol 32:2418–2426

Licht A, Pecht I, Schweitzer-Stenner R (2005) Regulation of mast cells' secretory response by co-clustering the Type I Fcε receptor with the mast cell function-associated antigen. Europ J Immunol 35

Lienard H, Bruhns P, Malbec O, Fridman WH, Daeron M (1999) Signal regulatory proteins negatively regulate immunoreceptor-dependent cell activation. J Biol Chem 274:32493–32499

Lin S, Cicala C, Scharenberg AM, Kinet JP (1996) The Fc(epsilon)RIbeta subunit functions as an amplifier of Fc(epsilon)RIgamma-mediated cell activation signals. Cell 85:985–995

Liu Q, Oliveira-Dos-Santos AJ, Mariathasan S, Bouchard D, Jones J, Sarao R, Kozieradzki I, Ohashi PS, Penninger JM, Dumont DJ (1998) The inositol polyphosphate 5-phosphatase SHIP is a crucial negative regulator of B cell antigen receptor signaling. J Exp Med 188: 1333–1342

Livingstone C, Patel G, Jones N (1995) ATF-2 contains a phosphorylation-dependent transcriptional activation domain. EMBO J 14:1785–1797

Lock P, Casagranda F, Dunn AR (1999) Independent SH2-binding sites mediate interaction of Dok-related protein with RasGTPase-activating protein and Nck. J Biol Chem 274:22775–22784

Lu TT, Barreuther M, Davis S, Madri JA (1997) Platelet endothelial cell adhesion molecule-1 is phosphorylatable by c-Src, binds Src-Src homology 2 domain, and exhibits immunoreceptor tyrosine-based activation motif-like properties. J Biol Chem 272:14442–14446

Lu-Kuo JM, Joyal D, Chambers J, Xu NN, Austen KF, Katz HR (1998) gp49B1 associates with the SHP-1 tyrosine phosphatase and inhibits activating signal transduction in mouse mast cells. FASEB J 12:A1061

Lu-Kuo JM, Joyal DM, Austen KF, Katz HR (1999) gp49B1 inhibits IgE-initiated mast cell activation through both immunoreceptor tyrosine-based inhibitory motifs, recruitment of src homology 2 domain-containing phosphatase-1, and suppression of early and late calcium mobilization. J Biol Chem 274:5791–5796

Maeda A, Kurosaki M, Ono M, Takai T, Kurosaki T (1998) Requirement of SH2-containing protein tyrosine phosphatases SHP-1 and SHP-2 for paired immunoglobulin-like receptor B (PIR-B)-mediated inhibitory signal. J Exp Med 187:1355–1360

Maeda A, Scharenberg AM, Tsukada S, Bolen JB, Kinet JP, Kurosaki T (1999) Paired immunoglobulin-like receptor B (PIR-B) inhibits BCR-induced activation of Syk and Btk by SHP-1. Oncogene 18:2291–2297

Maehama T, Dixon JE (1998) The tumor suppressor, PTEN/MMAC1, dephosphorylates the lipid second messenger, phosphatidylinositol 3,4,5-trisphosphate. J Biol Chem 273:13375–13378

Malaviya R, Abraham SN (2001) Mast cell modulation of immune responses to bacteria. Immunol Rev 179:16–24

Malbec O, Fong DC, Turner M, Tybulewicz VL, Cambier JC, Fridman WH, Daeron M (1998) Fc epsilon receptor I-associated lyn-dependent phosphorylation of Fc gamma receptor IIB during negative regulation of mast cell activation. J Immunol 160:1647–1658

Malbec O, Fridman WH, Daeron M (1999) Negative regulation of c-kit-mediated cell proliferation by Fc gamma RIIB. J Immunol 162:4424–4429

Manetz TS, Gonzalez-Espinosa C, Arudchandran R, Xirasagar S, Tybulewicz V, Rivera J (2001) Vav1 regulates phospholipase cgamma activation and calcium responses in mast cells. Mol Cell Biol 21:3763–3774

Marmor MD, Yarden Y (2004) Role of protein ubiquitylation in regulating endocytosis of receptor tyrosine kinases. Oncogene 23:2057–2070

Martelli MP, Boomer J, Bu M, Bierer BE (2001) T cell regulation of p62(dok) (Dok1) association with Crk-L. J Biol Chem 276:45654–45661

Martin A, Tsui HW, Shulman MJ, Isenman D, Tsui FWL (1999) Murine SHP-1 splice variants with altered Src homology 2 (SH2) domains – implications for the SH2-mediated intramolecular regulation of SHP-1. J Biol Chem 274:21725–21734

Masuda A, Yoshikai Y, Aiba K, Matsuguchi T (2002) Th2 cytokine production from mast cells is directly induced by lipopolysaccharide and distinctly regulated by c-Jun N-terminal kinase and p38 pathways. J Immunol 169:3801–3810

Matsumoto Y, Wang LL, Yokoyama WM, Aso T (2001) Uterine macrophages express the gp49B inhibitory receptor in midgestation. J Immunol 166:781–786

Matthews RJ, Bowne DB, Flores E, Thomas ML (1992) Characterization of hematopoietic intracellular protein tyrosine phosphatases – description of a phosphatase containing an Sh2 domain and another enriched in proline-rich, glutamic acid-rich, serine-rich, and threonine-rich sequences. Mol Cell Biol 12:2396–2405

Maurer D, Rieger A, Kilgus O, Reininger B, Ochiai K, Kinet JP, Stingl G (1993) Human peripheral-blood monocytes express the high-affinity IgE receptor Fc-epsilon-RI. J Immunol 150:A222

Maurer D, Ebner C, Reininger B, Fiebiger E, Kraft D, Kinet JP, Stingl G (1995) The high-affinity IgE receptor (Fc-epsilon-RI) mediates IgE-dependent allergen presentation. J Immunol 154: 6285–6290

Maurer D, Ebner C, Reininger B, Fiebiger E, Fischer GF, Wichlas S, Kinet JP, Stingl G (1996) Peripheral blood dendritic cells express Fc epsilon RI as a complex composed of Fc epsilon RI alpha- and Fc epsilon RI gamma-chains and use this receptor for IgE-mediated allergen presentation. J Invest Dermatol 106:739

McCormick MJ, Castells MC, Austen KF, Katz HR (1999) The gp49A gene has extensive sequence conservation with the gp49B gene and provides gp49A protein, a unique member of a large family of activating and inhibitory receptors of the immunoglobulin superfamily. Immunogenetics 50:286–294

Metzger H (1992) The receptor with high affinity for IgE. Immunol Rev 125:37–48

Miura K, MacGlashan DW (2000) Phosphatidylinositol-3 kinase regulates p21ras activation during IgE-mediated stimulation of human basophils. Blood 96:2199–2205

Miura K, Schroeder JT, Hubbard WC, MacGlashan DW Jr (1999) Extracellular signal-regulated kinases regulate leukotriene C4 generation, but not histamine release or IL-4 production from human basophils. J Immunol 162:4198–4206

Muraille E, Bruhns P, Pesesse X, Daeron M, Erneux C (2000) The SH2 domain containing inositol 5-phosphatase SHIP2 associates to the immunoreceptor tyrosine-based inhibition motif of Fc gammaRIIB in B cells under negative signaling. Immunol Lett 72:7–15

Murakami K, Sato S, Nagasawa S, Yamashita T (2000) Regulation of mast cell signaling through high-affinity IgE receptor by CD45 protein tyrosine phosphatase. Int Immunol 12:169–176

Nada S, Okada M, Macauley A, Cooper JA, Nakagawa H (1991) Cloning of a complementary-DNA for a protein-tyrosine kinase that specifically phosphorylates a negative regulatory site of P60C-Src. Nature 351:69–72

Nadler MJ, Matthews SA, Turner H, Kinet JP (2000) Signal transduction by the high-affinity immunoglobulin E receptor Fc epsilon RI: coupling form to function. Adv Immunol 76:325–355

Navarro F, Llano M, Bellon T, Colonna M, Geraghty DE, Lopez-Botet M (1999) The ILT2(LIR1) and CD94/NKG2A NK cell receptors respectively recognize HLA-G1 and HLA-E molecules co-expressed on target cells. Eur J Immunol 29:277–283

Nishizumi H, Yamamoto T (1997) Impaired tyrosine phosphorylation and Ca^{2+} mobilization, but not degranulation, in lyn-deficient bone marrow-derived mast cells. J Immunol 158:2350–2355

Numahata K, Komagata T, Hirasawa N, Someya K, Xiao YQ, Ohuchi K (2003) Analysis of the mechanism regulating the stability of rat macrophage inflammatory protein-2 mRNA in RBL-2H3 cells. J Cell Biochem 90:976–986

Nutku E, Aizawa H, Hudson SA, Bochner BS (2003) Ligation of Siglec-8: a selective mechanism for induction of human eosinophil apoptosis. Blood 101:5014–5020

Ohtake H, Ichikawa N, Okada M, Yamashita T (2002) Cutting edge: transmembrane phosphoprotein Csk-binding protein/phosphoprotein associated with glycosphingolipid-enriched microdomains as a negative feedback regulator of mast cell signaling through the FcepsilonRI. J Immunol 168:2087–2090

Oka T, Hori M, Tanaka A, Matsuda H, Karaki H, Ozaki H (2004) IgE alone-induced actin assembly modifies calcium signaling and degranulation in RBL-2H3 mast cells. Am J Physiol Cell Physiol 286:C256–C263

Ono M, Bolland S, Tempst P, Ravetch JV (1996) Role of the inositol phosphatase SHIP in negative regulation of the immune system by the receptor Fc(gamma)RIIB. Nature 383:263–266

Ortega SE, Pecht I (1988) A monoclonal antibody that inhibits secretion from rat basophilic leukemia cells and binds to a novel membrane component. J Immunol 141:4324–4332

Osborne MA, Zenner G, Lubinus M, Zhang XL, Songyang Z, Cantley LC, Majerus P, Burn P, Kochan JP (1996) The inositol 5′-phosphatase SHIP binds to immunoreceptor signaling motifs and responds to high affinity IgE receptor aggregation. J Biol Chem 271:29271–29278

Oshima K, Amin ARMR, Suzuki A, Hamaguchi M, Matsuda S (2002) SHPS-1, a multifunctional transmembrane glycoprotein. FEBS Lett 519:1–7

Ota Y, Samelson LE (1997) The product of the proto-oncogene c-Cbl: a negative regulator of the Syk tyrosine kinase. Science 276:418–420

Ota Y, Beitz LO, Scharenberg AM, Donovan JA, Kinet JP, Samelson LE (1996) Characterization of Cbl tyrosine phosphorylation and a Cbl–Syk complex in RBL-2H3 cells. J Exp Med 184: 1713–1723

Ott VL, Tamir I, Niki M, Pandolfi PP, Cambier JC (2002) Downstream of kinase, p62(dok), is a mediator of Fc gamma RIIB inhibition of Fc epsilon RI signaling. J Immunol 168:4430–4439

Ozawa K, Yamada K, Kazanietz MG, Blumberg PM, Beaven MA (1993) Different isozymes of protein kinase C mediate feedback inhibition of phospholipase C and stimulatory signals for exocytosis in rat RBL-2H3 cells. J Biol Chem 268:2280–2283

Pandey V, Mihara S, Fensome-Green A, Bolsover S, Cockcroft S (2004) Monomeric IgE stimulates NFAT translocation into the nucleus, a rise in cytosol Ca($^{2+}$), degranulation, and membrane ruffling in the cultured rat basophilic leukemia-2H3 mast cell line. J Immunol 172:4048–4058

Paolini R, Jouvin MH, Kinet JP (1991) Phosphorylation and dephosphorylation of the high-affinity receptor for immunoglobulin-e immediately after receptor engagement and disengagement. Nature 353:855–858

Paolini R, Molfetta R, Beitz LO, Zhang J, Scharenberg AM, Piccoli M, Frati L, Siraganian R, Santoni A (2002) Activation of Syk tyrosine kinase is required for c-Cbl-mediated ubiquitination of Fcepsilon RI and Syk in RBL cells. J Biol Chem 277:36940–36947

Parravicini V, Gadina M, Kovarova M, Odom S, Gonzalez-Espinosa C, Furumoto Y, Saitoh S, Samelson LE, O'Shea JJ, Rivera J (2002) Fyn kinase initiates complementary signals required for IgE-dependent mast cell degranulation. Nat Immunol 3:741–748

Pesesse X, Deleu S, De SF, Drayer L, Erneux C (1997) Identification of a second SH2-domain-containing protein closely related to the phosphatidylinositol polyphosphate 5-phosphatase SHIP. Biochem Biophys Res Commun 239:697–700

Pesesse X, Moreau C, Drayer AL, Woscholski R, Parker P, Erneux C (1998) The SH2 domain containing inositol 5-phosphatase SHIP2 displays phosphatidylinositol 3,4,5-trisphosphate and inositol 1,3,4,5-tetrakisphosphate 5-phosphatase activity. FEBS Lett 437:301–303

Phillips NE, Parker DC (1984) Cross-linking of B lymphocyte Fc gamma receptors and membrane immunoglobulin inhibits anti-immunoglobulin-induced blastogenesis. J Immunol 132:627–632

Piali L, Hammel P, Uherek C, Bachmann F, Gisler RH, Dunon D, Imhof BA (1995) Cd31/Pecam-1 is a ligand for alpha(v)beta(3) integrin involved in adhesion of leukocytes to endothelium. J Cell Biol 130:451–460

Pivniouk VI, Martin TR, Lu-Kuo JM, Katz HR, Oettgen HC, Geha RS (1999) SLP-76 deficiency impairs signaling via the high-affinity IgE receptor in mast cells. J Clin Invest 103:1737–1743

Pratt JC, Igras VE, Maeda H, Baksh S, Gelfand EW, Burakoff SJ, Neel BG, Gu HH (2000) Cutting edge: Gab2 mediates an inhibitory phosphatidylinositol 3′-kinase pathway in T cell antigen receptor signaling. J Immunol 165:4158–4163

Pribluda VS, Pribluda C, Metzger H (1994) Transphosphorylation as the mechanism by which the high-affinity receptor for IgE is phosphorylated upon aggregation. Proc Natl Acad Sci USA 91:11246–11250

Prieschl EE, Pendl GG, Harrer NE, Baumruker T (1995) P21(Ras) links Fc-epsilon-RI to NF-AT family member in mast-cells – the Ap3-like factor in this cell-type is an NF-AT family member. J Immunol 155:4963–4970

Pulverer BJ, Kyriakis JM, Avruch J, Nikolakaki E, Woodgett JR (1991) Phosphorylation of c-Jun mediated by MAP kinases. Nature 353:670–674

Qu X, Sada K, Kyo S, Maeno K, Miah SM, Yamamura H (2004) Negative regulation of FcepsilonRI-mediated mast cell activation by a ubiquitin-protein ligase Cbl-b. Blood 103:1779–1786

Rao N, Dodge I, Band H (2002) The Cbl family of ubiquitin ligases: critical negative regulators of tyrosine kinase signaling in the immune system. J Leuk Biol 71:753–763

Rawlings DJ, Scharenberg AM, Park H, Wahl MI, Lin SQ, Kato RM, Fluckiger AC, Witte ON, Kinet JP (1996) Activation of BTK by a phosphorylation mechanism initiated by SRC family kinases. Science 271:822–825

Razin E, Mencia-Huerta JM, Stevens RL, Lewis RA, Liu FT, Corey E, Austen KF (1983) IgE-mediated release of leukotriene C4, chondroitin sulfate E proteoglycan, beta-hexosaminidase, and histamine from cultured bone marrow-derived mouse mast cells. J Exp Med 157:189–201

Razin E, Szallasi Z, Kazanietz MG, Blumberg PM, Rivera J (1994) Protein-kinases C-beta and C-epsilon link the mast-cell high-affinity receptor for IgE to the expression of C-Fos and C-Jun. Proc Natl Acad Sci USA 91:7722–7726

Reth M (1989) Antigen receptor tail clue. Nature 338:383–384

Roach TI, White L, Thomas ML, Lowell C, Lagenaur C, Brown EJ (1999) Myeloid src family kinases are required for PIR-B and BIT phosphorylation in primary macrophages. Mol Biol Cell 10:319A

Robbins SH, Nguyen KB, Takahashi N, Mikayama T, Biron CA, Brossay L (2002) Cutting edge: inhibitory functions of the killer cell lectin-like receptor G1 molecule during the activation of mouse NK cells. J Immunol 168:2585–2589

Rojo S, Burshtyn DN, Long EO, Wagtmann N (1997) Type I transmembrane receptor with inhibitory function in mouse mast cells and NK cells. J Immunol 158:9–12

Rolli V, Gallwitz M, Wossning T, Flemming A, Schamel WWA, Zurn C, Reth M (2002) Amplification of B cell antigen receptor signaling by a Syk/ITAM positive feedback loop. Mol Cell 10:1057–1069

Rudge EU, Cutler AJ, Pritchard NR, Smith KGC (2002) Interleukin 4 reduces expression of inhibitory receptors on B cells and abolishes CD22 and Fc gamma RII-mediated B cell suppression. J Exp Med 195:1079–1085

Sagawa K, Kimura T, Swieter M, Siraganian RP (1997a) The protein-tyrosine phosphatase SHP-2 associates with tyrosine-phosphorylated adhesion molecule PECAM-1 (CD31). J Biol Chem 272:31086–31091

Sagawa K, Swaim W, Zhang J, Unsworth E, Siraganian RP (1997b) Aggregation of the high affinity IgE receptor results in the tyrosine phosphorylation of the surface adhesion protein PECAM-1 (CD31). J Biol Chem 272:13412–13418

Sagi-Eisenberg R, Lieman H, Pecht I (1985) Protein kinase-C regulation of the receptor-coupled calcium signal in histamine-secreting rat basophilic leukemia-cells. Nature 313:59–61

Saito K, Tolias KF, Saci A, Koon HB, Humphries LA, Scharenberg A, Rawlings DJ, Kinet JP, Carpenter CL (2003) BTK regulates PtdIns-4,5-P-2 synthesis: importance for calcium signaling and PI3 K activity. Immunity 19:669–678

Saitoh S, Arudchandran R, Manetz TS, Zhang W, Sommers CL, Love PE, Rivera J, Samelson LE (2000) LAT is essential for Fc(epsilon)RI-mediated mast cell activation. Immunity 12: 525–535

Salim K, Bottomley MJ, Querfurth E, Zvelebil MJ, Gout I, Scaife R, Margolis RL, Gigg R, Smith CIE, Driscoll PC, Waterfield MD, Panayotou G (1996) Distinct specificity in the recognition of phosphoinositides by the pleckstrin homology domains of dynamin and Bruton's tyrosine kinase. EMBO J 15:6241–6250

Sayama K, Diehn M, Matsuda K, Lunderius C, Tsai M, Tam SY, Botstein D, Brown PO, Galli SJ (2002) Transcriptional response of human mast cells stimulated via the Fc(epsilon)RI and identification of mast cells as a source of IL-11. Bio Med Central Immunol 3:5

Scharenberg AM, Lin S, Cuenod B, Yamamura H, Kinet JP (1995) Reconstitution of interactions between tyrosine kinases and the high affinity IgE receptor which are controlled by receptor clustering. EMBO J 14:3385–3394

Scheid MP, Woodgett JR (2003) Unravelling the activation mechanisms of protein kinase B/Akt. FEBS Lett 546:108–112

Schneider H, Cohen-Dayag A, Pecht I (1992) Tyrosine phosphorylation of phospholipase C gamma 1 couples the Fc epsilon receptor mediated signal to mast cells secretion. Int Immunol 4:447–453

Schweitzer-Stenner R, Pecht I (2005) Death of a dogma or enforcing the artificial: monomeric IgE binding may initiate mast cell response by inducing its receptor aggregation. J Immunol 174:4461–4464

Seiffert M, Brugger W, Kanz L, Ullrich A, Buhring HJ (1999) Human SIRP alpha 1 is an extracellular ligand for CD47 and is involved in transendothelial and transepithelial migration of neutrophils. Exp Hematol 27:41

Siraganian RP, Urata C, McGivney A (1983) Arachidonic acid release during IgE- and Ca^{2+}-ionophore activation of rat basophilic leukemia cells. Monogr Allergy 18:120–123

Siraganian RP, Zhang J, Suzuki K, Sada K (2002) Protein tyrosine kinase Syk in mast cell signaling. Mol Immunol 38:1229–1233

Sloane DE, Tedla N, Awoniyi M, MacGlashan DW, Borges L, Austen KF, Arm JP (2004) Leukocyte immunoglobulin-like receptors: novel innate receptors for human basophil activation and inhibition. Blood 104:2832–2839

Smeal T, Binetruy B, Mercola DA, Birrer M, Karin M (1991) Oncogenic and transcriptional cooperation with Ha-Ras requires phosphorylation of c-Jun on serines 63 and 73. Nature 354:494–496

Sohn HW, Gu H, Pierce SK (2003) Cbl-b negatively regulates B cell antigen receptor signaling in mature B cells through ubiquitination of the tyrosine kinase Syk. J Exp Med 197:1511–1524

Song JS, Haleem-Smith H, Arudchandran R, Gomez J, Scott PM, Mill JF, Tan TH, Rivera J (1999) Tyrosine phosphorylation of Vav stimulates IL-6 production in mast cells by a Rac/c-Jun N-terminal kinase-dependent pathway. J Immunol 163:802–810

Stingl G, Maurer D (1997) IgE-mediated allergen presentation via Fc epsilon RI on antigen-presenting cells. Int Arch Allergy Immunol 113:24–29

Sun J, Williams J, Yan HC, Amin KM, Albelda SM, DeLisser HM (1996) Platelet endothelial cell adhesion molecule-1 (PECAM-1) hemophilic adhesion is mediated by immunoglobulin-like domains 1 and 2 and depends on the cytoplasmic domain and the level of surface expression. J Biol Chem 271:18561–18570

Swieter M, Berenstein EH, Swaim WD, Siraganian RP (1995) Aggregation of IgE receptors in rat basophilic leukemia 2H3 cells induces tyrosine phosphorylation of the cytosolic protein-tyrosine phosphatase HePTP. J Biol Chem 270:21902–21906

Takai T, Ono M, Hikida M, Ohmori H, Ravetch JV (1996) Augmented humoral and anaphylactic responses in Fc gamma RII-deficient mice. Nature 379:346–349

Tamir I, Stolpa JC, Helgason CD, Nakamura K, Bruhns P, Daeron M, Cambier JC (2000) The Ras-GAP-binding protein p62dok is a mediator of inhibitory FcgammaRIIB signals in B cells. Immunity 12:347–358

Taniguchi T, Kobayashi T, Kondo J, Takahashi K, Nakamura H, Suzuki J, Nagai K, Yamada T, Nakamura S, Yamamura H (1991) Molecular-cloning of a porcine gene Syk that encodes a 72-kDa protein-tyrosine kinase showing high susceptibility to proteolysis. J Biol Chem 266:15790–15796

Tedla NT, Tassinari P, Samplaski M, Austen FK, Borges L, Cosman D, Arm JP (2001) Leukocyte immunoglobulin-like receptors expressed in eosinophils and basophils. FASEB J 15: A701

Teramoto H, Salem P, Robbins KC, Bustelo XR, Gutkind JS (1997) Tyrosine phosphorylation of the vav proto-oncogene product links FcepsilonRI to the Rac1-JNK pathway. J Biol Chem 272:10751–10755

Tkaczyk C, Beaven MA, Brachman SM, Metcalfe DD, Gilfillan AM (2003) The phospholipase C gamma 1-dependent pathway of Fc epsilon RI-mediated mast cell activation is regulated independently of phosphatidylinositol 3-kinase. J Biol Chem 278:48474–48484

Toker A, Newton AC (2000) Akt/protein kinase B is regulated by autophosphorylation at the hypothetical PDK-2 site. J Biol Chem 275:8271–8274

Turner H, Cantrell DA (1997) Distinct Ras effector pathways are involved in Fc epsilon R1 regulation of the transcriptional activity of Elk-1 and NFAT in mast cells. J Exp Med 185:43–53

Turner H, Kinet JP (1999) Signalling through the high-affinity IgE receptor Fc epsilonRI. Nature 402:24–30

Uehara T, Blery M, Kang DW, Chen CC, Ho LH, Gartland GL, Liu FT, Vivier E, Cooper MD, Kubagawa H (2001) Inhibition of IgE-mediated mast cell activation by the paired Ig-like receptor PIR-B. J Clin Invest 108:1041–1050

Valent P, Ashman LK, Hinterberger W, Eckersberger F, Majdic O, Lechner K, Bettelheim P (1989) Mast-cell typing – demonstration of a distinct hematopoietic-cell type and evidence for immunophenotypic relationship to mononuclear phagocytes. Blood 73:1778–1785

Veillette A, Thibaudeau E, Latour S (1998) High expression of inhibitory receptor SHPS-1 and its association with protein-tyrosine phosphatase SHP-1 in macrophages. J Biol Chem 273: 22719–22728

Vernon-Wilson EF, Kee WJ, Willis AC, Barclay AN, Simmons DL, Brown MH (2000) CD47 is a ligand for rat macrophage membrane signal regulatory protein SIRP (OX41) and human SIRP alpha 1. Eur J Immunol 30:2130–2137

Voehringer D, Blaser C, Brawand P, Raulet DH, Hanke T, Pircher H (2001) Viral infections induce abundant numbers of senescent CD8 T cells. J Immunol 167:4838–4843

Voehringer D, Koschella M, Pircher H (2002) Lack of proliferative capacity of human effector and memory T cells expressing killer cell lectinlike receptor G1 (KLRG1). Blood 100:3698–3702

Vonakis BM, Chen HX, Haleem-Smith H, Metzger H (1997) The SH4 domain of lyn kinase is constitutively associated with the high affinity receptor for IgE (Fc epsilon RI). J Allergy Clin Immunol 99:1653

Vonakis BM, Haleem-Smith H, Benjamin P, Metzger H (2001) Interaction between the unphosphorylated receptor with high affinity for IgE and Lyn kinase. J Biol Chem 276:1041–1050

Wang B, Rieger A, Kilgus O, Ochiai K, Maurer D, Fodinger D, Kinet JP, Stingl G (1992) Epidermal Langerhans cells from normal human skin bind monomeric IgE via Fc-epsilon-RI. J Exp Med 175:1353–1365

Wang JY, Koizumi T, Watanabe T (1996) Altered antigen receptor signaling and impaired Fas-mediated apoptosis of B cells in Lyn-deficient mice. J Exp Med 184:831–838

Wang LL, Mehta IK, LeBlanc PA, Yokoyama WM (1997) Mouse natural killer cells express gp49B1, a structural homologue of human killer inhibitory receptors. J Immunol 158:13–17

Wang LL, Chu DT, Dokun AO, Yokoyama WM (2000) Inducible expression of the gp49B inhibitory receptor on NK cells. J Immunol 164:5215–5220

Wang X, Huynh H, Gjorloff-Wingren A, Monosov E, Stridsberg M, Fukuda M, Mustelin T (2002) Enlargement of secretory vesicles by protein tyrosine phosphatase PTP-MEG2 in rat basophilic leukemia mast cells and Jurkat T cells. J Immunol 168:4612–4619

Wen R, Jou ST, Chen Y, Hoffmeyer A, Wang D (2002) Phospholipase C gamma 2 is essential for specific functions of Fc epsilon R and Fc gamma R. J Immunol 169:6743–6752

Whitmarsh AJ, Shore P, Sharrocks AD, Davis RJ (1995) Integration of MAP kinase signal transduction pathways at the serum response element. Science 269:403–407

Wilson BS, Kapp N, Lee RJ, Pfeiffer JR, Martinez AM, Platt Y, Letourneur F, Oliver JM (1995) Distinct functions of the Fc epsilon R1 gamma and beta subunits in the control of Fc epsilon R1-mediated tyrosine kinase activation and signaling responses in RBL-2H3 mast cells. J Biol Chem 270:4013–4022

Wodnar-Filipowicz A, Heusser CH, Moroni C (1989) Production of the haemopoietic growth factors GM-CSF and interleukin-3 by mast cells in response to IgE receptor-mediated activation. Nature 339:150–152

Wong MX, Roberts D, Bartley PA, Jackson DE (2002) Absence of platelet endothelial cell adhesion molecule-1 (CD31) leads to increased severity of local and systemic IgE-mediated anaphylaxis and modulation of mast cell activation. J Immunol 168:6455–6462

Xie ZH, Zhang J, Siraganian RP (2000) Positive regulation of c-Jun N-terminal kinase and TNF-alpha production but not histamine release by SHP-1 in RBL-2H3 mast cells. J Immunol 164:1521–1528

Xie ZH, Ambudkar I, Siraganian RP (2002) The adapter molecule Gab2 regulates Fc epsilon RI-mediated signal transduction in mast cells. J Immunol 168:4682–4691

Xu R, Pecht I (2001) The protein tyrosine kinase syk activity is reduced by clustering the mast cell function-associated antigen. Eur J Immunol 31:1571–1581

Xu R, Abramson J, Fridkin M, Pecht I (2001) SH2 domain-containing inositol polyphosphate 5′-phosphatase is the main mediator of the inhibitory action of the mast cell function-associated antigen. J Immunol 167:6394–6402

Yamaguchi M, Lantz CS, Oettgen HC, Katona IM, Fleming T, Miyajima I, Kinet JP, Galli SJ (1997a) IgE enhances mouse mast cell Fc(epsilon)RI expression in vitro and in vivo: evidence for a novel amplification mechanism in IgE-dependent reactions. J Exp Med 185:663–672

Yamaguchi M, Lantz CS, Oettgen HC, Katona IM, Fleming T, Yano K, Miyajima I, Kinet JP, Galli SJ (1997b) IgE enhances mouse and human mast cell Fc epsilon RI expression, which in turn enhances IgE-dependent mast cell mediator release. J Allergy Clin Immunol 99:422

Yamamoto T, Yumioka T, Sekine Y, Sato N, Minoguchi M, Yoshimura A, Matsuda T (2003) Regulation of Fc epsilon RI-mediated signaling by an adaptor protein STAP-2/BSK in rat basophilic leukemia RBL-2H3 cells. Biochem Biophys Res Comm 306:767–773

Yamanashi Y, Baltimore D (1997) Identification of the Abl- and rasGAP-associated 62 kDa protein as a docking protein, Dok. Cell 88:205–211

Yamasaki S, Nishida K, Hibi M, Sakuma M, Shiina R, Takeuchi A, Ohnishi H, Hirano T, Saito T (2001) Docking protein Gab2 is phosphorylated by ZAP-70 and negatively regulates T cell receptor signaling by recruitment of inhibitory molecules. J Biol Chem 276: 45175–45183

Yano H, Nakanishi S, Kimura K, Hanai N, Saitoh Y, Fukui Y, Nonomura Y, Matsuda Y (1993) Inhibition of histamine secretion by wortmannin through the blockade of phosphatidylinositol 3-kinase in RBL-2H3 cells. J Biol Chem 268:25846–25856

Yotsumoto K, Okoshi Y, Shibuya K, Yamazaki S, Tahara-Hanaoka S, Honda S, Osawa M, Kuroiwa A, Matsuda Y, Tenen DG, Iwama A, Nakauchi H, Shibuya A (2003) Paired activating and inhibitory immunoglobulin-like receptors, MAIR-I and MAIR-II, regulate mast cell and macrophage activation. J Exp Med 198:223–233

Yuasa T, Kubo S, Yoshino T, Ujike A, Matsumura K, Ono M, Ravetch JV, Takai T (1999) Deletion of fcgamma receptor IIB renders H-2(b) mice susceptible to collagen-induced arthritis. J Exp Med 189:187–194

Zhang J, Berenstein EH, Evans RL, Siraganian RP (1996) Transfection of Syk protein tyrosine kinase reconstitutes high affinity IgE receptor-mediated degranulation in a Syk-negative variant of rat basophilic leukemia RBL-2H3 cells. J Exp Med 184:71–79

Zhang C, Baumgartner RA, Yamada K, Beaven MA (1997a) Mitogen-activated protein (MAP) kinase regulates production of tumor necrosis factor-alpha and release of arachidonic acid in mast cells. Indications of communication between p38 and p42 MAP kinases. J Biol Chem 272:13397–13402

Zhang C, Hirasawa N, Beaven MA (1997b) Antigen activation of mitogen-activated protein kinase in mast cells through protein kinase C-dependent and independent pathways. J Immunol 158:4968–4975

Zhang J, Finnegan A, Glant TT (2003a) Cbl-b is a negative regulator of PI3-K and map kinases in T cells. Arthritis Rheum 48:S650

Zhang WY, Shao Y, Fang DY, Huang JY, Jeon MS, Liu YC (2003b) Negative regulation of T cell antigen receptor-mediated Crk-L-C3G signaling and cell adhesion by Cbl-b. J Biol Chem 278:23978–23983

Zhao CM, Yu DH, Shen R, Feng GS (1999) Gab2, a new pleckstrin homology domain-containing adapter protein, acts to uncouple signaling from ERK kinase to Elk-1. J Biol Chem 274: 19649–19654

Zhao RX, Fu XQ, Li QS, Krantz SB, Zhao ZZJ (2003) Specific interaction of protein tyrosine phosphatase-MEG2 with phosphatidylserine. J Biol Chem 278:22609–22614

Zhou JS, Friend DS, Feldweg AM, Daheshia M, Li L, Austen KF, Katz HR (2003) Prevention of lipopolysaccharide-induced microangiopathy by gp49B1: evidence for an important role for gp49B1 expression on neutrophils. J Exp Med 198:1243–1251

Zidovetzki R, Rost B, Armstrong DL, Pecht I (2003) Transmembrane domains in the functions of Fc receptors. Biophys Chem 100:555–575

Histamine Receptors and Signaling

Hargita Hegyesi, Zsuzsa Darvas, Robin L. Thurmond, András Falus

9.1 Introduction

Histamine is considered as one of the most important immunomodulator molecules owing to being a mediator of allergy and inflammation processes. Moreover, histamine is a chemical messenger and aminergic neurotransmitter, playing an important role in a multitude of physiological processes in the central nervous system and peripheral tissues. Histamine is synthesized in a restricted population of neurons located in the tuberomammillary nucleus of the posterior hypothalamus implicated in many brain functions (e. g., sleep/wakefulness, hormonal secretion, cardiovascular control, thermoregulation, food intake, and memory formation). In peripheral tissues, histamine is stored in mast cells, basophils and enterochromaffin cells. Mast cell histamine plays an important role in the pathogenesis of various allergic conditions, e. g., histamine release leads to various well-known symptoms of allergic conditions in the skin and the airway system. Histamine can influence the tumor cell proliferation and promotion directly as a growth factor, e. g., in colon tumor, gastric tumor and melanoma, and indirectly via the modification of, e. g., cytokine production in the tumor and in the tumor-surrounding environment.

The effects of histamine are mediated through four pharmacologically distinct subtypes of receptors, i. e., the H_1, H_2, H_3, and H_4 receptors, which are all members of the G-protein coupled receptor (GPCR) family. Histamine receptors display seven transmembrane (TM) domains, which are predicted to form α-helices that span the cell membrane, an extracellular N-terminus and a cytoplasmic C-terminus of variable length. The third and fifth TM domains of the receptors appear to be responsible for ligand binding, while the third intracellular loop is responsible for signaling pathway connection (Fig. 9.1).

Interestingly, the genes encoding the H_1, H_2, and H_3 receptors share less protein sequence identity with each other than with other biogenic amine receptor family members, e. g., with M2 muscarinergic receptor. Their overall homology is low (average 38%), suggesting that these histamine receptors evolved from different ancestor sequences (Zhu et al. 2001). Although lacking significant overall sequence homology, the histamine receptors apparently acquired crucial elements for the recognition of histamine during their evolution, e. g., in their ligand binding region. Nucleotide sequences of human H_3, H_4, H_1, and H_2 receptors are shown in Fig. 9.2.

Springer Series in Biophysics
S. Damjanovich (Ed.)
Biophysical Aspects of Transmembrane Signaling

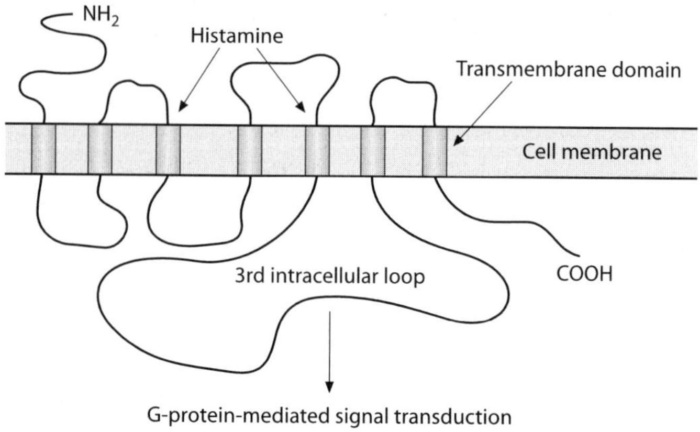

Fig. 9.1. General structure of histamine receptors. The third intracellular loop of receptors is mainly involved in the signal transduction pathways

9.2 H_1 Receptor

Human H_1 receptor (H_1R) is localized on the third chromosome (3p21-p14), composed of 487 amino acids and showed ~75–85% interspecies homology. Mouse and rat H_1R contains 488 and 486 amino acids, respectively. The human H_1R gene contains an intron in the 5′ untranslated region, immediately upstream of the start codon, but the coding sequence of this receptor is intronless. The promoter region of the gene appears to share some similarities with other GPCR genes, such as absence of the TATA and CAAT box at the appropriate localization (Smit et al. 1999; Bakker et al. 2004). Some nucleotide polymorphisms have been described in the H_1R gene, but these do not seem to be associated with atopic asthma (Sasaki et al. 2000).

9.2.1 The H_1 Receptor Signaling Pathway

H_1 receptor preferentially couples to the Gq/11 family of G-proteins and causes mobilization of intracellular Ca^{2+} in a pertussis toxin (PTX)-insensitive fashion. Histamine activates phospholipase C (PLC), which mediates the cleavage of the membrane phosphatidyl inositol diphosphate (PIP2), which results in formation of inositol triphosphate (IP3) and 1,2-diacyl glycerol (DAG). IP3 in turn mediates Ca^{2+} release from endoplasmic reticulum, and also increases Ca^{2+} influx from the extracellular space as a secondary but longer lasting event. This latter effect can be inhibited by nifedipine, suggesting the involvement of an L-type voltage-dependent Ca^{2+} ion channel. As a consequence of Ca^{2+} ion influx, a secondary breakdown of membrane phosphoinositides occurs. IP3 can be phospho-

```
H3      MERAPPDGPLNASGALAGDAAAAGGARGFSAAWTAVLAALMALLIVATVLGNALVMLAFV 60
H4      -------------------MADTNSTINLSLSTRVTLAFFMSLVAFAIMLGNALVILAFV 41
H1      -------MSLPNSSCLLEDKMCEGNKTTMASPQLMPLVVVLSTICLVTVGLNLLVLYAVR 53
H2      ----------------MAPNGTASSFCLDSTACKITITVVLAVLILITVAGNVVVCLAVG 44
                               ..    : .    :. .:: : .  :  * :*  *.

H3      ADSSLRTQNNFFLLNLAISDFLVGAFCIPLYVPYVLTGRWTFGRGLCKLWLVVDYLLCTS 120
H4      VDKNLRHRSSYFFLNLAISDFFVGVISIPLYIPHTLF-EWDFGKEICVFWLTTDYLLCTA 100
H1      SERKLHTVGNLYIVSLSVADLIVGAVVMPMNILYLLMSKWSLGRPLCLFWLSMDYVASTA 113
H2      LNRRLRNLTNCFIVSLAITDLLLGLLVLPFSAIYQLSCKWSFGKVFCNIYTSLDVMLCTA 104
         :  *:   . :::.*:::*:::* . :*:   : *  .* :*: :* ::   * : .*:

H3      SAFNIVLISYDRFLSVTRAVSYRAQQGDTRRAVRKMLLVWVLAFLLYGPAILSWEYLSGG 180
H4      SVYNIVLISYDRYLSVSNAVSYRTQHTGVLKIVTLMVAVWVLAFLVNGPMILVSESWKDE 160
H1      SIFSVFILCIDRYRSVQQPLRYLKYRTKTR-ASATILGAWFLSFLWVIPILGWNHFMQQT 172
H2      SILNLFMISLDRYCAVMDPLRYPVLVTPVRVAISLVLIWVISITLSFLSIHLGWNSRNET 164
        *  .:.::. **: :*  .: *      .      ::   .   *   .     .  .

H3      SSIPEGHCYAEFFYNWYFLITASTLEFFTPFLSVTFFNLSIYLNIQRRTRLRLDGAREAA 240
H4      GSE----CEPGFFSEWYILAITSFLEFVIPVILVAYFNMNIYWSLWKRDHL--------- 207
H1      SVRREDKCETDFYDVTWFKVMTAIINFYLPTLLMLWFYAKIYKAVRQHCQHRELINRSLP 232
H2      SKGNHTTSKCKVQVNEVYGLVDGLVTFYLPLLIMCITYYRIFKVARDQAKRINHISSWKA 224
        .      .   .          . : *  * : :      *:     : :

H3      GPEPPPEAQPSPPPPPGCWGCWQKGHGEAMPLHRYGVGEAAVGAEAGEATLGG------- 293
H4      -----SRCQSHPG---------------LTAVSSNICGHSFRGRLSSRRSLS-------- 239
H1      SFSEIKLRPENPKGDAKKPGKESPWEVLKRKPKDAGGGSVLKSPSQTPKEMKSPVVFSQE 292
H2      ATIREHKATVTLAAVMGAFIICWFPYFTAFVYRGLRGDDAINEVLEAIVLWLG------- 277
                                             .

H3      -------------------GGGGGSVASPTSSSGSSSRGTERPRSLKRGSKPSASSASLE 334
H4      -----------------------ASTEVPASFHSERQR---RKSSLMFSSRTKMNSNTIA 273
H1      DDREVDKLYCFPLDIEHMQAAAEGSSRDYVAVNRSHGQLKTDEQGLNTHGASEISEDQML 352
H2      ---------------------YANSALNPILYAALNRDFRTGYQQLFCCRLANRNSHKTS 316
                                *                    *     .. ..

H3      KRMKMVSQS-------------------------------------FTQRFRLSRDRK 355
H4      SKMGSFSQSD---------------------------------SVALHQREHVELLRARR 300
H1      GDSQSFSRTDSDTTTETAPGKGKLRSGSNTGLDYIKFTWKRLRSHSRQYVSGLHMNRERK 412
H2      LRSNASQLSR---------------------------------------------TQSREPR 333
              . :                                               *  :

H3      VAKSLAVIVSIFGLCWAPYTLLMIIRAACHGHCVP-DYWYETSFWLLWANSAVNPVLYPL 414
H4      LAKSLAILLGVFAVCWAPYSLFTIVLSFYSSATGPKSVWYRIAFWLQWFNSFVNPLLYPL 360
H1      AAKQLGFIMAAFILCWIPYFIFFMVIAFCKNCCNE--HLHMFTIWLGYINSTLNPLIYPL 470
H2      QQEEKPLKLQVWSG---------TEVTAPQGATDRPWLCLPECWSVELTHSFIHLFIHSF 384
          :.  . :  :              :   .              :   :* :: .::.:

H3      CHHSFRRAFTKLLCPQKLKIQPHSSLEHCWK 445
H4      CHKRFQKAFLKIFCIKKQPLPSQHSRSVSS- 390
H1      CNENFKKTFKRILHIRS-------------- 487
H2      ANIHPIPTTCQEL------------------ 397
```

Fig. 9.2. Sequence alignment of histamine receptor proteins. Human histamine receptor polypeptide sequences deduced from the cloned cDNAs from H_3, H_4, H_1, and H_2 receptors were aligned. An alignment will display by default the following *symbols* denoting degree of conservation observed in each column: *asterisk* means that the residues or nucleotides in that column are identical in all sequences in the alignment; *colon* means that conserved substitutions have been observed; *dot* means that semi-conserved substitutions are observed

rylated to produce IP4 which further increases the intracellular Ca^{2+} level. DAG activates a serine/threonine kinase, the protein kinase C (PKC) that can phosphorylate and activate other effector proteins in the cells. Additional secondary signaling pathways can be induced by the increased intracellular Ca^{2+} level and DAG. Via calcium/calmodulin (Ca^{2+}/CAM)-dependent enzyme, nitric oxide (NO) activity is stimulated to produce an elevated NO level, which results in the activation of guanylyl cyclase enzyme to produce cyclic guanosine-3′,5′-monophosphate (cGMP). Besides PKC activation, DAG can stimulate another enzyme, namely, phospholipase A2 (PLA2), which can form active arachidonic acid metabolites, such as prostaglandin E2, prostacyclin, and thromboxane A2 (Leurs et al. 1994; Hill et al. 1997).

Recently, some of the downstream elements of the H_1R signaling pathway have been described (Bakker et al. 2001). One of them is a transcription factor, the nuclear factor kappa B (NF-κB). Using luciferase-based reporter gene assay in H_1R transfected CHO cells, it was found that histamine stimulates the c-fos promoter through the activation of a PKC-α isoform and the mitogen-activated protein kinase (MAPK) pathway (Megson et al. 2001).

Data indicate that both Gq/11- and G-subunits play a role in H_1R-mediated NF-κB activation similar to NF-κB activation mediated via the bradykinin B2 receptor or US28 (Xie et al. 2000; Casarosa et al. 2001). The constitutive H_1R-mediated activation of NF-κB is mediated via free G-subunits from Gq/11-proteins, whereas histamine-mediated NF-κB activation is most likely regulated via Gq/11-subunits. Although PKC is a proper candidate for Gq/11-mediated NF-κB activation (Shahrestanifar et al. 1999) and both phosphatidylinositol 3-kinase (PI3 K) and the serine/threonine protein kinase Akt (Akt) have been shown to be involved in the G-protein-mediated activation, the actual target(s) of the G-protein and G-protein subunits in the H_1R-mediated responses are not well known and are currently under investigation.

The MAPK family can be divided into three different subfamilies: the extracellular signal-regulated kinases (ERK), the stress-regulated kinases (p38 MAPK) and the Jun N-terminal kinases (JNK). These kinases can be activated by distinct intracellular cascades, which are able to induce their phosphorylation on threonin (Thr) and tyrosine (Tyr) residues. They are expressed abundantly in the adrenal gland, where they participate in regulation of calcium homeostasis, and the synthesis and secretion of catecholamines. It has been demonstrated that in bovine adrenal chromaffin cells (BACC), the Ca^{2+} ion-dependent upregulation of tyrosine hydroxylase (TOH) produced by nicotine and KCl requires the participation of MAPK kinase 1/2 (MEK1/2), the upstream kinase that is capable of activating ERK1/2 (Griffiths and Marley 2001). Recently, it has been shown that in pheochromocytoma cells, overexpression of glia maturation factor induces TOH phosphorylation through a mechanism dependent on p38 and its downstream substrate, MAP kinase-activated protein kinase 2 (MAPKAP kinase-2) (Zaheer and Lim 1998).

Histamine can induce the H_1R-dependent activation of ERK1/2 and p38 in the smooth muscle cell line DDT(1)MF-2 (Robinson and Dickenson 2001) as well as the upregulation of JNK in gastric parietal and peritoneal mast cells (Nagahara et al. 1998; Azzolina et al. 2002). Some known elements and their function in the H_1 receptor-mediated signaling pathway are summarized in Fig. 9.3.

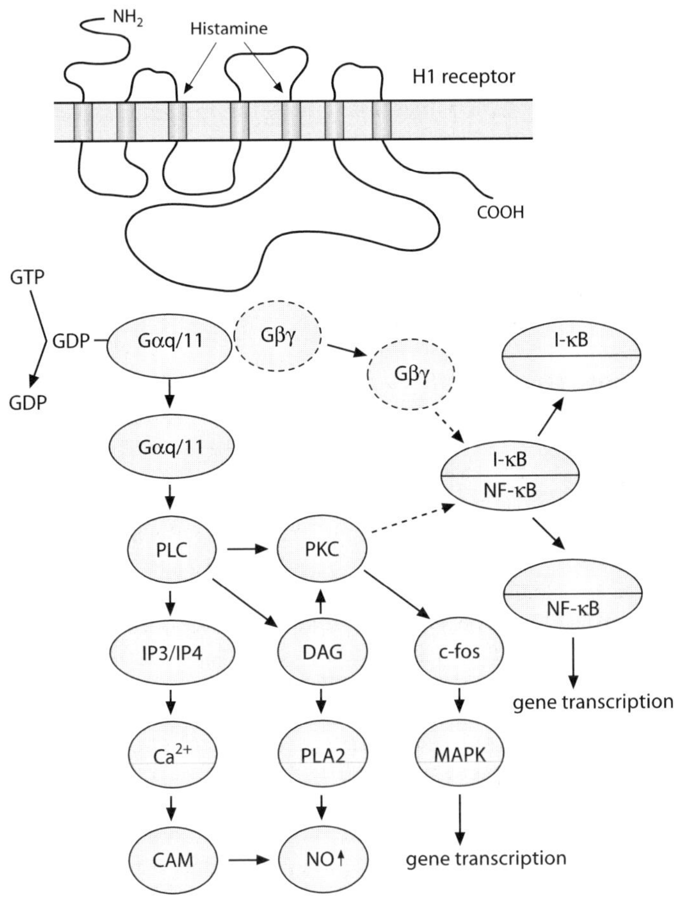

Fig. 9.3. Some known elements of H1 receptor signaling pathways. *Arrows with solid line* show well-characterized pathways and relations, while *arrows with dashed line* show connections that are not so well characterized, evidence coming from few experiments

9.2.2
Some Effects of Histamine Through H_1 Receptor

As mentioned above, H_1 receptor was the first membrane protein to be discovered as a mediator of histamine action. Its main activities include smooth muscle contraction and vascular permeability. It is the chief histamine receptor subtype involved in acute inflammatory and allergic disorders. H_1 receptors have been found in a wide variety of tissues including mammalian brain, airway smooth muscle, the gastrointestinal tract, genitourinary system, cardiovascular system, adrenal medulla, endothelial and various immune cells. Enhanced expression of H_1R was observed in the nasal mucosa of patients with allergic rhinitis, in cultured aortic intimal smooth muscle cells of patients suffering from atherosclerosis, in the inflamed joints of rheumatoid arthritis patients, etc. (Takagishi et al. 1995).

Histamine is widely distributed throughout the mammalian nervous system, e.g., in type 2 astrocytes where histamine stimulates phosphoinositide turnover and increases intracellular calcium concentration via H_1 receptors. In C6 glial cells, H_1R is involved in liberating calcium from internal stores by activation of PLC. The consequent increase in intracellular Ca^{2+} ions activates a Ca^{2+}-dependent K^+ ion channel, resulting in hyperpolarization of cell membrane. In addition, histamine H_1R activation has been shown to produce neuronal depolarization by blocking potassium conductance, which has a role in determining the membrane potential (Weiger et al. 1997).

H_1 receptor-deficient mice display abnormal circadian rhythm of locomotor activity, with increased activity during the light cycle and lower activity during the dark cycle, as well as decreased exploratory activity in a novel environment (Inoue et al. 1996).

Further reports indicate that H_1 receptor knockout ($H_1R^{-/-}$) mice demonstrate lower levels of anxiety and aggressiveness (Yanai et al. 1998); moreover, animals were less sensitive to pain, as measured by a variety of thermal, mechanical, and chemical nociceptive tests (Mobarakeh et al. 2000). Neurochemical analysis showed that cortical levels of dopamine and its metabolites were higher and the turnover rates of dopamine and serotonin were increased in $H_1R^{-/-}$ mice, indicating changes in monoamine activity. Fat deposition in response to a high fat diet was more rapid and the suppression of feeding behavior induced by leptin was attenuated in H1R mutants, demonstrating the involvement of H_1R in regulation of feeding behavior (Masaki et al. 2001).

H_1 receptor agonists do not appear to have any therapeutic potential according to our recent knowledge, but they are useful tools in research regarding the H_1 receptor function in different experimental conditions. New, highly potent and selective H_1R agonists seem to be the family members of histaprodifen. Its methylated derivative, Nα-methylhistaprodifen, is the most potent H_1 agonist known so far (Elz et al. 2000). As for H_1R antagonists, classical antihistamines included compounds such as mepyramine (also called pyrilamine) and triprolidine (Hill et al. 1977; van der Goot and Timmerman 2000). They are highly potent H_1 antagonists, and easily penetrate the brain, causing sedation. Many new so-called second-generation antihistamines are non-sedating H_1 antagonists (e.g., cetirizine, astemizole and loratadine), and they are used to treat allergic conditions (Curran et al. 2004).

9.3 H_2 Receptor

The sole existence of the H_1 receptor could not explain the effects of histamine on cardiac muscle and gastric acid secretion; therefore a second subtype was proposed, termed the H_2 receptor. It is involved in a wide array of physiological histamine actions, including the relaxation of airway and vascular smooth muscle, regulation of right atrial and ventricular muscle of the heart, inhibition of basophil chemotactic responsiveness, various actions on immune cells and inhibition of prostaglandin E2-stimulated duodenal epithelial bicarbonate secretion. Its principal action from a clinical point of view is related to its role in gastric acid secretion. H_2 receptor knockout mice have been recently developed, and appear to be viable and fertile, but show considerable alterations in the morphology and structure of the gastric mucosa (Kobayashi et al. 2000).

The H_2R gene located on chromosome 5 was cloned in 1991 (Gantz et al. 1991). The 5′ untranslated region of the gene contains some regulatory elements such as cAMP response element, GATA motifs and AP2 site, but apparently without TATA box-like sequence. In addition, multiple transcription initiation sites were revealed in this gene. Similarly to the H_1 receptor gene, the coding parts of the H_2 receptor gene lack intron sequences (Del Valle and Gantz 1997; Igaz et al. 2002). Several nucleotide sequence variants of the gene have been described. Among these, Lys207Arg, Asn217Asp and Lys231Arg could be particularly interesting since they are located in the third intracellular loop of protein involved in the signal transduction pathway (Orange et al. 1996). Val133Gly may also be relevant, since the second intracellular loop where it is situated appears to have some role also in the signal transduction pathway.

Human H_2 receptor protein contains 359 amino-acid residues, and its ligand binding site appears to be similar to the corresponding region of H_1R. The most notable difference between the two receptors is the comparatively much shorter third intracellular loop and the longer C-terminus in the H_2 receptor sequence (Del Valle and Gantz 1997).

9.3.1 The H_2 Receptor Signaling Pathway

Histamine H_2 receptors couple to adenyl cyclase via the G_s protein, and histamine stimulates cAMP production in many different cell types, e. g., the CNS and CNS-derived cells, gastric mucosa, cardiac myocytes, fat cells, vascular smooth cells, basophils and neutrophils (Del Valle and Gantz 1997). Elevated cAMP concentration activates protein kinase A (PKA), which is the downstream effector kinase of this pathway, phosphorylating a wide variety of proteins in the cells mentioned above.

However, the H_2R signaling pathway shows a dual face. In addition to the adenyl cyclase-mediated one, histamine through H_2 receptor can increase the intracellular Ca^{2+} ion level in some cell types, e. g., gastric parietal cells and HL-60 leukaemic cell line. This action of histamine seems to be a direct effect mediated by another G_q member of the PTX-sensitive G-protein family (Hill et al. 1997). In contrast to other receptors that stimulate the dual signaling pathway, H_2R activates each pathway directly and the required histamine concentration for the stimulation of both systems is identical.

9.3.2
Some Effects of Histamine Through H_2 Receptor

Dual coupling of H_2Rs to G_s and G_q was demonstrated for the first time in cardiac myocytes. It represents a novel mechanism to augment positive inotropic effects by activating two different signaling pathways via one type of histamine receptor. Activation of the Gs-cAMP-PKA pathway promotes Ca^{2+} influx through phosphorylation of L-type Ca^{2+} channels. Simultaneous activation of Gq-signaling pathways might result in phosphoinositide turnover and Ca^{2+} release from intracellular stores, thereby augmenting H_2-induced increases in the intracellular Ca^{2+} ion (Wellner-Kienitz et al. 2003).

Previously, it had been found that histamine H_1 and H_2 receptors are expressed in the atrium (Hattori et al. 1991). It was also known that the heart contains as much as 3 µg histamine/g tissue (Endou and Levi 1995). Effects of histamine on the regulation of cardiac function are species-specific and also tissue-specific, even in the same animal. Histamine has been known to increase force of contraction in rabbit and human atria via activation of adenyl-cyclase and to increase cAMP production (Longhurst and McNeill 1982).

It was found that activation of G-protein-coupled histamine H_2 receptor decreases atrial ANP release via cAMP-protein kinase signaling. A histamine-induced decrease in ANP release was a function of cAMP production. Because accentuation of cAMP production results in activation of L-type Ca^{2+} channels, and also an increase in Ca^{2+} influx via L-type channels inhibits ANP release in the beating atria and perfused heart, a role for increased Ca^{2+} influx in cAMP inhibition of ANP release by histamine was expected (Li et al. 2003).

Lie et al. (2003) found that both positive inotropic effect and an increase in cAMP production were blocked by cimetidine but not mepyramine, which indicates that the effects are related to the activation of histamine H_2 receptor. It was shown in rabbit atria that the increase in contractility and cAMP production by histamine is caused exclusively by an activation of H2 receptor. An increase in intracellular cAMP production may increase Ca^{2+} influx via cAMP-dependent protein kinase A (PKA), which in turn may result in an increase in contractility. It was shown that histamine-induced increase in Ca^{2+} current is blocked by H_2-receptor antagonist cimetidine and PKA inhibitor RP-adenosine 3′,5′-cyclic monophosphorothioate [RP-cAMP(S)].

H_2R agonists have been proven to be effective for acid peptic disorders of the gastrointestinal (GI) tract. H_2 receptor agonist, dimaprit, is a relatively selective H_2 receptor agonist, with minimal H_1R receptor agonism but considerable antagonistic potency on the H_3 and H_4 receptors. Amthamine [2-amino-5-(2-aminoethyl)-4-methylthiazole], a rigid dimaprit analogue, combines high H_2R selectivity and is equipotent to or slightly more potent than histamine itself (van der Goot and Timmerman 2000). One impromidine analog, arpromidine, due to its high agonistic potency at the H_2 receptor, can be considered a potential candidate for treatment of congestive heart failure.

Among H_2 receptor antagonists, cimetidine was the first to be clinically used for the treatment of gastric and duodenal ulcers, but it has become clear that cimetidine binds to non-H_2R imidazole recognition sites as well (van der Goot and Timmerman 2000). Besides cimetidine, tiotidin and ranitidine are potent and more selective H_2 receptor antagonists. Both compounds inhibits gastric acid secretion, but only ranitidine possesses therapeutic potential in gastric ulcer treatment because of the severe side effects of tiotidine.

Evidence has accumulated indicating that histamine exerts a variety of immunoregulatory actions. It inhibits CD4+ T cells and CD8+ T cells to produce histamine when activated with ConA (Kubo and Nakano 1999). Furthermore, histamine regulates cytokine production by T cells and macrophages (Sonobe et al. 2004). Histamine affects the balance of cytokines from T helper type 1 (Th1) cells and T helper type 2 (Th2) cells by shifting cytokine production from a Th1 to a Th2 pattern (Packard and Khan 2003). It is known that in certain circumstances, stress mediators, as well as histamine and adenosine, can be increased, and due to histamine effects the upregulated Th2 cytokines may also play a role in induction and progression of certain processes, e.g., in allergic/atopic reactions and in tumor growth. Modulation of IL-12 and IL-10 secretion by histamine involved the H_2 and H_3 receptors of antigen presenting cells. Histamine-dependent suppression of IL-2 mRNA expression and secretion from mouse splenocytes is abrogated by a protein kinase A-inhibitor, whereas the amine did not influence histaminase. In contrast to IFN-γ, neither α-FMH nor histaminase had any appreciable effect on the production of IL-4.

IL-10 production was enhanced by histamine through either an H_1R- or an H_2R-dependent fashion (Osna et al. 2001). The pretreatment of the splenocytes with histamine stimulates PMA- and ionomycin-induced IL-10 secretion from the cells, which was reversed by either H_1R or interfere with signal transduction pathways downstream of PKC, leading to production of IL-2 (Poluektova et al. 1999).

However, evidence has also accumulated to indicate cross talk between these receptors. For example, stimulation of the signal transducing system associated with PLC also activates H_2R-mediated reactions, leading to increased synthesis of cAMP (Garbarg and Schwartz 1988). In addition, histamine stimulates IL-10 production in Th2 cells which was reversed by either H_1R or H_2R antagonists, and the release of IFN-γ and IL-13 was enhanced by histamine in both H_1R- and H_2R-deficient spleen cells (Sonobe et al. 2004). Inhibition of IFN-γ synthesis by histamine was reversed by either cimetidine, an H_2R antagonist, or pyrilamine, an H_1R antagonist (Simons et al. 1995).

9.3.3
H_1 and H_2 Receptors on Human Neoplasias

In addition to the data available from studies of tumor cell lines derived from different human neoplasias and xenotransplanted models in nude mice, many investigations have also been carried out in biopsies of tumor tissues. Most reports are coincident in showing an increase in histamine synthesis and content in tumors as well as in the presence of histamine receptors.

In humans, several clinical trials have been carried out with H_2R antagonists as an adjunct to surgical resection, with conflicting results. While the H_2 antagonist cimetidine has been shown to inhibit the presence of tumor-infiltrating lymphocytes in colorectal cancer, this was not found in the case of breast cancer. Comparing breast cancer with colorectal cancer, no relationship between preoperative cimetidine administration and tumor cell proliferation was seen overall (Bowrey et al. 2000). A study with tissue samples obtained by surgery on 25 patients indicated that both H_1 and H_2 receptors are present in the human mammary gland, in benign lesions, and in breast carcinomas. In the different benign lesions, H_2R produced the increase in cAMP levels, while H_1R was coupled to PLC activation. On the other hand, the response observed in carcinomas was different: H_1R was invariably linked to the PLC pathway, but H_2R stimulated indistinctively both transductional pathways. In agreement with a previous report, approximately 25% of breast carcinomas were negative for H_2R. In all cases, the lack of expression of H_2R corresponded to tumors from patients with poor prognosis or evolution (Davio et al. 1993). According to a follow-up study to check risk factors for ovarian cancer development in females, long-time usage (5 years) of histamine receptor antagonists could be a risk factor in females (Lacey et al. 2004).

9.4 H_3 Receptor

As for H_1 and H_2 receptors, H_3R was initially identified on a pharmacological basis. The first report of the H_3 receptor came from Arrang et al. (1983) who found a new histamine receptor that acted as an autoreceptor which mediated histamine release from neurons. Lovenberg et al. (1999) cloned the receptor in 1999 after extensive searches through databases to identify sequences that might be GPCRs. They identified a sequence that encoded a 445-amino acid coding region with low homology to other biogenic amine receptors, which may have allowed it to elude discovery for so long. However, it did have an aspartic acid residue in transmembrane domain 3 which is a conserved residue for receptors that bind to primary amines. The cloned receptor, once expressed, clearly showed the pharmacological profile characteristic of the H_3 receptor (Lovenberg et al. 1999). The gene is located on human chromosome 20 (Coge et al. 2001a; Tardivel-Lacombe et al. 2001) and contains several introns, which is unusual for GPCRs. It has been reported that splice variants of the H_3 receptor exist in several species including humans, although their functional relevance remains somewhat controversial (Liu et al. 2000; Coge et al. 2001a; Drutel et al. 2001; Tardivel-Lacombe et al. 2001; Wellendorph et al. 2002; Wiedemann et al. 2002; Chen et al. 2003). Variants appear to either be truncated proteins that are not functional, or have partial deletions of the third intracellular loop, leading to varying pharmacology (Coge et al. 2001a; Drutel et al. 2001; Tardivel-Lacombe et al. 2001; Wellendorph et al. 2002). In addition, there are reports of polymorphisms, including one at position 280 (alanine to valine) in a patient with Shy-Drager syndrome (Wiedemann et al. 2002).

The H_3 receptor binds to histamine with a high affinity (K_d ~5 nM), which is consistent with its role as an autoreceptor. There has been a large body of work concerning the development of H_3 receptor ligands (see Stark 2003 for a review).

Most agonists of the receptor are imidazole derivatives related to histamine. The most commonly used reference agonists are R-α-methylhistamine and imetit, both of which bind to the H_3 receptor with subnanomolar affinity. However, both of these ligands have some cross-reactivity with the H_4 receptor (see below). Much more work has been done in developing H_3 receptor antagonists, since they appear to have more therapeutic utility. The early antagonists were also imidazoles like the agonists. This class includes thioperamide, which is a standard H_3 receptor antagonist used in many studies. As for the imidazole-based agonists, thioperamide also shows cross-reactivity to H_4 receptors in addition to cross-reactivity with 5-HT_3 and sigma receptors (Leurs et al. 1995). Recently, there has been much work to develop non-imidazole antagonists. Some examples of the different imidazole replacements are imidazopyridine (Shah et al. 2002; Chai et al. 2003), benzylamine (Apodaca et al. 2003), benzthiazole (Walczynski et al. 1999), aminoalkylguanidine (Linney et al. 2000), piperidine (Meier et al. 2002; Miko et al. 2003), quinoline (Turner et al. 2003) and piperazine amide (Faghih et al. 2002a,b; Vasudevan et al. 2002; Esbenshade et al. 2003). The piperazine amides have been shown to be efficacious in mouse obesity and dipsogenia models (Faghih et al. 2003; Gfesser et al. 2004).

The expression of the H_3 receptor based on mRNA appears to be mainly restricted to the central nervous system. The human receptor mRNA has been found in human brain samples, with the highest expression being in the caudate, cortex and thalamus and little expression seen in peripheral tissues (Lovenberg et al. 1999). The rat and mouse receptors show a similar pattern of expression in the brain and no expression in other tissues (Lovenberg et al. 2000; Morisset et al. 2000; Drutel et al. 2001; Chen et al. 2003). The distribution of the receptor can also be probed using radiolabeled ligands. This approach has shown the existence of the H_3 receptor protein in rat cerebral cortical membranes, striatum, hippocampus, substantia nigra and olfactory nucleus (Arrang et al. 1987; Pollard et al. 1993). Ligand binding has also been observed in guinea pig cerebral cortical membranes (Kilpatrick and Michel 1991), the lung (Arrang et al. 1987), ileum and large intestine (Korte et al. 1990). Human and non-human primate brains also have H_3 receptor ligand binding sites (Martinez-Mir et al. 1990).

9.4.1 The H_3 Receptor Signaling Pathway

In cells transfected with the human or rat H_3 receptor, histamine and R-α-methylhistamine could inhibit forskolin-stimulated cAMP production, and this inhibition is abolished by pretreatment of the cells with PTX, indicating that the receptor is coupled to $G_{i/o}$ G-proteins (Lovenberg et al. 1999, 2000; Drutel et al. 2001; Schwartz et al. 2003). The receptor can also signal via increases in intracellular calcium, as seen when the mouse receptor is cotransfected with chimeric G_{qi5} G-proteins (Chen et al. 2003). Furthermore, it has been shown that activation of rat H_3 receptor can increase p44/p42 MAP kinase phosphorylation and arachidonic acid release in transfected cells (Drutel et al. 2001; Schwartz et al. 2003). As for the cAMP response, both of these responses are PTX-sensitive. Finally, there is some evidence in transfected systems that the H_3 receptor can inhibit the activity of Na^+/H^+ exchangers (Silver et al. 2001).

In tissues, data are consistent with the results from transfected cells that the H_3 receptor is a GPCR that can couple to $G_{i/o}$ G-proteins. R-α-methylhistamine increases [^{35}S]GTPγS binding in rat cerebral cortical membranes which can be blocked by pretreatment with PTX (Clark and Hill 1996). It has been shown that histamine and R-α-methylhistamine can inhibit the release of noradrenaline in rat spinal cord slices and this effect can be blocked by PTX, but it appeared that cAMP did not play a role in the signaling (Celuch 1995). Poli et al. (1993) showed that R-α-methylhistamine reduced cholinergic neurotransmission in guinea pig duodenum that once again was PTX-sensitive but did not seem to involve changes in cAMP levels. The H_3 receptor in this system appeared to inhibit N-type Ca^{2+} channels, thereby reducing the intracellular calcium level (Poli et al. 1994). The H_3 receptor has been shown to inhibit Ca^{2+} channels in a PTX-sensitive fashion which led to inhibition of adrenergic responses in guinea pig heart (Endou et al. 1994), acetylcholine release in guinea pig ileum (Lee and Parsons 2000; Blandizzi et al. 2001), glutamate release in rat striatal synaptosomes (Molina-Hernandez et al. 2001), and GABA release from rat neurons (Jang et al. 2001). These effects do not seem to involve changes in cAMP or cGMP levels, or kinase activation.

One interesting aspect of H_3 receptor signaling is that it appears to possess a high degree of constitutive activity; that is, activity in the absence of agonists. In cells transfected with the H_3 receptor there is an inhibition of forskolin-mediated cAMP levels compared to untransfected cells, which can be further inhibited by the addition of agonists (Wieland et al. 2001). This opens up the possibility that H_3 receptor ligands that were initially characterized as antagonists may in fact be inverse agonists, i.e., they inhibit the constitutive activity of the receptor. Indeed, this has been found to be the case for thioperamide, clobenpropit, and iodophenpropit, whereas proxyfan acts as a neutral antagonist (Morisset et al. 2000; Wieland et al. 2001). Constitutive activity of the H_3 receptor has also been found in tissues and in vivo, where inverse agonists increase histamine neuronal activity (Morisset et al. 2000).

9.4.2 Some Effects of Histamine Through H_3 Receptors

The H_3 receptor was initially identified as an autoreceptor that mediated histamine release from neurons (Arrang et al. 1983). In histaminergic neurons, the H_3 receptor controls both the release and the synthesis of histamine (Arrang et al. 1987). The H_3 receptor has also been shown to act as a heteroreceptor where it affects the release of other neurotransmitters. Activation of the H_3 receptor has been shown to inhibit the release of acetylcholine (Clapham and Kilpatrick 1992; Blandina et al. 1996; Lee and Parsons 2000; Blandizzi et al. 2001), glutamate (Molina-Hernandez et al. 2001), noradrenaline (Schlicker et al. 1994), norepinephrine (Imamura et al. 1996; Levi and Smith 2000; Yamasaki et al. 2001; Koyama et al. 2003a), dopamine (Schlicker et al. 1993; Molina-Hernandez et al. 2000), serotonin (Schlicker et al. 1988; Fink et al. 1990), substance P (Ohkubo et al. 1995), and GABA (Arias-Montano et al. 2001; Jang et al. 2001). While the H_3 receptor appears to be involved in the induced release of many neurotransmitters, it may not play a role under steady-state conditions, since brain levels of dopamine, norepinephrine, and serotonin are not changed in H_3 receptor-deficient mice, whereas levels of histamine are decreased (Toyota et al. 2002).

This effect on neurotransmitter release has several physiological consequences. R-α-methylhistamine appears to increase slow-wave sleep and decrease wakefulness and REM sleep, whereas thioperamide can block these effects and on its own increase wakefulness (Lin et al. 1990; Monti et al. 1991). Studies on thioperamide-induced wakefulness in H_3-deficient mice proved that this effect was due to the H_3 receptor (Toyota et al. 2002). The H_3 receptor has also been linked to changes in memory and learning. H_3 agonists such as R-α-methylhistamine lead to impaired object recognition and passive avoidance responses in rats (Blandina et al. 1996). Two H_3 antagonists have been found to enhance the performance of rats in a repeat acquisition avoidance response, which is a model for attention deficit/hyperactivity disorder (Fox et al. 2002). In addition, thioperamide can partially block scopolamine-induced amnesia in a rat passive avoidance test (Giovannini et al. 1999). The H_3-deficient mice show normal learning behavior but are resistant to the effects of thioperamide on scopolamine-induced amnesia (Toyota et al. 2002). H_3R has also been implicated in food intake. Thioperamide and other H_3 receptor antagonists have been shown to decrease feeding and body weight in rats and mice (Attoub et al. 2001; Gfesser et al. 2004). Itoh et al. (1999) showed that thioperamide can inhibit peptide YY-induced hyperphagia in rats. There is one report (Takahashi et al. 2002) of the H_3 receptor knockout mice showing signs of mild obesity, but this may depend on the feeding conditions since it was not seen in another study (Toyota et al. 2002).

As a final note on the effects of histamine via the H_3 receptor, it must be pointed out that many of the studies have relied on the use of agonist/antagonist pairs such as R-α-methylhistamine and thioperamide as proof of H_3 involvement. However, it is now known that most of the imidazole-based H_3 ligands also have affinity for the H_4 receptor, and the findings in the literature should be judged accordingly.

9.4.3 Uses of H_3 Receptor Antagonists

As detailed above, animal studies indicate a role for H_3 receptors in numerous brain functions. The receptors clearly play a role in food intake, memory and sleep/wake cycles. This suggests that H_3 receptor ligands may be of use in treating a variety of conditions including sleep, memory, and cognitive disorders as well as obesity. In addition, there is evidence that H_3 antagonists may be useful in treating myocardial ischemia. There is a greater incidence and longer duration of ventricular fibrillation and increases in norepinephrine overflow in hearts subjected to ischemia when the H_3 receptor is absent (Koyama et al. 2003a,b). Currently there are no H_3 antagonists on the market, but several are reported to be in clinical trials for conditions including insomnia, cognitive disorders, anxiety, and obesity.

9.5 H_4 Receptor

Cloning of the H_3 receptor led directly to identification of the histamine H_4 receptor (Hough 2001; Liu et al. 2001a). Unlike the H_3 receptor, the H_4 receptor was largely unanticipated in the literature, except for a description by Raible et al. (1992, 1994) of a non-H_1, H_2, H_3 receptor on eosinophils. The primary amino acid sequence of the H_4 receptor clearly identifies it as a member of the GPCR family. The 390 amino acid protein is predicted to have seven transmembrane regions, DRY sequence at the end of transmembrane helix 3 and an aspartic acid residue in the second transmembrane domain, which indicate that it is a member of the biogenic amine GPCR family. The aspartic acid residue at position 94 in transmembrane helix 3, which is conserved among the other histamine receptors, is crucial for the binding of histamine (Shin et al. 2002). The H_4 receptor has highest sequence homology to the H_3 receptor, but its homology to H_1 and H_2 receptors is actually lower than that to other GPCRs (Nguyen et al. 2001). The H_4 receptor has been mapped to chromosome 18q11.2 and has a similar genomic structure as the H_3 receptor in that it contains two introns and three exons (Coge et al. 2001b; Liu et al. 2001a; Zhu et al. 2001). This suggests the possibility of the existence of splice variants, but to date none has been reported.

Most of the pharmacology of the H_4 receptor has been mapped out using transfected cells (see Jablonowski et al. 2004 for a review). Like the H_3 receptor, the H_4 receptor has a high affinity for histamine, with K_d values of approximately 8 nM, which is not surprising given that the sequence homology in the transmembrane region is 58% between the two receptors. This is also reflected in the fact that the standard imidazole-based ligands of H_3 also have affinity for the H_4 receptor (Nakamura et al. 2000; Oda et al. 2000; Liu et al. 2001a; Morse et al. 2001; Zhu et al. 2001). This includes R-α-methylhistamine and thioperamide, the latter of which binds to the H_4 receptor with an affinity comparable to that on H_3 receptors (Liu et al. 2001a). Most of the H_1 and H_2 receptor ligands, such as diphenhydramine, ranitidine and cimetidine, do not have affinity for the H_4 receptor. Recently, selective imidazole-based agonists (Hashimoto et al. 2003) and non-imidazole antagonists (Thurmond et al. 2004) for the H_4 receptor have been reported. The availability of selective ligands has been essential in probing the physiological role of the different histamine receptors (see above) and will be very important for the H_4 receptor as well.

Like the H_3 receptor, the H_4 receptor has a restricted expression pattern and seems to be limited to cells of the hematopoietic lineage (Oda et al. 2000; Liu et al. 2001a; Morse et al. 2001; Zhu et al. 2001). There are hints of expression on many hematopoietic cell types, but it is clearly expressed on eosinophils, mast cells, basophils, dendritic cells, and T cells (Liu et al. 2001a; Gantner et al. 2002; Hofstra et al. 2003). While this expression pattern differs from that of the H_3 receptor, there is significant overlap with H_1 and H_2 receptor, whose expression has also been reported on these cell types. However, expression of H_4 appears to be restricted to these hematopoietic cells, whereas the H_1 and H_2 receptors are more widely expressed. Eosinophils, mast cells, basophils, dendritic cells and T cells are all major players in the inflammatory response and presence of the H_4 receptor on these cell types suggest that it plays a role in inflammation. In support of this, it has been shown that H_4 mRNA levels can be modulated by IL-10 and IL-13 (Morse et al. 2001) and that the promoter region of the gene indicates that it may de regulated by other inflammatory stimuli such as interferon, TNFα and IL-6 (Coge et al. 2001b).

9.5.1 The H_4 Receptor Signaling Pathway

In transfected cells, the H_4 receptor can inhibit forskolin-stimulated cAMP increases; however, the levels of inhibition are much lower than those seen with the H_3 receptor under similar conditions (Nakamura et al. 2000; Oda et al. 2000; Liu et al. 2001a; Zhu et al. 2001). The effect on cAMP can be blocked by pretreating the cells with PTX, suggesting a role for $G\alpha_{i/o}$ proteins (Oda et al. 2000). Activation can also lead to increases in intracellular calcium, but this requires that it be cotransfected with promiscuous or chimeric G-proteins (Oda et al. 2000; Liu et al. 2001a,b; Morse et al. 2001; Zhu et al. 2001). There has been one report of activation of MAP kinases in transfected HEK-293 cells (Morse et al. 2001). The phosphorylation of the p44/p42 MAP kinase was sensitive to PTX treatment, implying that this effect is downstream of G-protein activation.

In primary cells, H_4 receptor signaling is mediated by increases in intracellular calcium levels. Raible et al. (1992, 1994) showed that histamine can induce calcium mobilization in human eosinophils and that this effect is not mediated by H_1, H_2, or H_3 receptors, implying a role for the H_4 receptor. In mouse primary mast cells, histamine induces calcium mobilization from intracellular stores which can be blocked by H_4 receptor antagonists but not by antagonists for other histamine receptors (Thurmond et al. 2004). In addition, histamine is unable to induce a calcium response in mast cells from H_4 receptor-deficient mice (Hofstra et al. 2003). The calcium response is similar to that seen via H_1 receptor activation in that it can be blocked by a phospholipase C inhibitor, U73122 (Hofstra et al. 2003). It is also blocked by PTX treatment, whereas the H_1R pathway is not affected, suggesting the involvement of different mechanisms. Thus, in primary cells, activation of H_4R leads to increases in intracellular calcium, which in mast cells is mediated by $G\alpha_{i/o}$ proteins and phospholipase C. It has been speculated that phospholipase C is activated by the released $G\beta\gamma$ subunits, and hydrolyzes phosphatidylinositol 4,5-biphosphate (PIP_2) to diacylglycerol and inositol-1,4,5-triphosphate (IP_3), leading to the release of calcium from intracellular stores such as the endoplasmic reticulum (Hofstra et al. 2003).

9.5.2 Some Effects of Histamine Through H_4 Receptors

The biological consequences of H_4R activation are also still being elucidated. Histamine can induce eosinophil chemotaxis, which can be blocked by thioperamide and the H_4 receptor-specific antagonist JNJ 7777120 (Clark et al. 1975, 1977; Wadee et al. 1980; O'Reilly et al. 2002; Ling et al. 2004). In addition, the chemotactic responses of eosinophils to other chemokines such as eotaxin can be enhanced by activation of the H_4 receptor (Buckland et al. 2003; Ling et al. 2004). Chemotaxis requires the polymerization of G-actin to F-actin and histamine has been shown to induce actin polymerization in eosinophils (Buckland et al. 2003). This effect can be blocked by thioperamide, suggesting that it is mediated by the H_4 receptor (Buckland et al. 2003). Adhesion molecules on the surface of eosinophils such as CD11b/CD18 (Mac-1) and CD54 (ICAM-1) can be upregulated by histamine (Buckland et al. 2003; Ling et al. 2004). This effect can be blocked by H_4 receptor antagonists, implying that the H_4 receptor is involved (Buckland et al. 2003; Ling et al. 2004). Thus, in eosinophils, the H_4 receptor mediates many of the events important for localization of eosinophils to sites of inflammation, like chemotaxis and adhesion molecule upregulation.

As for eosinophils, the H_4 receptor mediates histamine-induced chemotaxis in mast cells. This chemotaxis can be blocked by H_4 receptor antagonists and is absent in mast cells from H_4 receptor-deficient mice (Hofstra et al. 2003; Thurmond et al. 2004). Histamine-induced migration of mast cells can also be observed in vivo. Thurmond et al. (2004) showed that after histamine inhalation in mice there is an increase in the number of tracheal mast cells and a redistribution from connective tissues toward the epithelium. This effect can be blocked by the H_4 receptor antagonist, JNJ 7777120 (Thurmond et al. 2004). As seen with histamine inhalation, it is known that there is a redistribution of mast cells to the epithelial lining of the nasal mucosa in response to antigens (Fokkens et al. 1992; Slater et al. 1996). In addition, mast cell numbers increase at sites of inflammation in a number of different pathophysiological conditions such as asthma, rhinitis, rheumatoid arthritis, psoriasis, and cardiovascular disease (Kirby et al. 1987; Crimi et al. 1991; Marone et al. 1997; Peterson et al. 1998; Amin et al. 2000; Gauvreau et al. 2000; Kassel et al. 2001; Krishnaswamy et al. 2001; Woolley 2003). Histamine-induced chemotaxis via the H_4 receptor may play a role in the local migration of mast cells to sites of inflammation.

The H_4 receptor appears to be expressed on dendritic cells, as well as both CD4+ and CD8+ T cells, but little is known about its function in these cells. In T cells, the H_4 receptor may control cytokine release. Both cimetidine (an H_2 receptor antagonist) and thioperamide were able to block the histamine-induced IL-16 release from CD8+ T cells, which implied a role for both H_2 and H_4 receptors (Gantner et al. 2002).

H_4 receptor antagonists have shown anti-inflammatory properties in mice. Both thioperamide and JNJ 7777120 have been shown to reduce neutrophil infiltration in response to zymosan in vivo (Takeshita et al. 2003; Thurmond et al. 2004). The compounds are only effective in mast cell-mediated models, suggesting that these cells are the target cells for the antagonists. In one of the models

(Takeshita et al. 2003), the H_4 receptor antagonist reduced the levels of leukotriene B_4 presumably produced by mast cells. Since leukotriene B_4 is a potent neutrophil chemotactic agent, this may provide an explanation for the anti-inflammatory effects.

9.5.3 Uses of H_4 Receptor Antagonists

While no H_4 antagonists are currently used in the clinic, the current literature suggests several possible indications. Histamine is a major mediator in allergic rhinitis where H_1 receptor antagonists are useful but not completely effective. One indication that H_4 antagonists may have a therapeutic use in this condition comes from the fact that neither H_1 nor H_2 antagonists can inhibit histamine-induced congestion (Howarth et al. 2000). Furthermore, the major role that mast cells play in allergic rhinitis, coupled with the expression of H_4 on these cells, may also suggest that H_4 receptor antagonists will be useful for treating this condition. Another aspect where the H_4 receptor may be of benefit is in the allergic hyper-responsiveness developed upon repeated exposure to antigens that is characterized by increased sensitivity to allergens. An increase in the number of mast cells and other inflammatory cells in the nasal mucosa is thought to be responsible for this response. Therefore, if H_4 receptor antagonists can inhibit mast cell and eosinophil accumulation at sites of inflammation, they may be effective in reducing allergic hyper-responsiveness.

The expression and function of the H_4 receptor in mast cells, eosinophils, dendritic cells and T cells open up the possibility of use in a number of therapeutic areas. These cell types play a major role in the etiology of asthma. Histamine is known to be released in asthma; however, H_1 receptor antagonists have only modest effects (Larsen 2001; Lordan and Holgate 2002; Walsh 2002), leaving open the possibility that H_4 receptor antagonists may be useful.

Autoimmune diseases, such as rheumatoid arthritis, multiple sclerosis, type I diabetes and systemic lupus erythematosus, are other areas that deserve further study in order to elucidate the role of the H_4 receptor, since these conditions are driven by dendritic cells and T cells. Finally, itching associated with conditions like atopic dermatitis and urticaria is also thought to be partially mediated by histamine (Hagermark 1992), and, although H_1 antagonists are useful in the treatment of urticaria, they are ineffective in other conditions like atopic dermatitis (Henz et al. 1998). In support of the role of H_4 receptors in itch, Bell et al. (2004) implicated both H_1 and H_4 receptors in modulating histamine-induced itch in mice.

Table 9.1 summarizes the most important data of the four known and characterized histamine receptors.

Table 9.1. Histamine receptors and some elements of their signal pathways

Receptor	Chromosomal localization	Amino acids	G-protein type	Main elements of the signaling pathway
H1	3	487	Gq/11	PKC, PLC, PLA NOS, MAP kinases, e.g., ERK1/2, p38
H2	5	359	Gsα and Gq	PKA, PLC, c-fos
H3	20	445	$G_{i/0}$	MAP kinase, e.g., p44/42 kinase High constitutive activity
H4	18	390	$G_{i/0}$	PLC MAP kinase, e.g., p44/42 kinase

References

Amin K, Ludviksdottir D, Janson C, Nettelbladt O, Bjornsson E, Roomans GM, Boman G, Seveus L, Venge P (2000) Inflammation and structural changes in the airways of patients with atopic and nonatopic asthma. BHR Group. Am J Resp Crit Care Med 162:2295–2301

Apodaca R, Dvorak CA, Xiao W, Barbier AJ, Boggs JD, Wilson SJ, Lovenberg TW, Carruthers NI (2003) A new class of diamine-based human histamine H3 receptor antagonists: 4-(aminoalkoxy)benzylamines. J Med Chem 46:3938–3944

Arias-Montano JA, Floran B, Garcia M, Aceves J, Young JM (2001) Histamine H3 receptor-mediated inhibition of depolarization-induced, dopamine D1 receptor-dependent release of [3H]-γ-aminobutyric acid from rat striatal slices. Br J Pharmacol 133:165–171

Arrang JM, Garbarg M, Schwartz JC (1983) Autoinhibition of brain histamine release mediated by a novel class (H3) of histamine receptor. Nature 302:832–837

Arrang JM, Garbarg M, Lancelot JC, Lecomte JM, Pollard H, Robba M, Schunack W, Schwartz JC (1987) Highly potent and selective ligands for histamine H3-receptors. Nature 327:117–123

Attoub S, Moizo L, Sobhani I, Laigneau J-P, Lewin MJM, Bado A (2001) The H3 receptor is involved in cholecystokinin inhibition of food intake in rats. Life Sci 69:469–478

Azzolina A, Guarneri P, Lampiasi N (2002) Involvement of p38 and JNK MAPKs pathways in substance P-induced production of TNF-alpha by peritoneal mast cells. Cytokine 18:72–80

Bakker RA, Schoonus SB, Smit MJ, Timmerman H, Leurs R (2001) Histamine H(1)-receptor activation of nuclear factor-kappa B: roles for G beta gamma- and G alpha(q/11)-subunits in constitutive and agonist-mediated signalling. Mol Pharmacol 60:1133–1142

Bakker RA, Casarosa P, Timmerman H, Smit MJ, Leurs R (2004) Constitutively active Gq/11-coupled receptors enable signaling by co-expressed G(i/o)-coupled receptors. J Biol Chem 279:5152–5161

Bell JK, McQueen DS, Rees JL (2004) Involvement of histamine H4 and H1 receptors in scratching induced by histamine receptor agonists in Balb C mice. Br J Pharmacol 142(2):374–380

Blandina P, Giorgetti M, Bartolini L, Cecchi M, Timmerman H, Leurs R, Pepeu G, Giovannini MG (1996) Inhibition of cortical acetylcholine release and cognitive performance by histamine H3 receptor activation in rats. Br J Pharmacol 119:1656–1664

Blandizzi C, Colucci R, Tognetti M, De Paolis B, Del Tacca M (2001) H3 receptor-mediated inhibition of intestinal acetylcholine release: pharmacological characterization of signal transduction pathways. Naunyn-Schmiedeberg's Arch Pharmacol 363:193–202

Bowrey PF, King J, Magarey C, Schwartz P, Marr P, Bolton E, Morris DL (2000) Histamine, mast cells and tumour cell proliferation in breast cancer: does preoperative cimetidine administration have an effect? Br J Cancer 82:167–170

Buckland KF, Williams TJ, Conroy DM (2003) Histamine induces cytoskeletal changes in human eosinophils via the H4 receptor. Br J Pharmacol 140:1117–1127

Casarosa P, Bakker RA, Verzijl D, Navis M, Timmerman H, Leurs R, Smit MJ (2001) Constitutive signaling of the human cytomegalovirus-encoded chemokine receptor US28. J Biol Chem 276:1133–1137

Celuch SM (1995) Possible participation of histamine H3 receptors in the modulation of noradrenaline release from rat spinal cord slices. Eur J Pharmacol 287:127–133

Chai W, Breitenbucher JG, Kwok A, Li X, Wong V, Carruthers NI, Lovenberg TW, Mazur C, Wilson SJ, Axe FU, Jones TK (2003) Non-imidazole heterocyclic histamine H3 receptor antagonists. Bioorg Med Chem Lett 13:1767–1770

Chen J, Liu C, Lovenberg TW (2003) Molecular and pharmacological characterization of the mouse histamine H3 receptor. Eur J Pharmacol 467:57–65

Clapham J, Kilpatrick GJ (1992) Histamine H3 receptors modulate the release of [3H]acetylcholine from slices of rat entorhinal cortex: evidence for the possible existence of H3 receptor subtypes. Br J Pharmacol 107:919–923

Clark EA, Hill SJ (1996) Sensitivity of histamine H3 receptor agonist-stimulated [35S]GTP gamma[S] binding to pertussis toxin. Eur J Pharmacol 296:223–225

Clark RAF, Gallin JI, Kaplan AP (1975) Selective eosinophil chemotactic activity of histamine. J Exp Med 142:1462–1476

Clark RAF, Sandler JA, Gallin JI, Kaplan AP (1977) Histamine modulation of eosinophil migration. J Immunol 118:137–145

Coge F, Guenin S-P, Audinot V, Renouard-Try A, Beauverger P, Macia C, Ouvry C, Nagel N, Rique H, Boutin JA, Galizzi J-P (2001a) Genomic organization and characterization of splice variants of the human histamine H3 receptor. Biochem J 355:279–288

Coge F, Guenin S-P, Rique H, Boutin JA, Galizzi J-P (2001b) Structure and expression of the human histamine H4-receptor gene. Biochem Biophys Res Commun 284:301–309

Crimi E, Chiaramondia M, Milanese M, Rossi GA, Brusasco V (1991) Increased numbers of mast cells in bronchial mucosa after the late-phase asthmatic response to allergen. Am Rev Resp Dis 144:1282–1286

Curran MP, Scott LJ, Perry CM (2004) Cetirizine: a review of its use in allergic disorders. Drugs 64:523–561

Davio C, Cricco G, Bergoc R, Andrade N, Rivera E (1993) H1 and H2 histamine receptors in human mammary carcinomas. Agents Actions Spec Conf Issue 38:71–74

Del Valle J, Gantz I (1997) Novel insights into histamine H2 receptor biology. Am J Physiol 273:G987–G996

Drutel G, Peitsaro N, Karlstedt K, Wieland K, Smit MJ, Timmerman H, Panula P, Leurs R (2001) Identification of rat H3 receptor isoforms with different brain expression and signaling properties. Mol Pharmacol 59:1–8

Elz S, Kramer K, Pertz HH, Detert H, Laak AM, Kuhne R, Schunack W (2000) Histaprodifens: synthesis, pharmacological in vitro evaluation, and molecular modeling of a new class of highly active and selective histamine H(1)-receptor agonists. J Med Chem 43:1071–1084

Endou M, Levi R (1995) Histamine in the heart. Eur J Clin Invest 25(Suppl 1): 5–11

Endou M, Poli E, Levi R (1994) Histamine H3-receptor signaling in the heart: possible involvement of Gi/Go proteins and N-type Ca^{++} channels. J Pharmacol Exp Ther 269:221–229

Esbenshade TA, Krueger KM, Miller TR, Kang CH, Denny LI, Witte DG, Yao BB, Fox GB, Faghih R, Bennani YL, Williams M, Hancock AA (2003) Two novel and selective nonimidazole histamine H3 receptor antagonists A-304121 and A-317920: I. In vitro pharmacological effects. J Pharmacol Exp Ther 305:887–896

Faghih R, Dwight W, Black L, Liu H, Gentles R, Phelan K, Esbenshade TA, Ireland L, Miller TR, Kang C-H, Krueger KM, Fox GB, Hancock AA, Bennani YL (2002a) Structure–activity relationships of non-imidazole H3 receptor ligands. Part 2: binding preference for δ-amino acids motifs. Bioorg Med Chem Letters 12:2035–2037

Faghih R, Dwight W, Vasudevan A, Dinges J, Conner SE, Esbenshade TA, Bennani YL, Hancock AA (2002b) Aminoalkoxybiphenylnitriles as histamine-3 receptor ligands. Bioorg Med Chem Lett 12:3077–3079

Faghih R, Dwight W, Pan JB, Fox GB, Krueger KM, Esbenshade TA, McVey JM, Marsh K, Bennani YL, Hancock AA (2003) Synthesis and SAR of aminoalkoxy-biaryl-4-carboxamides: novel and selective histamine H3 receptor antagonists. Bioorg Med Chem Lett 13:1325–1328

Fink K, Schlicker E, Neise A, Gothert M (1990) Involvement of presynaptic H3 receptors in the inhibitory effect of histamine on serotonin release in the rat brain cortex. Naunyn-Schmiedeberg's Arch Pharmacol 342:513–519

Fokkens WJ, Godthelp T, Holm AF, Blom H, Mulder PG, Vroom TM, Rijntjes E (1992) Dynamics of mast cells in the nasal mucosa of patients with allergic rhinitis and non-allergic controls: a biopsy study. Clin Exp Allergy 22:701–710

Fox GB, Pan JB, Esbenshade TA, Bennani YL, Black LA, Faghih R, Hancock AA, Decker MW (2002) Effects of histamine H3 receptor ligands GT-2331 and ciproxifan in a repeated acquisition avoidance response in the spontaneously hypertensive rat pup. Behav Brain Res 131:151–161

Gantner F, Sakai K, Tusche MW, Cruikshank WW, Center DM, Bacon KB (2002) Histamine H4 and H2 receptors control histamine-induced interleukin-16 release from human CD8+ T cells. J Pharmacol Exp Ther 303:300–307

Gantz I, Munzert G, Tashiro T, Schaffer M, Wang L, DelValle J, Yamada T (1991) Molecular cloning of the human histamine H2 receptor. Biochem Biophys Res Commun 178:1386–1392

Garbarg M, Schwartz JC (1988) Synergism between histamine H1- and H2-receptors in the cAMP response in guinea pig brain slices: effects of phorbol esters and calcium. Mol Pharmacol 33:38–43

Gauvreau GM, Lee JM, Watson RM, Irani AM, Schwartz LB, O'Byrne PM (2000) Increased numbers of both airway basophils and mast cells in sputum after allergen inhalation challenge of atopic asthmatics. Am J Resp Crit Care Med 161:1473–1478

Gfesser GA, Zhang H, Dinges J, Fox GB, Pan JB, Esbenshade TA, Yao BB, Witte D, Miller TR, Kang C-H, Krueger KM, Bennani YL, Hancock AA, Faghih R (2004) Structure–activity relationships of non-imidazole H3 receptor ligands. Part 3: 5-substituted 3-phenyl-1,2,4-oxadiazoles as potent antagonists. Bioorg Med Chem Lett 14:673–676

Giovannini MG, Bartolini L, Bacciottini L, Greco L, Blandina P (1999) Effects of histamine H3 receptor agonists and antagonists on cognitive performance and scopolamine-induced amnesia. Behav Brain Res 104:147–155

Griffiths J, Marley PD (2001) Ca^{2+}-dependent activation of tyrosine hydroxylase involves MEK1. Neuroreport 12:2679–2683

Hagermark O (1992) Peripheral and central mediators of itch. Skin Pharmacol 5:1–8

Hashimoto T, Harusawa S, Araki L, Zuiderveld Obbe P, Smit Martine J, Imazu T, Takashima S, Yamamoto Y, Sakamoto Y, Kurihara T, Leurs R, Bakker Remko A, Yamatodani A (2003) A selective human H(4)-receptor agonist: (-)-2-cyano-1-methyl-3-[(2R,5R)-5-[1H-imidazol-4(5)-yl]tetrahydrofuran-2-y] methylguanidine. J Med Chem 46:3162–3165

Hattori Y, Gando S, Endou M, Kanno M (1991) Characterization of histamine receptors modulating inotropic and biochemical activities in rabbit left atria. Eur J Pharmacol 196:29–36

Henz BM, Metzenauer P, O'Keefe E, Zuberbier T (1998) Differential effects of new-generation H1-receptor antagonists in pruritic dermatoses. Allergy 53:180–183

Hill SJ, Young JM, Marrian DH (1977) Specific binding of 3H-mepyramine to histamine H1 receptors in intestinal smooth muscle. Nature 270:361–363

Hill SJ, Ganellin CR, Timmerman H, Schwartz JC, Shankley NP, Young JM, Schunack W, Levi R, Haas HL (1997) International Union of Pharmacology. XIII. Classification of histamine receptors. Pharmacol Rev 49:253–278

Hofstra CL, Desai PJ, Thurmond RL, Fung-Leung W-P (2003) Histamine H4 receptor mediates chemotaxis and calcium mobilization of mast cells. J Pharmacol Exp Ther 305:1212–1221

Hough LB (2001) Genomics meets histamine receptors: new subtypes, new receptors. Mol Pharmacol 59:415–419

Howarth PH, Salagean M, Dokic D (2000) Allergic rhinitis: not purely a histamine-related disease. Allergy 55(Suppl 64):7–16

Igaz P, Fitzimons CP, Szalai C, Falus A (2002) Histamine genomics in silico: polymorphisms of the human genes involved in the synthesis, action and degradation of histamine. Am J Pharmacogenom 2:67–72

Imamura M, Lander HM, Levi R (1996) Activation of histamine H3-receptors inhibits carrier-mediated norepinephrine release during protracted myocardial ischemia: comparison with adenosine A1-receptors and A2-adrenoceptors. Circ Res 78:475–481

Inoue I, Yanai K, Kitamura D, Taniuchi I, Kobayashi T, Niimura K, Watanabe T (1996) Impaired locomotor activity and exploratory behavior in mice lacking histamine H_1 receptors. Proc Nat Acad Sci USA 93:13316–13320

Itoh E, Fujimiya M, Inui A (1999) Thioperamide, a histamine H3 receptor antagonist, powerfully suppresses peptide YY-induced food intake in rats. Biol Psychiatry 45:475–481

Jablonowski JA, Carruthers NI, Thurmond RL (2004) Histamine H4 receptor and potential therapeutic uses for H4 ligands. Mini Rev Med Chem 4(9):993–1000

Jang I-S, Rhee J-S, Watanabe T, Akaike N, Akaike N (2001) Histaminergic modulation of GABAergic transmission in rat ventromedial hypothalamic neurons. J Physiol 534:791–803

Kassel O, De Blay F, Duvernelle C, Olgart C, Israel-Biet D, Krieger P, Moreau L, Muller C, Pauli G, Frossard N (2001) Local increase in the number of mast cells and expression of nerve growth factor in the bronchus of asthmatic patients after repeated inhalation of allergen at low-dose. Clin Exp Allergy 31:1432–1440

Kilpatrick GJ, Michel AD (1991) Characterization of the binding of the histamine H3 receptors agonist [3H](R)-alpha-methyl histamine to homogenates of rat and guinea pig cortex. Agents Actions Suppl 33:69–75

Kirby JG, Hargreave FE, Gleich GJ, O'Byrne PM (1987) Bronchoalveolar cell profiles of asthmatic and nonasthmatic subjects. Am Rev Resp Dis 136:379–383

Kobayashi T, Tonai S, Ishihara Y, Koga R, Okabe S, Watanabe T (2000) Functional and morphological regulation of the gastric mucosa in histamine H2 receptor-deficient mice. J Clin Invest 105:1741–1749

Korte A, Myers J, Shih NY, Egan RW, Clark MA (1990) Characterization and tissue distribution of H3 histamine receptors in guinea pigs by Na-methylhistamine. Biochem Biophys Res Commun 168:979–986

Koyama M, Heerdt PM, Levi R (2003a) Increased severity of reperfusion arrhythmias in mouse hearts lacking histamine H3-receptors. Biochem Biophys Res Commun 306:792–796

Koyama M, Seyedi N, Fung-Leung W-P, Lovenberg TW, Levi R (2003b) Norepinephrine release from the ischemic heart is greatly enhanced in mice lacking histamine H3 receptors. Mol Pharmacol 63:378–382

Krishnaswamy G, Kelley J, Johnson D, Youngberg G, Stone W, Huang SK, Bieber J, Chi DS (2001) The human mast cell: functions in physiology and disease. Front Biosci 6:1109–1127

Kubo Y, Nakano K (1999) Regulation of histamine synthesis in mouse CD4+ and CD8+ T lymphocytes. Inflamm Res 48:149–153

Lacey JV Jr, Sherman ME, Hartge P, Schatzkin A, Schairer C (2004) Medication use and risk of ovarian carcinoma: a prospective study. Int J Cancer 108:281–286

Larsen JS (2001) Do antihistamines have a role in asthma therapy? Pharmacotherapy 21(3 Pt 2): 28S–33S

Lee JJ, Parsons ME (2000) Signaling mechanisms coupled to presynaptic A1- and H3-receptors in the inhibition of cholinergic contractile responses of the guinea pig ileum. J Pharmacol Exp Ther 295:607–613

Leurs R, Traiffort E, Arrang JM, Tardivel-Lacombe J, Ruat M, Schwartz JC (1994) Guinea pig histamine H1 receptor. II. Stable expression in Chinese hamster ovary cells reveals the interaction with three major signal transduction pathways. J Neurochem 2:519–527

Leurs R, Smit MJ, Timmerman H (1995) Molecular pharmacological aspects of histamine receptors. Pharmacol Ther 66:413–463

Levi R, Smith NCE (2000) Histamine H3-receptors: a new frontier in myocardial ischemia. J Pharmacol Exp Ther 292:825–830

Li D, Wen JF, Jin JY, Jin H, Ann HS, Kim SZ, Kim SH, Lee HS, Cho KW (2003) Histamine inhibits atrial myocytic ANP release via H2 receptor-cAMP-protein kinase signalling Am J Physiol-Regul Integr Comp Physiol 285(2):R380–R393

Lin JS, Sakai K, Vanni-Mercier G, Arrang JM, Garbarg M, Schwartz JC, Jouvet M (1990) Involvement of histaminergic neurons in arousal mechanisms demonstrated with H3-receptor ligands in the cat. Brain Res 523:325–330

Ling P, Ngo K, Nguyen S, Thurmond RL, Edwards JP, Karlsson L, Fung-Leung W-P (2004) Histamine H4 receptor mediates eosinophil chemotaxis with cell shape change and adhesion molecule up-regulation. Br J Pharmacol 142(1):161–171

Linney ID, Buck IM, Harper EA, Kalindjian SB, Pether MJ, Shankley NP, Watt GF, Wright PT (2000) Design, synthesis, and structure–activity relationships of novel non-imidazole histamine H3 receptor antagonists. J Med Chem 43:2362–2370

Liu C, Ma XJ, Lovenberg TW (2000) Alternative splicing of the histamine H3 receptor mRNA at the third cytoplasmic loop is not detectable in humans. Mol Brain Res 83:145–150

Liu C, Ma X-J, Jiang X, Wilson SJ, Hofstra CL, Blevitt J, Pyati J, Li X, Chai W, Carruthers N, Lovenberg TW (2001a) Cloning and pharmacological characterization of a fourth histamine receptor (H4) expressed in bone marrow. Mol Pharmacol 59:420–426

Liu C, Wilson SJ, Kuei C, Lovenberg TW (2001b) Comparison of human, mouse, rat, and guinea pig histamine H4 receptors reveals substantial pharmacological species variation. J Pharmacol Exp Ther 299:121–130

Longhurst PA, McNeill JH (1982) Rabbit atrial histamine receptors and cyclic AMP. Can J Physiol Pharmacol 60:1730–1732

Lordan JL, Holgate ST (2002) H1-antihistamines in asthma. Clin Allergy Immunol 17:221–248

Lovenberg TW, Roland BL, Wilson SJ, Jiang X, Pyati J, Huvar A, Jackson MR, Erlander MG (1999) Cloning and functional expression of the human histamine H3 receptor. Mol Pharmacol 55:1101–1107

Lovenberg TW, Pyati J, Chang H, Wilson SJ, Erlander MG (2000) Cloning of rat histamine H3 receptor reveals distinct species pharmacological profiles. J Pharmacol Exp Ther 293:771–778

Marone G, Casolaro V, Patella V, Florio G, Triggiani M (1997) Molecular and cellular biology of mast cells and basophils. Int Arch Allergy Appl Immunol 114:207–217

Martinez-Mir MI, Pollard H, Moreau J, Arrang JM, Ruat M, Traiffort E, Schwartz JC, Palacios JM (1990) Three histamine receptors (H1, H2 and H3) visualized in the brain of human and non-human primates. Brain Res 526:322–327

Masaki TH, Yoshimatsu S, Chiba T, Watanabe T, Sakata T (2001) Targeted disruption of histamine H_1-receptor attenuates regulatory effects of leptin on feeding, adiposity, and UCP family in mice. Diabetes 50:385–391

Megson AC, Walker EM, Hill SJ (2001) Role of protein kinase Calpha in signaling from the histamine H(1) receptor to the nucleus. Mol Pharmacol 59:1012–1021

Meier G, Ligneau X, Pertz HH, Ganellin CR, Schwartz J-C, Schunack W, Stark H (2002) Piperidine–hydrocarbon compounds as novel non-imidazole histamine H3-receptor antagonists. Bioorg Med Chem 10:2535–2542

Miko T, Ligneau X, Pertz HH, Ganellin CR, Arrang J-M, Schwartz J-C, Schunack W, Stark H (2003) Novel nonimidazole histamine H3 receptor antagonists: 1-(4-(phenoxymethyl)benzyl) piperidines and related compounds. J Med Chem 46:1523–1530

Mobarakeh JI, Sakurada S, Katsuyama S, Kutsuwa M, Kuramasu A, Lin ZY, Watanabe T, Hashimoto Y, Yanai K (2000) Role of histamine H_1 receptor in pain perception: a study of the receptor gene knockout mice. Eur J Pharmacol 391:81–89

Molina-Hernandez A, Nunez A, Arias-Montano JA (2000) Histamine H3-receptor activation inhibits dopamine synthesis in rat striatum. Neuroreport 11:163–166

Molina-Hernandez A, Nunez A, Sierra J-J, Arias-Montano J-A (2001) Histamine H3 receptor activation inhibits glutamate release from rat striatal synaptosomes. Neuropharmacology 41:928–934

Monti JM, Jantos H, Boussard M, Altier H, Orellana C, Olivera S (1991) Effects of selective activation or blockade of the histamine H3 receptor on sleep and wakefulness. Eur J Pharmacol 205:283–287

Morisset S, Rouleau A, Lineau X, Gbahou F, Tardivel-Lacombe J, Stark H, Schunack W, Ganellin CR, Schwartz JC, Arrang J-M (2000) High constitutive activity of native H3 receptors regulates histamine neurons in brain. Nature 408:860–864

Morse KL, Behan J, Laz TM, West RE Jr, Greenfeder SA, Anthes JC, Umland S, Wan Y, Hipkin RW, Gonsiorek W, Shin N, Gustafson EL, Qiao X, Wang S, Hedrick JA, Greene J, Bayne M, Monsma

FJ Jr (2001) Cloning and characterization of a novel human histamine receptor. J Pharmacol Exp Ther 296:1058–1066

Nagahara A, Wang L, Del Valle J, Todisco A (1998) Regulation of c-Jun NH2-terminal kinases in isolated canine gastric parietal cells. Am J Physiol 275:G740–G748

Nakamura T, Itadani H, Hidaka Y, Ohta M, Tanaka K (2000) Molecular cloning and characterization of a new human histamine receptor, HH4R. Biochem Biophys Res Commun 279:615–620

Nguyen T, Shapiro DA, George SR, Setola V, Lee DK, Cheng R, Rauser L, Lee SP, Lynch KR, Roth BL, O'Dowd BF (2001) Discovery of a novel member of the histamine receptor family. Mol Pharmacol 59:427–433

Oda T, Morikawa N, Saito Y, Masuho Y, Matsumoto S-I (2000) Molecular cloning and characterization of a novel type of histamine receptor preferentially expressed in leukocytes. J Biol Chem 275:36781–36786

Ohkubo T, Shibata M, Inoue M, Kaya H, Takahashi H (1995) Regulation of substance P release mediated via prejunctional histamine H3 receptors. Eur J Pharmacol 273:83–88

Orange PR, Heath PR, Wright SR, Pearson RC (1996) Allelic variations of the human histamine H2 receptor gene. Neuroreport 7:1293–1296

O'Reilly M, Alpert R, Jenkinson S, Gladue RP, Foo S, Trim S, Peter B, Trevethick M, Fidock M (2002) Identification of a histamine H4 receptor on human eosinophils – role in eosinophil chemotaxis. J Recept Signal Tr R 22:431–448

Osna N, Elliott K, Khan MM (2001) Regulation of interleukin-10 secretion by histamine in TH2 cells and splenocytes. Int Immunopharmacol 1:85–96

Packard KA, Khan MM (2003) Effects of histamine on Th1/Th2 cytokine balance. Int Immunopharmacol 3:909–920

Peterson LJ, Hansen U, Kristensen JK, Nielsen H, Skov PS, Nielsen HJ (1998) Studies on mast cells and histamine release in psoriasis: the effect of ranitidine. Acta Dermatol Venereol (Stockholm) 78:190–193

Poli E, Pozzoli C, Coruzzi G, Bertaccini G (1993) Histamine H3-receptor-induced inhibition of duodenal cholinergic transmission is independent of intracellular cyclic AMP and GMP. Gen Pharmacol 24:1273–1278

Poli E, Pozzoli C, Coruzzi G, Bertaccini G (1994) Signal transducing mechanisms coupled to histamine H3 receptors and alpha-2 adrenoceptors in the guinea pig duodenum: possible involvement of N-type Ca^{++} channels. J Pharmacol Exp Ther 270:788–794

Pollard H, Moreau J, Arrang JM, Schwartz JC (1993) A detailed autoradiographic mapping of histamine H3 receptors in rat brain areas. Neuroscience 52:169–189

Poluektova LY, Huggler GK, Patterson EB, Khan MM (1999) Involvement of protein kinase A in histamine-mediated inhibition of IL-2mRNA expression in mouse splenocytes. Immunopharmacology 41:77–87

Raible DG, Schulman ES, DiMuzio J, Cardillo R, Post TJ (1992) Mast cell mediators prostaglandin-D2 and histamine activate human eosinophils. J Immunol 148:3536–3542

Raible DG, Lenahan T, Fayvilevich Y, Kosinski R, Schulman ES (1994) Pharmacologic characterization of a novel histamine receptor on human eosinophils. Am J Res Crit Care Med 149: 1506–1511

Robinson AJ, Dickenson JM (2001) Activation of the p38 and p42/p44 mitogen-activated protein kinase families by the histamine H(1) receptor in DDT(1)MF-2 cells. Br J Pharmacol 133:1378–1386

Sasaki Y, Ihara K, Ahmed S, Yamawaki K, Kusuhara K, Nakayama H, Nishima S, Hara T (2000) Lack of association between atopic asthma and polymorphisms of the histamine H1 receptor, histamine H2 receptor, and histamine N-methyltransferase genes. Immunogenetics 51: 238–240

Schlicker E, Betz R, Gothert M (1988) Histamine H3 receptor-mediated inhibition of serotonin release in the rat brain cortex. Naunyn-Schmiedeberg's Arch Pharmacol 337:588–590

Schlicker E, Fink K, Detzner M, Gothert M (1993) Histamine inhibits dopamine release in the mouse striatum via presynaptic H3 receptors. J Neural Transm Gen Sect 93:1–10

Schlicker E, Kathmann M, Detzner M, Exner HJ, Gothert M (1994) H3 receptor-mediated inhibition of noradrenaline release: an investigation into the involvement of Ca^{2+} and K^{+} ions, G protein and adenylate cyclase. Naunyn-Schmiedeberg's Arch Pharmacol 350:34–41

Schwartz JC, Morisset S, Rouleau A, Ligneau X, Gbahou F, Tardivel-Lacombe J, Stark H, Schunack W, Ganellin CR, Arrang JM (2003) Therapeutic implications of constitutive activity of receptors: the example of the histamine H3 receptor. J Neural Transm Suppl 64:1–16

Shah C, McAtee L, Breitenbucher JG, Rudolph D, Li X, Lovenberg TW, Mazur C, Wilson SJ, Carruthers NI (2002) Novel human histamine H3 receptor antagonists. Bioorg Med Chem Lett 12:3309–3312

Shahrestanifar M, Fan X, Manning DR (1999) Lysophosphatidic acid activates NF-kappaB in fibroblasts. A requirement for multiple inputs. J Biol Chem 274:3828–3833

Shin N, Coates E, Murgolo NJ, Morse KL, Bayne M, Strader CD, Monsma FJ Jr (2002) Molecular modeling and site-specific mutagenesis of the histamine-binding site of the histamine H4 receptor. Mol Pharmacol 62:38–47

Silver RB, Mackins CJ, Smith NCE, Koritchneva IL, Lefkowitz K, Lovenberg TW, Levi R (2001) Coupling of histamine H3 receptors to neuronal Na^{+}/H^{+} exchange: a novel protective mechanism in myocardial ischemia. Proc Natl Acad Sci USA 98:2855–2859

Simons FE, Sussman GL, Simons KJ (1995) Effect of H2-antagonist cimetidine on the pharmacokinetics and pharmacodynamics of the H1-receptor antagonists hydroxyzine and cetirizine in patients with chronic urticaria. J Allergy Clin Immun 95:685–693

Slater A, Smallman LA, Drake-Lee AB (1996) Increase in epithelial mast cell numbers in the nasal mucosa of patients with perennial allergic rhinitis. J Laryngol Otol 110:929–933

Smit MJ, Hoffmann M, Timmerman H, Leurs R (1999) Molecular properties and signalling pathways of the histamine H1 receptor. Clin Exp Allergy 29(Suppl 3):19–28

Sonobe Y, Nakane H, Watanabe T, Nakano K (2004) Regulation of Con A-dependent cytokine production from CD4+ and CD8+ T lymphocytes by autosecretion of histamine. Inflamm Res 53:87–92

Stark H (2003) Recent advances in histamine H3/H4 receptor ligands. Expert Opin Ther Pat 13:851–865

Takagishi T, Sasaguri Y, Nakano R, Arima N, Tanimoto A, Fukui H, Morimatsu M (1995) Expression of the histamine H1 receptor gene in relation to atherosclerosis. Am J Pathol 146:981–988

Takahashi K, Suwa H, Ishikawa T, Kotani H (2002) Targeted disruption of H3 receptors results in changes in brain histamine tone leading to an obese phenotype. J Clin Invest 110:1791–1799

Takeshita K, Sakai K, Bacon KB, Gantner F (2003) Critical role of histamine H4 receptor in leukotriene B4 production and mast cell-dependent neutrophil recruitment induced by zymosan in vivo. J Pharmacol Exp Ther 307:1072–1078

Tardivel-Lacombe J, Morisset S, Gbahou F, Schwartz JC, Arrang J-M (2001) Chromosomal mapping and organization of the human histamine H3 receptor gene. Neuroreport 12:321–324

Thurmond RL, Desai PJ, Dunford PJ, Fung-Leung W-P, Hofstra CL, Jiang W, Nguyen S, Riley JP, Sun S, Williams KN, Edwards JP, Karlsson L (2004) A potent and selective histamine H4 receptor antagonist with anti-inflammatory properties. J Pharmacol Exp Ther 309:404–413

Toyota H, Dugovic C, Koehl M, Laposky AD, Weber C, Ngo K, Wu Y, Lee DH, Yanai K, Sakurai E, Watanabe T, Liu C, Chen J, Barbier AJ, Turek FW, Fung-Leung WP, Lovenberg TW (2002) Behavioral characterization of mice lacking histamine H3 receptors. Mol Pharmacol 62:389–397

Turner SC, Esbenshade TA, Bennani YL, Hancock AA (2003) A new class of histamine H3-receptor antagonists: synthesis and structure–activity relationships of 7,8,9,10-tetrahydro-6H-cyclohepta[β]quinolines. Bioorg Med Chem Lett 13:2131–2135

van der Goot H, Timmerman H (2000) Selective ligands as tools to study histamine receptors. Eur J Med Chem 35:5–20

Vasudevan A, Conner SE, Gentles RG, Faghih R, Liu H, Dwight W, Ireland L, Kang CH, Esbenshade TA, Bennani YL, Hancock AA (2002) Synthesis and evaluation of potent pyrrolidine H3 antagonists. Bioorg Med Chem Lett 12:3055–3058

Wadee AA, Anderson R, Sher R (1980) In vitro effects of histamine on eosinophil migration. Int Arch Allergy Appl Immunol 63:322–329

Walczynski K, Guryn R, Zuiderveld OP, Timmerman H (1999) Non-imidazole histamine H3 ligands. Part 2. New 2-substituted benzothiazoles as histamine H3 antagonists. Arch Pharm 332:389–398

Walsh GM (2002) Second-generation antihistamines in asthma therapy: is there a protective effect? Am J Resp Med 1:27–34

Weiger T, Stevens DR, Wunder L, Haas HL (1997) Histamine H1 receptors in C6 glial cells are coupled to calcium-dependent potassium channels via release of calcium from internal stores. Naunyn-Schmiedeberg's Arch Pharmacol 355:559–565

Wellendorph P, Goodman MW, Burstein ES, Nash NR, Brann MR, Weiner DM (2002) Molecular cloning and pharmacology of functionally distinct isoforms of the human histamine H3 receptor. Neuropharmacology 42:929–940

Wellner-Kienitz MC, Bender K, Meyer T, Pott L (2003) Coupling to Gs and G(q/11) of histamine H2 receptors heterologously expressed in adult rat atrial myocytes. Biochim Biophys Acta 1642:67–77

Wiedemann P, Boenisch H, Oerters F, Bruess M (2002) Structure of the human histamine H3 receptor gene (HRH3) and identification of naturally occurring variations. J Neural Transm 109:443–453

Wieland K, Bongers G, Yamamoto Y, Hashimoto T, Yamatodani A, Menge WMBP, Timmerman H, Lovenberg TW, Leurs R (2001) Constitutive activity of histamine H3 receptors stably expressed in SK-N-MC cells: display of agonism and inverse agonism by H3 antagonists. J Pharmacol Exp Ther 299:908–914

Woolley DE (2003) The mast cell in inflammatory arthritis. New Engl J Med 348:1709–1711

Xie P, Browning DD, Hay N, Mackman N, Ye RD (2000) Activation of NF-kappa B by bradykinin through a Galpha(q)- and Gbeta gamma-dependent pathway that involves phosphoinositide 3-kinase and Akt. J Biol Chem 275:24907–24914

Yamasaki T, Tamai I, Matsumura Y (2001) Activation of histamine H3 receptors inhibits renal noradrenergic neurotransmission in anesthetized dogs. Am J Physiol 280:R1450–R1456

Yanai K, Son LZ, Endou M, Sakurai E, Nakagawasai O, Tadano T, Kisara K, Inoue I, Watanabe T (1998) Behavioural characterization and amounts of brain monoamines and their metabolites in mice lacking histamine H_1 receptors. Neuroscience 87:479–487

Zaheer A, Lim R (1998) Overexpression of glia maturation factor (GMF) in PC12 pheochromocytoma cells activates p38 MAP kinase, MAPKAP kinase-2, and tyrosine hydroxylase. Biochem Biophys Res Commun 250:278–282

Zhu Y, Michalovich D, Wu H-L, Tan KB, Dytko GM, Mannan IJ, Boyce R, Alston J, Tierney LA, Li X, Herrity NC, Vawter L, Sarau HM, Ames RS, Davenport CM, Hieble JP, Wilson S, Bergsma DJ, Fitzgerald LR (2001) Cloning, expression, and pharmacological characterization of a novel human histamine receptor. Mol Pharmacol 59:434–441

Signal Transduction by Ion Channels in Lymphocytes

Zoltán Varga, Rezső Gáspár

10.1 Introduction

The best-known function of ion channels is the generation of action potentials in excitable cells; however, these pore-forming transmembrane proteins play crucial roles in non-excitable cells, such as lymphocytes, as well. Among other essential cellular processes, many ion channels are involved in signal transduction pathways. In lymphocytes, various ion channels affect these pathways mainly via the control of the membrane potential and calcium signaling.

This chapter deals with the role of ion channels in cellular responses to different stimuli, such as T cell activation following antigenic challenge, regulatory volume decrease in lymphocytes in hypoosmotic media and apoptosis induced by drugs, ligands, or ionizing radiation. The distribution of lymphocyte ion channels in the plasma membrane and its functional role are also discussed.

10.2 The Role of Ion Channels in T Cell Activation

For an efficient generalized immune response upon antigenic challenge, the clonal expansion of lymphocytes is necessary. T lymphocytes must first recognize foreign antigenic peptide fragments presented by antigen-presenting cells. The binding of the peptide-loaded major histocompatibility complex (MHC) to the T cell receptor complex (TCR/CD3) initiates the signal transduction pathway, which ultimately leads to the activation of T cells (Fig. 1). After a few hours of T cell receptor stimulation, lymphocytes are committed to proliferate even if the activating stimulus is removed (McCardy et al. 1988). The concentration of intracellular ions (e.g. Ca^{2+}, Na^{+}, pH; reviewed in Grinstein and Dixon 1989; Cahalan et al. 2001) and metabolites (glucose, amino acids) changes due to the alteration in membrane-transport processes during this period (Jacobs et al. 1989; Bental and Deutsch 1993). Then, the transcription of a variety of activation-associated genes occurs, including that of the autocrine growth factor interleukin-2, which drives cell-cycle progression (Shaw et al. 1988).

The stimulation of T cells via the T cell receptor complex activates protein tyrosine kinases, such as members of the Src, Zap-70/Syk, Tec and Csk families of non-receptor tyrosine kinases, and consequently phospholipase C-γ (PLCγ). Following activation, PLCγ cleaves phosphatidylinositol 4,5-biphosphate (PiP_2) into

Springer Series in Biophysics
S. Damjanovich (Ed.)
Biophysical Aspects of Transmembrane Signaling

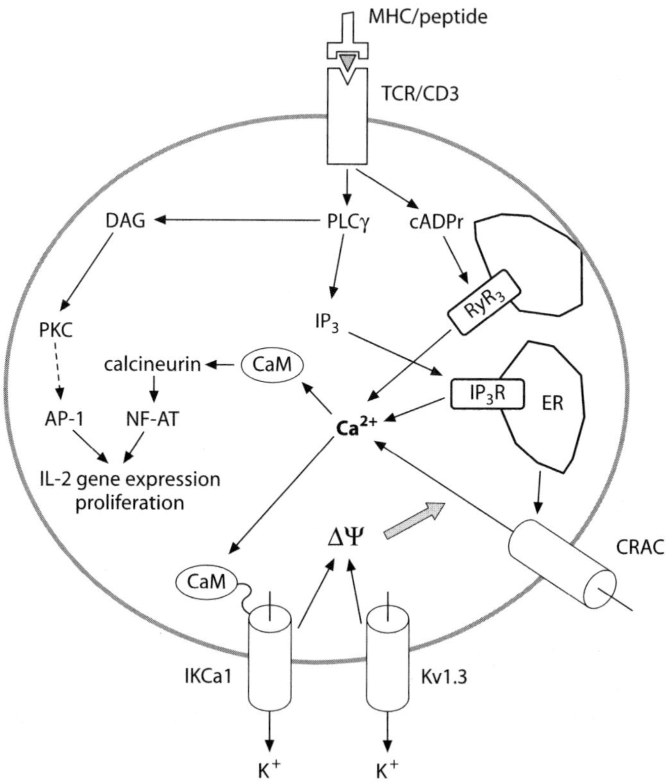

Fig. 10.1. Signal transduction steps during T cell activation. Stimulation of the T cell receptor/CD3 (*TCR/CD3*) complex by the peptide-loaded major histocompatibility complex of an antigen-presenting cell activates tyrosine kinases and phospholipase C-γ, which cleaves phosphatidylinositol 4,5-biphosphate into 1,2-diacylglycerol (*DAG*) and inositol 1,4,5-trisphosphate (*IP_3*). DAG activates the protein kinase C (*PKC*) pathway that ultimately leads to assembly of the Fos/Jun transcription factor complex (*AP-1*). IP_3 binds to its receptor in the endoplasmic reticulum (*ER*) membrane, causing Ca^{2+} release. TCR/CD3 stimulation also leads to generation of cyclic ADP-ribose (*cADPr*), which releases Ca^{2+} from internal stores as well. IP_3-dependent store-emptying activates Ca^{2+} release-activated Ca^{2+} (*CRAC*) channels that provide a sustained Ca^{2+} influx from the extracellular space. The Ca^{2+} binding protein calmodulin (*CaM*) activates phosphatase calcineurin, which then dephosphorylates the nuclear factor of activated T cells (*NF-AT*), enabling it to accumulate in the nucleus and together with AP-1 promote interleukin-2 gene expression. Sustained Ca^{2+} influx through CRAC channels requires sufficient driving force provided by the adequately negative membrane potential. This is generated by Ca^{2+} and voltage-activated K^+ channels. *RyR3* Type 3 ryanodine receptor; *IKCa1* intermediate conductance KCa ion channel in human lymphocytes

1,2-diacylglycerol (DAG) and inositol 1,4,5-trisphosphate (IP_3), and at this point two signaling pathways of lymphocyte activation diverge. In the protein kinase C (PKC) pathway, membrane-associated DAG activates PKC, which phosphorylates several intracellular substrates and triggers transcription via the assembly of the Fos/Jun transcription factor complex on AP1 elements in several genes (Northrop et al. 1993; Acuto and Cantrell 2000). In the first phase of the Ca^{2+}-dependent pathway, IP_3 binds to its receptor located in the membrane of the endoplasmic reticulum, which results in the release of Ca^{2+} stored in the lumen of the ER. This causes an increase in the free intracellular Ca^{2+} concentration ($[Ca^{2+}]_i$) from a resting level of ~100 nM to a peak concentration of ~500 nM.

Along with IP_3, cyclic ADP-ribose (cADPr) is also capable of inducing Ca^{2+} release from internal stores other than the endoplasmic reticulum in certain T lymphocyte cell lines (Bourguignon et al. 1995; Guse et al. 1995). These stores were found to be insensitive to thapsigargin and localized close to the plasma membrane. Sustained increased levels of cADPr are generated in Jurkat cells by the activation of the soluble ADP-ribosyl cyclase following the stimulation of the TCR/CD3 receptor complex (Guse et al. 1999). Thus, cADPr acts as a second messenger, which is generated in response to the stimulation of the T cell receptor and produces Ca^{2+} release from intracellular stores. The molecular target of cADPr seems to be the type-3 ryanodine receptor/calcium-release channel expressed in Jurkat T cells (Guse et al. 1999).

The relationship between the IP_3-dependent and cADPr-dependent signaling pathways was investigated by Guse and coworkers (1999) in Jurkat T cells. They showed that IP_3-mediated Ca^{2+} release is necessary for further Ca^{2+} signaling, but cADPr is essential for its sustained phase. Thus, the two systems seem to act in a temporal sequence rather than in an alternative manner.

For efficient signal transduction, the sustained elevation of cytosolic free Ca^{2+} concentration is required, which is realized by Ca^{2+} influx from the extracellular space. This Ca^{2+} influx relies on the operation of the ion channels in T cells, especially the Ca^{2+} release-activated Ca^{2+} (CRAC) channel, and two potassium channels that help control the membrane potential during the sustained Ca^{2+} signal, the voltage-gated Kv1.3 and the Ca^{2+}-activated IKCa1 potassium channels. The Ca^{2+}-calmodulin-dependent phosphatase calcineurin is activated by the calcium signal and it dephosphorylates the transcription factor NF-AT (nuclear factor of activated T cells), which then accumulates in the nucleus and binds to the promoter element of the interleukin-2 gene (Rao et al. 1997; Crabtree 1999). At this point the IP_3-dependent and PKC pathways of activation converge as NF-AT and AP-1 proteins act synergistically by forming complexes to regulate gene expression (Macian et al. 2001). In order to keep NF-AT in the nucleus in the transcriptionally active state, a sustained Ca^{2+} signal is required. Activation of the IL-2 gene and IL-2 expression are critical commitment points beyond which further T cell activation becomes antigen independent (Crabtree 1989).

The Ca^{2+} entry from the extracellular space, which underlies the sustained Ca^{2+} signal essential for the function of the immune system, is triggered by the depletion of intracellular Ca^{2+} stores, rather than by the direct action of the increased IP_3 or cADPr concentration generated during signal transduction. This was proven by demonstrating Ca^{2+} influx in response to the depletion of Ca^{2+}

stores without a specific IP_3 or cADPr signal induced by, for example, thapsigargin, a SERCA Ca^{2+} pump inhibitor or the use of intracellular Ca^{2+} chelators, such as BAPTA.

10.2.1 Ca^{2+} Release-Activated Ca^{2+} (CRAC) Channels

The ion channel that responds to store depletion and generates Ca^{2+} influx in T cells is the Ca^{2+} release-activated Ca^{2+} (CRAC) channel. Zweifach and Lewis (1993) provided the first electrophysiological evidence for the existence of these channels in Jurkat cells. They found that the thapsigargin-activated Ca^{2+} current was indistinguishable from the one activated by the cross-linking of the T cell receptor.

Likewise, the cross-linking of the T cell receptor–CD3 complex and intracellular store depletion by intracellular perfusion of inositol triphosphate or thapsigargin or extracellular perfusion of ionomycin elicited similar low-amplitude Ca^{2+}-selective currents in human T cells (Partiseti et al. 1994). This implies that the store depletion-activated CRAC channels control the main Ca^{2+} influx pathway upon mitogenic stimulation in human T cells.

CRAC channels are not easily identified and characterized due to their very low Ca^{2+} conductance and the enormous number of low conductance cation channels with similar properties. For a few years there was confusion regarding the conductance of the CRAC channel because a Mg^{2+}-inhibited cation channel (MIC) with somewhat similar properties but much greater conductance was mistaken for CRAC (Kerschbaum and Cahalan 1999). Later, the two channels were distinguished on the basis of several differing characteristics.

The molecular identity of CRAC channels is still a mystery, although many candidates have been suggested from the superfamily of transient receptor potential (TRP) cation channels. The ion channels encoded by this gene family are related to the *Drosophila trp* gene. They are expressed in a variety of tissues and cell types and perform a wide range of functions such as temperature sensor, taste receptor, mechano- and osmosensor; and they are also involved in Ca^{2+} transport mechanisms (Minke and Cook 2002). To date, none of these candidates have been indisputably confirmed as the real CRAC channel.

The search for the molecular identity of this channel has been hampered by the lack of potent and highly specific blockers. Capsaicin and SKF 96365, two compounds reported to have inhibitory effects on CRAC current, do not fulfill these requirements (Fischer et al. 2001; Kozak et al. 2002; Prakriya and Lewis 2002). However, a recently discovered bistrifluoromethyl pyrazole derivative, termed BTP2, has been shown to specifically inhibit CRAC channels in the nanomolar range, thereby becoming a very promising candidate for probing CRAC channel structure and function (Zitt et al. 2004).

Based on the results of several laboratories, the hallmarks of the CRAC current that distinguish it from other similar inward cation currents have been established (Parekh and Penner 1997; Hermosura et al. 2002; Kozak et al. 2002; Prakriya and Lewis 2002). It has high selectivity for Ca^{2+}, but also conducts some other divalent ions, e.g. Ba^{2+}; it is not blocked by internal Mg^{2+}; it has very low single channel conductance for Ca^{2+} (~21 fS), but in divalent-free solutions conducts Na^+ ions well (~0.2 pS); it displays strong inward rectification and strong inhibition by 20 μM SKF96365.

The Ca^{2+} current through CRAC channels is regulated by various feedback mechanisms. In different cell types the modulation of I_{CRAC} was reported by mechanisms such as direct Ca^{2+}-induced inactivation, inhibition by calmodulin or indirect effects via membrane potential changes generated by Ca^{2+}-sensitive ion channel activities (Zweifach and Lewis 1995; Vaca 1996). Additional mechanisms that were identified as modulators of I_{CRAC} include PKC-dependent phosphorylation and sphingomyelinase-mediated production of sphingosine (Parekh and Penner 1995; Mathes et al. 1998).

There seem to be some missing links in the Ca^{2+} entry phase of the signal transduction cascade, since besides the CRAC channels' molecular identity and structure, the mechanism by which they sense the depletion of internal Ca^{2+} stores is also ambiguous. Several distinctly different models have been suggested to explain how store depletion leads to the opening of CRAC channels. One of these models attributes the transmission of the signal to a diffusible molecule, a "calcium influx factor" (CIF). Several groups isolated such factors that could generate Ca^{2+} influx, but later most of these proved to lack one or more criteria that would be required of a "true" CIF (Bird and Putney 1993; Randriamampita and Tsien 1993; Su et al. 2001).

Another model proposes that following store depletion, functional CRAC channels are inserted into the membrane from vesicles. This hypothesis is supported by experiments in which single channels seemed to appear one by one during the whole-cell measurement of CRAC current (Fasolato et al. 1993; Kerschbaum and Cahalan 1999). Despite some supporting results (Yao et al. 1999), no decisive proof has been found for this proposition either.

Yet another theory hypothesizes that there is direct physical coupling between the IP_3 receptor (IP_3R) and the CRAC channel such that store depletion induces a conformational change of IP_3R, which in turn directly activates CRAC channels. This theory is as controversial as the other models. Thus, it is possible that several different mechanisms such as conformational coupling or signaling by a diffusible factor exist for initiating capacitative Ca^{2+} entry and the actual mechanism used may vary according to cell type or may even occur by the combination of different mechanisms (Putney et al. 2001).

10.2.2 Voltage-Gated and Ca^{2+}-Activated K^+ Channels

The opening probability of CRAC channels, unlike that of voltage-gated channels, does not depend on the membrane potential; it is linked to the emptying of the internal Ca^{2+} stores (Zweifach and Lewis 1993). When CRAC channels are opened by the appropriate stimulus, the magnitude of the Ca^{2+} current through these channels depends on the electrochemical driving force for Ca^{2+}. A more negative membrane potential results in larger Ca^{2+} current.

The resting membrane potential of unstimulated T lymphocytes is between –50 and –70 mV (reviewed in Grinstein and Dixon 1989; Lewis and Cahalan 1995). Since the membrane potential of T cells is mainly set by potassium channels, the modulation of Ca^{2+} influx through CRAC channels heavily relies on the operation of these channels.

The dominant voltage-gated K^+ channel in human T cells, the *n*-type potassium channel, was first characterized in 1984 (Decoursey et al. 1984; Matteson and Deutsch 1984). This channel is encoded by the Kv1.3 gene, and therefore it has been referred to as the Kv1.3 channel within the *Shaker* family of Kv channels.

The structure as well as the biophysical properties and mode of assembly of this channel have been well characterized using the whole-cell (Cahalan et al. 1985; Pahapill and Schlichter 1990; Panyi et al. 1995; Panyi and Deutsch 1996) and cell-attached patch modes of the patch-clamp technique (Pahapill and Schlichter 1992; Verheugen et al. 1995). As Kv1.3 is a voltage-gated channel, its open probability increases with increasing voltage, reaching saturation at –10 to 0 mV; the activation threshold is about –60 mV. The single channel conductance is 10–12 pS.

Kv1.3 channels can effectively control the membrane potential, because their activation threshold is close to the resting potential of the cells and the open probability increases very steeply with depolarizations exceeding the activation threshold. Furthermore, at depolarized membrane potentials, the channels enter an inactivated state, but at the resting potential this steady-state inactivation is not complete. Thus, through feedback loops a range of membrane potentials is determined at which these channels are active. It is also important that the lymphocyte membrane has extremely high electrical resistance (10–20 GΩ). As a result, the activity of a small number of ion channels generates detectable changes in the membrane potential (Maltsev 1990; Verheugen et al. 1995).

In summary, the biophysical properties of Kv1.3 channels and that of the lymphocyte membrane provide the basis for effectively clamping the membrane potential close to the activation threshold of the channels.

In addition to the ones gated by voltage (Kv), potassium channels exist in a wide variety of forms, such as ATP-gated, G-protein-regulated or mechanical stress-activated. Even two-pore potassium channels have been described (Hajdu et al. 2003a). Ca^{2+}-activated K^+ channels (K_{Ca}) are also present in the plasma membrane of various non-excitable cells (Weidema et al. 1993; Hajas et al. 2004) including human and rodent T cells. Human T lymphocytes express an intermediate conductance Ca^{2+}-activated K^+ channel with ~11 pS unitary conductance. This channel is encoded by the hKCa4 gene (Logsdon et al. 1997) and referred to as the hIKCa1 channel (or $K_{Ca}3.1$).

This channel is considered to be voltage-independent, since the opening of the channels is not affected significantly by the membrane potential (Grissmer et al. 1993). In contrast, the opening of IKCa1 channels is highly sensitive to the intracellular Ca^{2+} concentration. The activation threshold of the channels is between 200 and 300 nM $[Ca^{2+}]_i$, half-maximal activation occurs between 300 and 450 nM $[Ca^{2+}]_i$ and maximal activation of the conductance requires 1 μM $[Ca^{2+}]_i$ (Grissmer et al. 1993; Verheugen et al. 1995). These channels are silent in intact resting T lymphocytes because the resting $[Ca^{2+}]_i$ is below the activation threshold of the IKCa1 channels.

The Ca^{2+} sensor of the IKCa1 channel is the Ca^{2+}-binding protein calmodulin (CaM). CaM binds to the C-terminus of IKCa1 subunits and mediates the changes in $[Ca^{2+}]_i$ to the channel (Fanger et al. 1999). Each of the channel subunits binds a CaM molecule and the concerted action of the four molecules leads to channel opening.

10.2.3
Ion Channels, Membrane Potential and Lymphocyte Activation

The key role of K^+ channels during lymphocyte activation was demonstrated by the use of classical blockers of K^+ channels, such as quinine, 4-aminopyridine, and tetraethylammonium, which were shown to inhibit T cell activation. These blockers inhibited PHA-induced proliferation and interleukin-2 production of human peripheral blood T cells at concentrations comparable to those required to block the K^+ current (Chandy et al. 1984; Decoursey et al. 1984; Deutsch et al. 1986). Even some of the classical immunosuppressive agents, such as cyclosporin A, were found to affect the membrane potential and block K^+ currents in lymphocytes (Vereb et al. 1990; Panyi et al. 1996). However, these compounds were not specific for Kv1.3 or IKCa1 channels, and blocked these channels with low affinity. Peptide toxins isolated from scorpion venoms proved to be much more efficient and specific blockers, and, as such, facilitated the determination of the role of K^+ channels in lymphocyte physiology.

Charybdotoxin (ChTx), for example, blocks voltage-gated and Ca^{2+}-activated K^+ channels of lymphocytes (Sands et al. 1989; Grissmer et al. 1993), and inhibits lymphocyte proliferation and IL-2 synthesis at mRNA level at nanomolar concentrations (Price et al. 1989; Freedman et al. 1992). The discovery of peptide toxins, which were selective for Kv1.3 over IKCa1, allowed the separation of the contribution of these channels to the control of membrane potential. Selective block of Kv1.3 channels by Margatoxin (MgTx) and Noxiustoxin (NxTx) depolarized the membrane potential of resting T cells to a similar extent as ChTx, which blocked both IKCa1 and Kv1.3 channels (Leonard et al. 1992). This suggested that the membrane potential in resting T cells is determined by Kv1.3 but not by IKCa1 channels.

The critical link between K^+ channel activity, membrane potential and lymphocyte activation is the modulation of the Ca^{2+} influx by the membrane potential (Sarkadi et al. 1990); thus, interfering with K^+ channel activity should affect Ca^{2+} signaling during T cell activation. This was demonstrated by Lin et al. (1993) who showed that ChTx inhibits specifically T cell activation induced by signals that elicit a rise in $[Ca^{2+}]_i$, whereas alternative activation pathways (e.g. anti-CD28/phorbol ester-dependent proliferation) remained unaltered. They also found that the Kv1.3-specific blockers NxTx and MgTx were as effective in inhibiting the Ca^{2+} signal as ChTx, strengthening the hypothesis that Kv1.3 is the K^+ channel responsible for the maintenance of a proper Ca^{2+} signal. Several recent experiments with more potent and selective Kv1.3 inhibitors have confirmed these observations (Kalman et al. 1998).

Based on these results, the following model has been proposed for the involvement of K^+ channels in lymphocyte activation (reviewed in Lewis and Cahalan

1995; Cahalan and Chandy 1997; Cahalan et al. 2001; Panyi et al. 2004b). During the Ca^{2+} signal, Ca^{2+} current through CRAC channels causes depolarization (Fanger et al. 2001), which would decrease further Ca^{2+} entry due to the decreased electrochemical driving force for Ca^{2+} in the absence of counterbalancing cation efflux. This cation efflux is provided by Kv1.3 channels that are activated by depolarization. Their steep voltage dependence allows these channels to respond promptly to a small Ca^{2+}-induced depolarization; the outward K^+ current through Kv1.3 channels can clamp the membrane at a hyperpolarized potential and thus allow sustained Ca^{2+} influx through CRAC channels required for gene expression (Negulescu et al. 1994; Kerschbaum et al. 1997).

Although the initial rise in the Ca^{2+} concentration following TCR/CD3 stimulation should be sufficient for the activation of Ca^{2+}-activated K^+ channels, their very limited number in resting T cells makes the contribution of these channels minimal to membrane potential control. In contrast, preactivated human T cells upregulate IKCa1 channels, so their influence on membrane potential and, consequently, on Ca^{2+} signaling, increases dramatically.

Accordingly, anti-CD3 and PMA/ionomycin-induced proliferation of human peripheral blood lymphocytes that were preactivated by similar treatments was suppressed by nanomolar concentrations of the IKCa1 selective blocker TRAM-34. In contrast, much higher concentrations of TRAM-34 were required to inhibit the mitogen-induced proliferation of resting T cells (Ghanshani et al. 2000; Wulff et al. 2000). Conversely, the mitogen-induced proliferation of resting T cells was effectively suppressed by selective Kv1.3 inhibitors, whereas selective IKCa1 inhibitors were ineffective (Ghanshani et al. 2000). Moreover, chronically activated T cells, such as those involved in autoimmune responses, express very high numbers of Kv1.3 and low numbers of IKCa1 channels, and consequently they are effectively suppressed by low concentrations of Kv1.3-specific blockers (Beeton et al. 2001b).

Thus, high-affinity blockers of T cell K^+ channels can interfere with the signal transduction pathways of T cell activation in vitro. Several recent publications have reported that K^+ channel blockers could inhibit immune responses in vivo as well (Koo et al. 1997; Beeton et al. 2001a,b), thereby emphasizing the potential use of these blockers as immunosuppressive agents in autoimmune diseases. The latest discovery of T cell subset-specific expression K^+ channels holds even more promise for specific and well-targeted modulation of certain immune processes.

10.2.4 Distribution of Kv1.3 Channels in the Lymphocyte Membrane

The plasma membrane isolates cellular components from the environment and supports membrane proteins, but, in addition to these basic functions, it also plays an active role in signal transduction. Specialized membrane domains, lipid rafts, are involved in numerous signaling pathways. There is a growing body of evidence for the participation of such rafts in the T cell activation signaling pathway and the role of the association of potassium channels with these rafts.

The in vivo activation of T cells requires interaction with professional antigen-presenting cells. During this interaction, a specialized contact area, called the immunological synapse (IS), is formed between the cells, where functionally important proteins segregate into microdomains (Davis 2002). The redistribution of TCR/CD3, CD28 and CD2 membrane proteins to the center and that of the lymphocyte function-associated antigen 1 (LFA-1) to the periphery of the supramolecular activation clusters (SMACs) have been reported, along with the selective enrichment of cytosolic protein talin in the periphery, and Lck or PKCθ in the center of the IS (Monks et al. 1998; Grakoui et al. 1999; Bromley et al. 2001; Holdorf et al. 2002; van der Merwe 2002). Protein redistribution during IS formation might be guided by lipid raft reorganization and by physical coupling mediated by adaptor proteins or by cytoskeletal attachments.

Ion channels are embedded in the plasma membrane and therefore have intimate contact with the surrounding lipid environment. Consequently, changes in the composition or physicochemical properties of this lipid environment may modify the operation of the channels (Martens et al. 2000; Romanenko et al. 2002; Hajdu et al. 2003b; Tillman and Cascio 2003).

Several recent studies have demonstrated the uneven distribution of ion channels in the membrane of non-excitable cells (Martens et al. 2000; Szucs et al. 2004). Voltage-gated potassium channels in particular were shown to localize to specialized membrane microdomains (reviewed in Martens et al. 2004). Changes in lipid raft structure or redistribution of ion channels to different rafts are expected to modulate channel activity. This was demonstrated for Kv1.3 channels as well, by showing that biophysical parameters of the channel changed as the cholesterol content of the membrane was modified. Moreover, increasing the cholesterol content of the membrane resulted in whole-cell currents with biphasic activation kinetics, suggesting the existence of two channel populations in different lipid microenvironments (Hajdu et al. 2003b). Another study reported that Kv1.3 channels localize to small sphingolipid- and cholesterol-enriched membrane rafts and that the transformation of these rafts to large ceramide-enriched platforms by various means resulted in the clustering of Kv1.3 within these platforms and the inhibition of the channel's activity (Bock et al. 2003).

As detailed above, Kv1.3 channels have a pivotal role in the signal transduction leading to T cell activation. The localization of these channels to lipid rafts may facilitate the redistribution of channels to the proximity of signaling molecules, for example kinases, which in turn can modify channel activity to fine tune certain signaling steps.

This notion is supported by electron microscope images that showed non-random distribution of Kv1.3 channels in the plasma membrane of Jurkat T cells, indicating the preferential localization of these channels in certain areas of the membrane. Also, a high degree of overlap was found in the distribution of Kv1.3 channels and CD3 molecules in the membrane, and energy transfer measurements showed the molecular proximity of these two molecules (Panyi et al. 2003). Moreover, fluorescently labeled Kv1.3 channels showed a patchy but nearly even distribution in the membrane of solitary allogen-activated human cytotoxic T lymphocytes (CTLs), whereas channels clustered in the IS in CTLs engaged with target cells (Panyi et al. 2004a).

The proximity of Kv1.3 to several other molecules important in signaling was also demonstrated, indicating the involvement of Kv1.3 channels in the signal transduction machinery of T cells. Direct interaction of the human homologue of the *Drosophila* discs-large tumor suppressor protein (hDlg) and $p56^{lck}$, an Src-family tyrosine kinase with distinguished role in early T cell activation, was shown, in addition to the in vivo association of hDlg with Kv1.3 in T lymphocytes (Hanada et al. 1997). These results suggest that hDlg may function as an adaptor protein coupling $p56^{lck}$ to Kv1.3, facilitating efficient tyrosine phosphorylation of the channel.

Similarly, ZIP1 and ZIP2 adaptor proteins were shown to bind to both potassium channel $K_v\beta_2$ subunits, which are cytoplasmic auxiliary proteins modifying channel activity, and protein kinase C-ζ (PKC-ζ), acting as a physical link between the ion channel and the kinase (Gong et al. 1999; McCormack et al. 1999). Since Kv1.3 channel activity is known to be modulated by phosphorylation (Jonas and Kaczmarek 1996; Bowlby et al. 1997), the proximity of kinases involved in signaling allows for the precise regulation of Kv1.3 activity within the signaling complex.

The reported molecular proximity and possible functional coupling between Kv1.3 channels and β_1-integrin in the plasma membrane of T cells may be important in the regulation of T cell activation, since β_1-integrin is known to stabilize the IS (Levite et al. 2000).

Thus, Kv1.3, at least at certain stages of the signaling pathway, appears to be part of a large signaling complex containing TCR/CD3, CD4, β_1-integrin, intracellular adaptor proteins and kinases. The physical closeness of Kv1.3 to these signaling molecules seems to be essential for efficient and well-controlled signal transduction. However, further work is needed to characterize and understand the significance of Kv1.3 compartmentalization in the membrane of T cells and its functional consequences.

10.3 The Role of Ion Channels in Volume Regulation by Lymphocytes

The earliest observations on the volume regulation of lymphocytes described the response of human peripheral blood mononuclear cells (PBM) to the application of hypotonic media (Cheung et al. 1982). Electronic cell sizing, electron microscopy and cellular water determinations showed that hypotonic swelling was followed by a regulatory volume decrease (RVD) phase. The experimental findings suggested that RVD occurred largely by passive loss of cellular K^+, resulting from a selective increase in permeability to this ion. In addition, the Na^+-K^+ pump appeared to be activated upon cell swelling by a mechanism other than Na^+ entry into the cell, but this was not essential for RVD.

Patch-clamp recording soon led to the characterization of ion channels expressed in several types of cells of the immune system, including T lymphocytes, B lymphocytes and macrophages (Cahalan et al. 1987; Gallin and Sheehy 1988). Besides the role of these ion channels in lymphocyte activation, their role in the regulation of cell volume was implicated (Grinstein et al. 1984). T cells placed in hypotonic solutions initially swell as water enters, but then lose K^+, Cl^-

and water as the cell resumes its original volume. The ability of lymphocytes to regulate their volume in the face of osmotic gradients may be essential to their proper function in regions of the kidney, and may also reflect a homeostatic mechanism responsive during cell adhesion and growth.

At least two independent ionic transport pathways, one for K^+ and one for Cl^-, become activated during RVD (Sarkadi et al. 1984a,b). The discovery of osmotically activated Cl^- channels promoted a hypothesis for RVD, in which the activation of chloride channels is the triggering event. This results in depolarization and subsequent activation of K^+ channels and causes the loss of K^+, Cl^- and water, as the cell returns to its original volume following osmotic swelling (Cahalan and Lewis 1988).

The basic idea of this hypothesis has not been ruled out; a wide variety of lymphocyte ion channels have been characterized and the role of some in homeostatic volume regulation has been established (Deutsch and Lee 1989; Cahalan and Lewis 1990; Schlichter et al. 1990; Gardner et al. 1991; Lee et al. 1992; Deutsch and Chen 1993; Lewis et al. 1993). In one line of investigations the properties of the K^+ pathway underlying RVD in human blood lymphocytes were further studied by using an animal toxin, charybdotoxin (ChTX). At least two types of channels, one ChTX-sensitive and the other ChTX-insensitive, were observed in human T lymphocytes (Grinstein and Smith 1990; Deutsch et al. 1991). The observations also suggest that K^+ loss during RVD is mediated by Ca^{2+}-independent, ChTX-sensitive voltage-gated ion channels in human blood lymphocytes.

It is the Kv1.3 voltage-gated K^+ channel that plays a role in RVD of T lymphocytes in response to hypotonic shock. A mouse cytotoxic T lymphocyte cell line, CTLL-2, devoid of voltage-dependent K^+ channels, is not capable of volume regulation. Transient transfection of these cells with Kv1.3 reconstituted their ability to volume regulate (Deutsch and Chen 1993). As predicted from the experiments, the ability of RVD depends critically on the ability of the cells to respond to volume-induced changes in membrane potential.

Parallel to the identification of the role of K^+ channels in RVD, another line of investigations focused on the role of Cl^- channels in RVD of lymphocytes. As mentioned above, a special type of chloride channel has been described in T cells that could be activated by a transmembrane osmotic gradient (Cahalan and Lewis 1988; Lewis et al. 1993). The unitary conductance of this Cl^- channel is around 2 pS and each cell contains on the order of 10^4 such channels that can be activated, making these osmotically activated chloride channels the most abundant and widespread ion channels in lymphocytes. In addition, a large (maximum unitary conductance 365 pS), multiple conductance Cl^- channel was discovered in human T lymphocytes (Schlichter et al. 1990).

The existence of the mini-Cl^- channels together with the recognition of the importance of K^+ channels have been essential in formulating the hypothesis of RVD described above (Cahalan and Lewis 1988). Details of this hypothesis are as follows. The triggering event of RVD is the activation of the mini-Cl^- channels. RVD is associated with membrane depolarization as a result of the chloride conductance increase. The increase in Cl^- conductance induced by cell swelling is expected to depolarize the cells from the normal resting potential (normally fluctuating) of −50 to −70 mV. The depolarization itself would increase the driving force for K^+ efflux through open K^+ channels, but more importantly would increase activation of the *n*-type K^+ channels (Kv1.3), further increasing K^+ efflux. The membrane potential would thereby take up a value between the K^+ and Cl^- equilibrium potentials, driving both K^+ and Cl^- ions out of the cell through their respective channels. As the cell loses KCl, water would follow, and the cell would change size until the reduction in cell volume is sufficient to close the Cl^- channels. According to this hypothesis, osmotically activated Cl^- channels act as sensors for cell volume increase and their influence controls the activation of K^+ channels and RVD.

Although mini-Cl^- channels are favored in this model, the role of maxi-Cl^- channels cannot definitely be excluded. A small number of highly conductive anion channels might be switched on during RVD (Sarkadi et al. 1984b).

The model is also valid for RVD of lymphocytes, although a calcium-permeable cation channel activated by cell swelling may complement this normally Ca^{2+}-independent process for RVD in thymocytes (Cahalan and Chandy 1997; Cahalan et al. 2001). It has been shown that the outward-rectifying chloride conductance activated by osmotic swelling can also be regulated by an increase in extracellular K^+. This effect seems to be general since it has been observed in osteoblasts, as well as in human leukemic T lymphocytes (Steinert and Grissmer 1997). The volume regulation of human lymphocytes can also be affected by PKC due to the enhancement of Kv1.3 activity in T cells. The PKC-induced increase in Kv1.3 conductance may increase the ability of lymphocytes to volume regulate (Chung and Schlichter 1997). It is also an interesting observation that an increase neither in cytosolic Ca^{2+} nor in membrane area is necessary for activation of volume-sensitive anion channels in lymphocytes (Ross et al. 1994; Ross and Cahalan 1995); however, the participation of cytoskeletal elements in channel activation has been demonstrated. Most interestingly, the presence of intracellular ATP is necessary for maintenance and repeated activation of the swelling-activated anion current (Levitan et al. 1995).

The osmotic swelling of lymphocytes was shown to open outwardly rectifying chloride channels (ORCC) through the Src-like kinase $p56^{lck}$. Osmotic activation of ORCC could be blocked by tyrosine kinase inhibitors; furthermore, ORCC activation could be achieved not only by osmotic swelling, but also by intracellular purified $p56^{lck}$. In addition to osmoregulation in lymphocytes, these mechanisms may play an important role when cells actively change their volume, i.e. during proliferation and apoptosis (Lepple-Wienhues et al. 2000).

Cell volume regulation has further implications in connection with lymphocyte apoptosis, because the loss of cell volume is an early and fundamental feature of programmed cell death or apoptosis. One of the most interesting questions relating apoptotic cell shrinkage to volume regulation and ion channels is how the cell death pathway is coupled to the activation of cell shrinkage and whether cellular shrinkage utilizes cell volume regulatory paths commonly employed in osmotic responses.

The loss of cell volume is not a passive component of apoptosis because a number of experimental findings highlight the importance of this process as an early and necessary regulatory event in the signaling of the cell death cycle. Additionally, the loss of intracellular ions has been shown to play a primary role in cell shrinkage, caspase activation, and nuclease activity during apoptosis. Thus, an understanding of the role that ion channels and plasma membrane transporters play in cellular signaling during apoptosis may have important physiological implications for immune cells, especially lymphocyte function (Gomez-Angelats et al. 2000).

10.4 The Role of Ion Channels in the Apoptosis of Lymphocytes

Lymphocyte ion channels have been implicated not only in the activation, proliferation and volume regulation of lymphocytes but also in the control of apoptosis in these cells.

Lymphoid cells can undergo apoptosis in response to a wide variety of stimuli, including exposure to glucocorticoids (Wyllie 1980; Compton et al. 1987), T cell receptor stimulation (Smith et al. 1989), ionizing radiation (Klassen et al. 1993), Fas/CD95 receptor ligation (Nagata 1994; Matiba et al. 1997), chemotherapeutic drugs (Friesen et al. 1996), Ca^{2+} influx (Caron-Leslie and Cidlowski 1991) and extracellular ATP (Nagy et al. 2000).

An early event in cells undergoing apoptosis is the loss of cell volume or cell shrinkage. Primary thymocytes, for example, which represent one of the most extensively studied model systems for apoptosis, exhibit cell shrinkage following the administration of glucocorticoids (Wyllie 1980; Thomas and Bell 1981; Compton et al. 1987). Similarly, thymocytes irradiated with γ-rays underwent a quick loss of cell volume in the first stage of apoptotic cell shrinkage, which was followed by a more gradual shrinking phase (Klassen et al. 1993). Both stages were attributed to a loss of both water and ions. The loss of intracellular ions, particularly potassium, was shown to play a primary role in cell shrinkage, caspase activation and nuclease activity during apoptosis, and the prevention of K^+ efflux inhibited the execution of the apoptotic program (Bortner et al. 1997; Hughes et al. 1997). The significant K^+ efflux in the cell death process was found to be independent of the apoptosis-inducing stimulus (Dallaporta et al. 1998).

Several studies reported a connection between apoptotic stimuli and Kv1.3 channel activity. However, some of these results are contradicting and a clear role of Kv1.3 channels in the cell death program is yet to emerge. A suppression of the activity of voltage-gated Kv1.3 channels was described during Fas-induced apoptosis in Jurkat T cells, which was mediated by the Src-like tyrosine kinase $p56^{lck}$ (Szabo et al. 1996). Similarly, treatment with ceramide, a lipid metabolite synthesized following Fas ligation, also inhibited Kv1.3 channels in a $p56^{lck}$-dependent manner (Gulbins et al. 1997). In contrast, a Fas-associated death domain protein and caspase 8-dependent increase in Kv1.3 channel activity following Fas ligation was also reported in Jurkat cells (Storey et al. 2003). The opposing results may arise from the use of different solutions, the tenfold difference in the Fas ligand concentrations used or the fact that one group preincubated the cells with the ligand and the other added it only after achieving whole-cell configuration. Certainly, the enhanced activity of Kv1.3 channels and the concomitant K^+ efflux would be a plausible explanation for the loss of K^+ during apoptosis observed by several other groups. However, the presence of the potent Kv1.3 blocker Shk-Dap[22] did not prevent cell shrinkage (Storey et al. 2003), which implies the involvement of other channels in potassium efflux, possibly that of two-pore domain potassium channels (Trimarchi et al. 2002). Nevertheless, the compatibility of Kv1.3 inhibition by Fas ligation with K^+ loss and cell shrinkage cannot be ruled out if other possible pathways are considered.

The requirement of Kv1.3 activity was also demonstrated in actinomycin D-induced apoptosis in CTLL-2 T lymphocytes (Bock et al. 2002). These cells, which genetically lack Kv1.3 channels, proved apoptosis-resistant upon the administration of the cytostatic drug. Expression of Kv1.3 channels in these cells following transfection rendered them actinomycin D-sensitive. Furthermore, the same study observed an enhancement of Kv1.3 activity in Jurkat T cells after 16 h of incubation with actinomycin D. Thus, the role of Kv1.3 channels in the apoptotic program may vary by cell type or according to the initiating stimulus, or may even consist of various phases in which channel activity is modulated differentially.

Other ions, as well as different osmolytes, were also shown to play an important role in Fas-induced apoptosis in Jurkat T cells, similarly to Kv1.3 channels, the stimulation of the Fas receptor in Jurkat cells or the administration of ceramide-induced activation of the outwardly rectifying chloride channel via a mechanism that involves Src-like kinase-dependent phosphorylation (Szabo et al. 1998). Pharmacological inhibition of this chloride conductance resulted in the inhibition of apoptosis.

Earlier studies provided evidence indicating a role for Ca^{2+} in the apoptotic pathway (Story et al. 1992; Oshimi and Miyazaki 1995). More recently, the block of CRAC channels following CD95 stimulation has been reported. The block occurred via sphingomyelinase and resulted in the prevention of Ca^{2+} influx, NF-AT activation and IL-2 synthesis.

Thus, stimulation of lymphocytes through CD95 exerts multiple effects on ion channels. The CD95-associated apoptotic pathway seems to activate the outwardly rectifying chloride channels and inhibit CRAC channels, and its effect on Kv1.3 is controversial and may be biphasic.

As mentioned earlier, following treatment with ceramide or the stimulation of acid sphyngomyelinase, Kv1.3 channels localized to ceramide-rich membrane platforms, which inhibited their activity (Bock et al. 2003). Similarly, ceramide was shown to be essential for CD95 clustering which appears to be a necessary step in apoptotic signaling (Grassme et al. 2001). In addition, the coupling of $p56^{lck}$ tyrosine kinase to Kv1.3 by hDlg was also demonstrated. These results suggest that the fusion of membrane rafts and the concomitant redistribution of Kv1.3, $p56^{lck}$ and CD95 molecules to large ceramide-rich platforms create a signaling complex in which the proximity of these molecules allows for efficient phosphorylation and modulation of Kv1.3 in the apoptotic pathway, as is the case during the antigenic activation of T cells.

In the immune system, apoptosis is involved in intrathymic elimination of self-reactive thymocytes. The role of a nonselective cation channel P_{2X1}, activated by ATP, was described in the dexamethasone-induced apoptosis of thymocytes. These channels are upregulated during apoptosis and their antagonists inhibit cell death. However, P_{2X1} channels do not play a role in the apoptosis of peripheral

T cells, providing a unique mechanism that contributes to thymocyte cell death (Chvatchko et al. 1996). Extracellular ATP was shown to affect the membrane potential and induce apoptosis in murine thymocytes in a dose- and cytosolic pH-dependent manner. Ca^{2+}-dependent hyperpolarization and cytosolic acidification triggered by low extracellular ATP doses are essential early signals in the apoptosis of murine thymocytes, and are probably mediated by P_{2X1}-type ATP-gated ion channels (Matkó et al. 1993; Nagy et al. 2000).

Besides plasma membrane ion channels, the involvement of intracellular ion channels has been implicated in lymphocyte apoptosis. T lymphocytes that were made deficient in type-1 IP_3 receptors could not produce IP_3-induced Ca^{2+} release, exhibited defective signaling via the TCR (Jayaraman et al. 1995) and became resistant to apoptosis induced by stimuli such as dexamethasone, TCR stimulation, ionizing radiation and Fas (Jayaraman and Marks 1997). Conversely, an increase in the number of functional IP_3 receptors resulted in enhanced cell death (Boehning et al. 2004). Thus, IP_3 receptors are required for the completion of the apoptotic program; in turn, type-1 IP_3 receptors are degraded by caspases during apoptosis of Jurkat T cells (Hirota et al. 1999).

Mitochondrial channels are also known to have a central role in apoptosis. The involvement of mitochondrial ATP and Ca^{2+}-activated potassium channels was shown in various cell types (Inoue et al. 1991; Holmuhamedov et al. 1998; Siemen et al. 1999; Xu et al. 2002) and that of a putative chloride intracellular channel (CLIC) was also suggested (Ashley 2003). Similar mitochondrial channels may also exist in lymphocytes and function in the cell death pathway.

References

Acuto O, Cantrell D (2000) T cell activation and the cytoskeleton. Annu Rev Immunol 18:165–184

Ashley RH (2003) Challenging accepted ion channel biology: p64 and the CLIC family of putative intracellular anion channel proteins (Review). Mol Membr Biol 20:1–11

Beeton C, Barbaria J, Giraud P, Devaux J, Benoliel AM, Gola M, Sabatier JM, Bernard D, Crest M, Beraud E (2001a) Selective blocking of voltage-gated K^+ channels improves experimental autoimmune encephalomyelitis and inhibits T cell activation. J Immunol 166:936–944

Beeton C, Wulff H, Barbaria J, Clot-Faybesse O, Pennington M, Bernard D, Cahalan MD, Chandy KG, Beraud E (2001b) Selective blockade of T lymphocyte $K^{(+)}$ channels ameliorates experimental autoimmune encephalomyelitis, a model for multiple sclerosis. Proc Natl Acad Sci USA 98:13942–13947

Bental M, Deutsch C (1993) Metabolic changes in activated T cells: an NMR study of human peripheral blood lymphocytes. Magn Reson Med 29:317

Bird GS, Putney JW Jr (1993) Inhibition of thapsigargin-induced calcium entry by microinjected guanine nucleotide analogues. Evidence for the involvement of a small G-protein in capacitative calcium entry. J Biol Chem 268:21486–21488

Bock J, Szabo I, Jekle A, Gulbins E (2002) Actinomycin D-induced apoptosis involves the potassium channel Kv1.3. Biochem Biophys Res Commun 295:526–531

Bock J, Szabo I, Gamper N, Adams C, Gulbins E (2003) Ceramide inhibits the potassium channel Kv1.3 by the formation of membrane platforms. Biochem Biophys Res Commun 305:890–897

Boehning D, Patterson RL, Snyder SH (2004) Apoptosis and calcium: new roles for cytochrome *c* and inositol 1,4,5-trisphosphate. Cell Cycle 3:252–254

Bortner CD, Hughes FM Jr, Cidlowski JA (1997) A primary role for K^+ and Na^+ efflux in the activation of apoptosis. J Biol Chem 272:32436–32442

Bourguignon LY, Chu A, Jin H, Brandt NR (1995) Ryanodine receptor–ankyrin interaction regulates internal Ca^{2+} release in mouse T-lymphoma cells. J Biol Chem 270:17917–17922

Bowlby MR, Fadool DA, Holmes TC, Levitan IB (1997) Modulation of the Kv1.3 potassium channel by receptor tyrosine kinases. J Gen Physiol 110:601–610

Bromley SK, Burack WR, Johnson KG, Somersalo K, Sims TN, Sumen C, Davis MM, Shaw AS, Allen PM, Dustin ML (2001) The immunological synapse. Annu Rev Immunol 19:375–396

Cahalan MD, Chandy KG (1997) Ion channels in the immune system as targets for immunosuppression. Curr Opin Biotechnol 8:749–756

Cahalan MD, Lewis RS (1988) Role of potassium and chloride channels in volume regulation by T lymphocytes. In: Gunn RB, Parker JC (eds) Cell physiology of blood. Rockefeller University Press, New York, pp 281–302

Cahalan MD, Lewis RS (1990) Functional roles of ion channels in lymphocytes. Semin Immunol 2:107–117

Cahalan MD, Chandy KG, Decoursey TE, Gupta S (1985) A voltage-gated potassium channel in human T lymphocytes. J Physiol (Lond) 358:197–237

Cahalan MD, Chandy KG, Decoursey TE, Gupta S, Lewis RS, Sutro JB (1987) Ion channels in T lymphocytes. Adv Exp Med Biol 213:85–101

Cahalan MD, Wulff H, Chandy KG (2001) Molecular properties and physiological roles of ion channels in the immune system. J Clin Immunol 21:235–252

Caron-Leslie LA, Cidlowski JA (1991) Similar actions of glucocorticoids and calcium on the regulation of apoptosis in S49 cells. Mol Endocrinol 5:1169–1179

Chandy KG, Decoursey TE, Cahalan MD, McLaughlin C, Gupta S (1984) Voltage-gated potassium channels are required for human T lymphocyte activation. J Exp Med 160:369–385

Cheung RK, Grinstein S, Dosch HM, Gelfand EW (1982) Volume regulation by human lymphocytes: characterization of the ionic basis for regulatory volume decrease. J Cell Physiol 112: 189–196

Chung I, Schlichter LC (1997) Native Kv1.3 channels are upregulated by protein kinase C. J Membr Biol 156:73–85

Chvatchko Y, Valera S, Aubry JP, Renno T, Buell G, Bonnefoy JY (1996) The involvement of an ATP-gated ion channel, P(2X1), in thymocyte apoptosis. Immunity 5:275–283

Compton MM, Caron LA, Cidlowski JA (1987) Glucocorticoid action on the immune system. J Steroid Biochem 27:201–208

Crabtree GR (1989) Contingent genetic regulatory events in T lymphocyte activation. Science 243:355–361

Crabtree GR (1999) Generic signals and specific outcomes: signaling through Ca^{2+}, calcineurin, and NF-AT. Cell 96:611–614

Dallaporta B, Hirsch T, Susin SA, Zamzami N, Larochette N, Brenner C, Marzo I, Kroemer G (1998) Potassium leakage during the apoptotic degradation phase. J Immunol 160:5605–5615

Davis DM (2002) Assembly of the immunological synapse for T cells and NK cells. Trends Immunol 23:356–363

Decoursey TE, Chandy KG, Gupta S, Cahalan MD (1984) Voltage-gated K^+ channels in human T lymphocytes: a role in mitogenesis? Nature 307:465–468

Deutsch C, Chen L-Q (1993) Heterologous expression of specific K^+ channels in T lymphocytes: functional consequences for volume regulation. Proc Natl Acad Sci USA 90:10036–10040

Deutsch C, Lee SC (1989) Modulation of K^+ currents in human lymphocytes by pH. J Physiol (Lond) 413:399–413

Deutsch C, Krause D, Lee SC (1986) Voltage-gated potassium conductance in human T lymphocytes stimulated with phorbol ester. J Physiol (Lond) 372:405–423

Deutsch C, Price M, Lee S, King VF, Garcia ML (1991) Characterization of high affinity binding sites for charybdotoxin in human T lymphocytes. Evidence for association with the voltage-gated K^+ channel. J Biol Chem 266:3668–3674

Fanger CM, Ghanshani S, Logsdon NJ, Rauer H, Kalman K, Zhou J, Beckingham K, Chandy KG, Cahalan MD, Aiyar J (1999) Calmodulin mediates calcium-dependent activation of the intermediate conductance KCa channel, IKCa1. J Biol Chem 274:5746–5754

Fanger CM, Rauer H, Neben AL, Miller MJ, Wulff H, Rosa JC, Ganellin CR, Chandy KG, Cahalan MD (2001) Calcium-activated potassium channels sustain calcium signaling in T lymphocytes. Selective blockers and manipulated channel expression levels. J Biol Chem 276:12249–12256

Fasolato C, Hoth M, Penner R (1993) A GTP-dependent step in the activation mechanism of capacitative calcium influx. J Biol Chem 268:20737–20740

Fischer BS, Qin D, Kim K, McDonald TV (2001) Capsaicin inhibits Jurkat T-cell activation by blocking calcium entry current I (CRAC). J Pharmacol Exp Ther 299:238–246

Freedman BD, Price M, Deutsch C (1992) Evidence for voltage modulation of IL2 production by human blood lymphocytes. J Immunol 149:3784–3794

Friesen C, Herr I, Krammer PH, Debatin KM (1996) Involvement of the CD95 (APO-1/FAS) receptor/ligand system in drug-induced apoptosis in leukemia cells. Nat Med 2:574–577

Gallin EK, Sheehy PA (1988) Leukocyte ion channels and their functional implications. In: Gallin JI, Goldstein IM, Snyderman R (eds) Inflammation: basic principles and clinical correlates. Raven Press, New York, pp 651–664

Gardner P, McDonald T, Nishimoto I, Wagner J, Schumann M, Chen J, Schulman H (1991) Regulation of lymphocyte chloride channels. Adv Exp Med Biol 290:319–326

Ghanshani S, Wulff H, Miller MJ, Rohm H, Neben A, Gutman GA, Cahalan MD, Chandy KG (2000) Up-regulation of the IKCa1 potassium channel during T-cell activation. Molecular mechanism and functional consequences. J Biol Chem 275:37137–37149

Gomez-Angelats M, Bortner CD, Cidlowski JA (2000) Cell volume regulation in immune cell apoptosis. Cell Tissue Res 301:33–42

Gong J, Xu J, Bezanilla M, van Huizen R, Derin R, Li M (1999) Differential stimulation of PKC phosphorylation of potassium channels by ZIP1 and ZIP2. Science 285:1565–1569

Grakoui A, Bromley SK, Sumen C, Davis MM, Shaw AS, Allen PM, Dustin ML (1999) The immunological synapse: a molecular machine controlling T cell activation. Science 285:221–227

Grassme H, Jekle A, Riehle A, Schwarz H, Berger J, Sandhoff K, Kolesnick R, Gulbins E (2001) CD95 signaling via ceramide-rich membrane rafts. J Biol Chem 276:20589–20596

Grinstein S, Dixon SJ (1989) Ion transport, membrane potential, and cytoplasmic pH in lymphocytes: changes during activation. Physiol Rev 69:417–481

Grinstein S, Smith JD (1990) Calcium-independent cell volume regulation in human T lymphocytes. Inhibition by charybdotoxin. J Gen Physiol 95:97–120

Grinstein S, Rothstein A, Sarkadi B, Gelfand EW (1984) Responses of lymphocytes to anisotonic media: volume-regulating behavior. Am J Physiol 246:C204–C215

Grissmer S, Nguyen AN, Cahalan MD (1993) Calcium-activated potassium channels in resting and activated human T lymphocytes. Expression levels, calcium dependence, ion selectivity, and pharmacology. J Gen Physiol 102:601–630

Gulbins E, Szabo I, Baltzer K, Lang F (1997) Ceramide-induced inhibition of T lymphocyte voltage-gated potassium channel is mediated by tyrosine kinases. Proc Natl Acad Sci USA 94:7661–7666

Guse AH, da Silva CP, Emmrich F, Ashamu GA, Potter BV, Mayr GW (1995) Characterization of cyclic adenosine diphosphate-ribose-induced Ca^{2+} release in T lymphocyte cell lines. J Immunol 155:3353–3359

Guse AH, da Silva CP, Berg I, Skapenko AL, Weber K, Heyer P, Hohenegger M, Ashamu GA, Schulze-Koops H, Potter BV, Mayr GW (1999) Regulation of calcium signalling in T lymphocytes by the second messenger cyclic ADP-ribose. Nature 398:70–73

Hajas G, Zsiros E, Laszlo T, Hajdu P, Somodi S, Rethi B, Gogolak P, Ludanyi K, Panyi G, Rajnavolgyi E (2004) New phenotypic, functional and electrophysiological characteristics of KG-1 cells. Immunol Lett 92:97–106

Hajdu P, Ulens C, Panyi G, Tytgat J (2003a) Drug- and mutagenesis-induced changes in the selectivity filter of a cardiac two-pore background K^+ channel. Cardiovasc Res 58:46–54

Hajdu P, Varga Z, Pieri C, Panyi G, Gaspar R Jr (2003b) Cholesterol modifies the gating of Kv1.3 in human T lymphocytes. Pflugers Arch 445:674–682

Hanada T, Lin L, Chandy KG, Oh SS, Chishti AH (1997) Human homologue of the *Drosophila* discs large tumor suppressor binds to p56lck tyrosine kinase and Shaker type Kv1.3 potassium channel in T lymphocytes. J Biol Chem 272:26899–26904

Hermosura MC, Monteilh-Zoller MK, Scharenberg AM, Penner R, Fleig A (2002) Dissociation of the store-operated calcium current I (CRAC) and the Mg-nucleotide-regulated metal ion current MagNuM. J Physiol 539:445–458

Hirota J, Furuichi T, Mikoshiba K (1999) Inositol 1,4,5-trisphosphate receptor type 1 is a substrate for caspase-3 and is cleaved during apoptosis in a caspase-3-dependent manner. J Biol Chem 274:34433–34437

Holdorf AD, Lee KH, Burack WR, Allen PM, Shaw AS (2002) Regulation of Lck activity by CD4 and CD28 in the immunological synapse. Nat Immunol 3:259–264

Holmuhamedov EL, Jovanovic S, Dzeja PP, Jovanovic A, Terzic A (1998) Mitochondrial ATP-sensitive K^+ channels modulate cardiac mitochondrial function. Am J Physiol 275:H1567–H1576

Hughes FM Jr, Bortner CD, Purdy GD, Cidlowski JA (1997) Intracellular K^+ suppresses the activation of apoptosis in lymphocytes. J Biol Chem 272:30567–30576

Inoue I, Nagase H, Kishi K, Higuti T (1991) ATP-sensitive K^+ channel in the mitochondrial inner membrane. Nature 352:244–247

Jacobs DB, Lee T-P, Jung CY, Mookerjee BK (1989) Mechanism of mitogen-induced stimulation of glucose transport in human peripheral blood mononuclear cells. J Clin Invest 83:437

Jayaraman T, Marks AR (1997) T cells deficient in inositol 1,4,5-trisphosphate receptor are resistant to apoptosis. Mol Cell Biol 17:3005–3012

Jayaraman T, Ondriasova E, Ondrias K, Harnick DJ, Marks AR (1995) The inositol 1,4,5-trisphosphate receptor is essential for T-cell receptor signaling. Proc Natl Acad Sci USA 92:6007–6011

Jonas EA, Kaczmarek LK (1996) Regulation of potassium channels by protein kinases. Curr Opin Neurobiol 6:318–323

Kalman K, Pennington MW, Lanigan MD, Nguyen A, Rauer H, Mahnir V, Paschetto K, Kem WR, Grissmer S, Gutman GA, Christian EP, Cahalan MD, Norton RS, Chandy KG (1998) ShK-Dap22, a potent Kv1.3-specific immunosuppressive polypeptide. J Biol Chem 273:32697–32707

Kerschbaum HH, Cahalan MD (1999) Single-channel recording of a store-operated Ca^{2+} channel in Jurkat T lymphocytes. Science 283:836–839

Kerschbaum HH, Negulescu PA, Cahalan MD (1997) Ion channels, Ca^{2+} signaling, and reporter gene expression in antigen-specific mouse T cells. J Immunol 159:1628–1638

Klassen NV, Walker PR, Ross CK, Cygler J, Lach B (1993) Two-stage cell shrinkage and the OER for radiation-induced apoptosis of rat thymocytes. Int J Radiat Biol 64:571–581

Koo GC, Blake JT, Talento A, Nguyen M, Lin S, Sirotina A, Shah K, Mulvany K, Hora D, Cunningham P, Wunderler DL, McManus OB, Slaughter R, Bugianesi R, Felix J, Garcia ML, Williamson J, Kaczorowski GJ, Sigal NH, Springer MS, Feeny W (1997) Blockade of the voltage-gated potassium channel Kv1.3 inhibits immune responses in vivo. J Immunol 158:5120–5128

Kozak JA, Kerschbaum HH, Cahalan MD (2002) Distinct properties of CRAC and MIC channels in RBL cells. J Gen Physiol 120:221–235

Lee SC, Levy DI, Deutsch C (1992) Diverse K^+ channels in primary human T lymphocytes. J Gen Physiol 99:771–793

Leonard RJ, Garcia ML, Slaughter RS, Reuben JP (1992) Selective blockers of voltage-gated K^+ channels depolarize human T lymphocytes: mechanism of the antiproliferative effect of charybdotoxin. Proc Natl Acad Sci USA 89:10094–10098

Lepple-Wienhues A, Szabo I, Wieland U, Heil L, Gulbins E, Lang F (2000) Tyrosine kinases open lymphocyte chloride channels. Cell Physiol Biochem 10:307–312

Levitan I, Almonte C, Mollard P, Garber SS (1995) Modulation of a volume-regulated chloride current by F-actin. J Membr Biol 147:283–294

Levite M, Cahalon L, Peretz A, Hershkoviz R, Sobko A, Ariel A, Desai R, Attali B, Lider O (2000) Extracellular $K^{(+)}$ and opening of voltage-gated potassium channels activate T cell integrin function: physical and functional association between Kv1.3 channels and beta1 integrins. J Exp Med 191:1167–1176

Lewis RS, Cahalan MD (1995) Potassium and calcium channels in lymphocytes. Annu Rev Immunol 13:623–653

Lewis RS, Ross PE, Cahalan MD (1993) Chloride channels activated by osmotic stress. J Gen Physiol 101:801–826

Lin CS, Boltz RC, Blake JT, Nguyen M, Talento A, Fischer PA, Springer MS, Sigal NH, Slaughter RS, Garcia ML, Kaczorowski GJ, Koo GC (1993) Voltage-gated potassium channels regulate calcium-dependent pathways involved in human T lymphocyte activation. J Exp Med 177: 637–645

Logsdon NJ, Kang J, Togo JA, Christian EP, Aiyar J (1997) A novel gene, hKCa4, encodes the calcium-activated potassium channel in human T lymphocytes. J Biol Chem 272:32723–32726

Macian F, Lopez-Rodriguez C, Rao A (2001) Partners in transcription: NFAT and AP-1. Oncogene 20:2476–2489

Maltsev VA (1990) Oscillating and triggering properties of T cell membrane potential. Immunol Lett 26:277–282

Martens JR, Navarro-Polanco R, Coppock EA, Nishiyama A, Parshley L, Grobaski TD, Tamkun MM (2000) Differential targeting of Shaker-like potassium channels to lipid rafts. J Biol Chem 275:7443–7446

Martens JR, O'Connell K, Tamkun M (2004) Targeting of ion channels to membrane microdomains: localization of KV channels to lipid rafts. Trends Pharmacol Sci 25:16–21

Mathes C, Fleig A, Penner R (1998) Calcium release-activated calcium current (ICRAC) is a direct target for sphingosine. J Biol Chem 273:25020–25030

Matiba B, Mariani SM, Krammer PH (1997) The CD95 system and the death of a lymphocyte. Semin Immunol 9:59–68

Matkó J, Nagy P, Panyi G, Vereb G, Bene L, Mátyus L, Damjanovich S (1993) Biphasic effect of extracellular ATP on the membrane potential of mouse thymocytes. Biochem Biophys Res Commun 191:378–384

Matteson DR, Deutsch C (1984) K^+ channels in T lymphocytes: a patch clamp study using monoclonal antibody adhesion. Nature 307:468–471

McCardy CW, Ely CM, Westin E, Carchman RA (1988) Coordination and reversibility of signs for proliferative activation and IL2 mRNA production in human T lymphocytes by phorbol ester and calcium ionophore. J Biol Chem 263:18537–18544

McCormack T, McCormack K, Nadal MS, Vieira E, Ozaita A, Rudy B (1999) The effects of Shaker beta-subunits on the human lymphocyte K^+ channel Kv1.3. J Biol Chem 274:20123–20126

Minke B, Cook B (2002) TRP channel proteins and signal transduction. Physiol Rev 82:429–472

Monks CR, Freiberg BA, Kupfer H, Sciaky N, Kupfer A (1998) Three-dimensional segregation of supramolecular activation clusters in T cells. Nature 395:82–86

Nagata S (1994) Apoptosis regulated by a death factor and its receptor: Fas ligand and Fas. Philos Trans R Soc Lond B Biol Sci 345:281–287

Nagy PV, Feher T, Morga S, Matko J (2000) Apoptosis of murine thymocytes induced by extracellular ATP is dose- and cytosolic pH-dependent. Immunol Lett 72:23–30

Negulescu PA, Shastri N, Cahalan MD (1994) Intracellular calcium dependence of gene expression in single T lymphocytes. Proc Natl Acad Sci USA 91:2873–2877

Northrop JP, Ullman KS, Crabtree GR (1993) Characterization of the nuclear and cytoplasmic components of the lymphoid-specific nuclear factor of activated T cells (NF-AT) complex. J Biol Chem 268:2917–2923

Oshimi Y, Miyazaki S (1995) Fas antigen-mediated DNA fragmentation and apoptotic morphologic changes are regulated by elevated cytosolic Ca^{2+} level. J Immunol 154:599–609

Pahapill PA, Schlichter LC (1990) Modulation of potassium channels in human T lymphocytes: effects of temperature. J Physiol (Lond) 422:103–126

Pahapill PA, Schlichter LC (1992) Modulation of potassium channels in intact human T lymphocytes. J Physiol (Lond) 445:407–430

Panyi G, Deutsch C (1996) Assembly and suppression of endogenous Kv1.3 channels in human T cells. J Gen Physiol 107:409–420

Panyi G, Sheng Z-F, Tu L-W, Deutsch C (1995) C-type inactivation of a voltage-gated K^+ channel occurs by a cooperative mechanism. Biophys J 69:896–904

Panyi G, Gáspár R, Krasznai Z, ter Horst JJ, Ameloot M, Aszalós A, Steels P, Damjanovich S (1996) Immunosuppressors inhibit voltage-gated potassium channels in human peripheral blood lymphocytes. Biochem Biophys Res Commun 221:254–258

Panyi G, Bagdany M, Bodnar A, Vamosi G, Szentesi G, Jenei A, Matyus L, Varga S, Waldmann TA, Gaspar R, Damjanovich S (2003) Colocalization and nonrandom distribution of Kv1.3 potassium channels and CD3 molecules in the plasma membrane of human T lymphocytes. Proc Natl Acad Sci USA 100:2592–2597

Panyi G, Vamosi G, Bacso Z, Bagdany M, Bodnar A, Varga Z, Gaspar R, Matyus L, Damjanovich S (2004a) Kv1.3 potassium channels are localized in the immunological synapse formed between cytotoxic and target cells. Proc Natl Acad Sci USA 101:1285–1290

Panyi G, Varga Z, Gaspar R (2004b) Ion channels and lymphocyte activation. Immunol Lett 92:55–66

Parekh AB, Penner R (1995) Depletion-activated calcium current is inhibited by protein kinase in RBL-2H3 cells. Proc Natl Acad Sci USA 92:7907–7911

Parekh AB, Penner R (1997) Store depletion and calcium influx. Physiol Rev 77:901–930

Partiseti M, Le Deist F, Hivroz C, Fischer A, Korn H, Choquet D (1994) The calcium current activated by T cell receptor and store depletion in human lymphocytes is absent in a primary immunodeficiency. J Biol Chem 269:32327–32335

Prakriya M, Lewis RS (2002) Separation and characterization of currents through store-operated CRAC channels and Mg^{2+}-inhibited cation (MIC) channels. J Gen Physiol 119:487–507

Price M, Lee SC, Deutsch C (1989) Charybdotoxin inhibits proliferation and interleukin 2 production in human peripheral blood lymphocytes. Proc Natl Acad Sci USA 86:10171–10175

Putney JW Jr, Broad LM, Braun FJ, Lievremont JP, Bird GS (2001) Mechanisms of capacitative calcium entry. J Cell Sci 114:2223–2229

Randriamampita C, Tsien RY (1993) Emptying of intracellular Ca^{2+} stores releases a novel small messenger that stimulates Ca^{2+} influx. Nature 364:809–814

Rao A, Luo C, Hogan PG (1997) Transcription factors of the NFAT family: regulation and function. Annu Rev Immunol 15:707–747

Romanenko VG, Rothblat GH, Levitan I (2002) Modulation of endothelial inward-rectifier K^+ current by optical isomers of cholesterol. Biophys J 83:3211–3222

Ross PE, Cahalan MD (1995) Ca^{2+} influx pathways mediated by swelling or stores depletion in mouse thymocytes. J Gen Physiol 106:415–444

Ross PE, Garber SS, Cahalan MD (1994) Membrane chloride conductance and capacitance in Jurkat T lymphocytes during osmotic swelling. Biophys J 66:169–178

Sands SB, Lewis RS, Cahalan MD (1989) Charybdotoxin blocks voltage-gated K^+ channels in human and murine T lymphocytes. J Gen Physiol 93:1061–1074

Sarkadi B, Mack E, Rothstein A (1984a) Ionic events during the volume response of human peripheral blood lymphocytes to hypotonic media. I. Distinctions between volume-activated Cl^- and K^+ conductance pathways. J Gen Physiol 83:497–512

Sarkadi B, Mack E, Rothstein A (1984b) Ionic events during the volume response of human peripheral blood lymphocytes to hypotonic media. II. Volume- and time-dependent activation and inactivation of ion transport pathways. J Gen Physiol 83:513–527

Sarkadi B, Tordai A, Gárdos G (1990) Membrane depolarization selectively inhibits receptor-operated calcium channels in human (Jurkat) T lymphocytes. Biochim Biophys Acta 1027:130–140

Schlichter LC, Grygorczyk R, Pahapill PA, Grygorczyk C (1990) A large, multiple-conductance chloride channel in normal human T lymphocytes. Pflugers Arch 416:413–421

Shaw JP, Utz PJ, Durand DB, Toole JJ, Emmel EA, Crabtree GR (1988) Identification of a putative regulator of early T cell activation genes. Science 241:202–205

Siemen D, Loupatatzis C, Borecky J, Gulbins E, Lang F (1999) Ca^{2+}-activated K channel of the BK-type in the inner mitochondrial membrane of a human glioma cell line. Biochem Biophys Res Commun 257:549–554

Smith CA, Williams GT, Kingston R, Jenkinson EJ, Owen JJ (1989) Apoptosis. Nature 338:10

Steinert M, Grissmer S (1997) Novel activation stimulus of chloride channels by potassium in human osteoblasts and human leukaemic T lymphocytes. J Physiol 500(3):653–660

Storey NM, Gomez-Angelats M, Bortner CD, Armstrong DL, Cidlowski JA (2003) Stimulation of Kv1.3 potassium channels by death receptors during apoptosis in Jurkat T lymphocytes. J Biol Chem 278:33319–33326

Story MD, Stephens LC, Tomasovic SP, Meyn RE (1992) A role for calcium in regulating apoptosis in rat thymocytes irradiated in vitro. Int J Radiat Biol 61:243–251

Su Z, Csutora P, Hunton D, Shoemaker RL, Marchase RB, Blalock JE (2001) A store-operated nonselective cation channel in lymphocytes is activated directly by $Ca^{(2+)}$ influx factor and diacylglycerol. Am J Physiol Cell Physiol 280:C1284–C1292

Szabo I, Gulbins E, Apfel H, Zhang X, Barth P, Busch AE, Schlottmann K, Pongs O, Lang F (1996) Tyrosine phosphorylation-dependent suppression of a voltage-gated K^+ channel in T lymphocytes upon Fas stimulation. J Biol Chem 271:20465–20469

Szabo I, Lepple-Wienhues A, Kaba KN, Zoratti M, Gulbins E, Lang F (1998) Tyrosine kinase-dependent activation of a chloride channel in CD95-induced apoptosis in T lymphocytes. Proc Natl Acad Sci USA 95:6169–6174

Szucs A, Szappanos H, Toth A, Farkas Z, Panyi G, Csernoch L, Sziklai I (2004) Differential expression of purinergic receptor subtypes in the outer hair cells of the guinea pig. Hearing Res 196(1–2):2–7

Thomas N, Bell PA (1981) Glucocorticoid-induced cell-size changes and nuclear fragility in rat thymocytes. Mol Cell Endocrinol 22:71–84

Tillman TS, Cascio M (2003) Effects of membrane lipids on ion channel structure and function. Cell Biochem Biophys 38:161–190

Trimarchi JR, Liu L, Smith PJ, Keefe DL (2002) Apoptosis recruits two-pore domain potassium channels used for homeostatic volume regulation. Am J Physiol Cell Physiol 282:C588–C594

Vaca L (1996) Calmodulin inhibits calcium influx current in vascular endothelium. FEBS Lett 390:289–293

van der Merwe PA (2002) Formation and function of the immunological synapse. Curr Opin Immunol 14:293–298

Vereb G, Panyi G, Balázs M, Mátyus L, Matkó J, Damjanovich S (1990) Effect of cyclosporin A on the membrane potential and Ca^{2+} level of human lymphoid cell lines and mouse thymocytes. Biochim Biophys Acta 1019:159–165

Verheugen JA, Vijverberg HP, Oortgiesen M, Cahalan MD (1995) Voltage-gated and $Ca^{(2+)}$-activated K^+ channels in intact human T lymphocytes. Noninvasive measurements of membrane currents, membrane potential, and intracellular calcium. J Gen Physiol 105:765–794

Weidema AF, Ravesloot JH, Panyi G, Nijweide P, Ypey DL (1993) A Ca^{2+}-dependent K^+-channel in freshly isolated and cultured chick osteoclasts. Biochim Biophys Acta 1149:63–72

Wulff H, Miller MJ, Hansel W, Grissmer S, Cahalan MD, Chandy KG (2000) Design of a potent and selective inhibitor of the intermediate-conductance Ca^{2+}-activated K^+ channel, IKCa1: a potential immunosuppressant. Proc Natl Acad Sci USA 97:8151–8156

Wyllie AH (1980) Glucocorticoid-induced thymocyte apoptosis is associated with endogenous endonuclease activation. Nature 284:555–556

Xu W, Liu Y, Wang S, McDonald T, Van Eyk JE, Sidor A, O'Rourke B (2002) Cytoprotective role of Ca^{2+}- activated K^+ channels in the cardiac inner mitochondrial membrane. Science 298:1029–1033

Yao Y, Ferrer-Montiel AV, Montal M, Tsien RY (1999) Activation of store-operated Ca^{2+} current in *Xenopus* oocytes requires SNAP-25 but not a diffusible messenger. Cell 98:475–485

Zitt C, Strauss B, Schwarz EC, Spaeth N, Rast G, Hatzelmann A, Hoth M (2004) Potent inhibition of Ca^{2+} release-activated Ca^{2+} channels and T-lymphocyte activation by the pyrazole derivative BTP2. J Biol Chem 279:12427–12437

Zweifach A, Lewis RS (1993) Mitogen-regulated Ca^{2+} current of T lymphocytes is activated by depletion of intracellular Ca^{2+} stores. Proc Natl Acad Sci USA 90:6295–6299

Zweifach A, Lewis RS (1995) Rapid inactivation of depletion-activated calcium current (ICRAC) due to local calcium feedback. J Gen Physiol 105:209–226

Subject Index

V

Printing: Krips bv, Meppel
Binding: Stürtz, Würzburg